T0215091

Inventing Accuracy

Inside Technology

Wiebe E. Bijker, W. Bernard Carlson, and Trevor Pinch, editors

Artificial Experts: Social Knowledge and Intelligent Machines, H. M. Collins, 1990
Viewing the Earth: The Social Construction of the Landsat Satellite System, Pamela E. Mack, 1990
Inventing Accuracy: A Historical Sociology of Nuclear Missile Guidance, Donald MacKenzie, 1990

Inventing Accuracy
A Historical Sociology of Nuclear Missile Guidance

Donald MacKenzie

The MIT Press
Cambridge, Massachusetts
London, England

First MIT Press paperback edition, 1993

© 1990 Massachusetts Institute of Technology

All rights reserved. No part of this book may be reproduced in any form by any electronic or mechanical means (including photocopying, recording, or information storage and retrieval) without permission in writing from the publisher.

This book was set in Baskerville by DEKR Corporation

Library of Congress Cataloging-in-Publication Data

MacKenzie, Donald A.
  Inventing accuracy : an historical sociology of nuclear missile guidance / Donald MacKenzie.
    p.   cm. — (Inside technology)
  Includes bibliographical references.

  ISBN-13 978-0-262-13258-9 (HB), 978-0-262-63147-1 (PB)
  1. Ballistic missiles—United States—Guidance systems—History.
2. Nuclear weapons—United States—History. 3. United States—Military policy. 4. Nuclear warfare. I. Title. II. Series.
UG1312.B34M33   1990
358′.174—dc20   90-5915
                                                                      CIP

To Alice, that you may grow up into a world in which the subject of this book is in every sense a matter of history

Contents

Acknowledgments

It is impossible to study a topic like this and to write a book about it without incurring many debts of gratitude. It would not have been possible at all without the financial help of three grants from the Nuffield Foundation. Almost any other funding agency would have found the idea of a sociologist studying missile guidance bizarre. The Nuffield Foundation did not— or if it did, decided to gamble by supporting it anyway. The Economic and Social Research Council also helped, in a less intrepid but still useful way, with a small grant for research on "The Management of Technical Uncertainty" and through its support for the Edinburgh University work under the Programme in Information and Communication Technologies (PICT).

Throughout the work I was also encouraged by finding colleagues who regarded it as a worthwhile enterprise. Thomas P. Hughes did not just support me intellectually, he also arranged some graduate teaching for me at the University of Pennsylvania at a point when the collapse of the pound sterling against the U.S. dollar threatened to leave me hitchhiking from interview to interview. Bruno Latour's characteristic enthusiasm was always a joy, and his sympathetic but critical reading of the entire draft manuscript was an invaluable help in reworking it.

More generally, the international network of the social studies of science and technology has been a constant source of both ideas and practical help. At the start of the research, the Program in Science, Technology, and Society at the Massachusetts Institute of Technology (MIT), for example, gave me a welcome, an office, a telephone, and access to a photocopier, without all four of which life would have been very difficult. Alex Roland, Secretary of the Society for the History of Technology, also read the full manuscript and gave me the benefit of his considerable knowledge of the history of U.S. aerospace. The editors of this series, Wiebe Bijker, Trevor Pinch, and Bernie Carlson, were unfailingly encouraging.

I began by knowing nothing about inertial navigation or missile guidance. Phil Odor and David Holloway were the only people I knew who did, from the quite different perspectives of former Royal Air Force instrument fitter and scholar of the Soviet intercontinental ballistic missile (ICBM) program, respectively. They got me started. The field's technical literature is sparse in Edinburgh University Library, as in most university libraries, so it was my good fortune that one of the main European suppliers of inertial navigators, Ferranti plc, was both local and willing to help. In the United States the archivists at MIT, the Library of Congress, the Office of Air Force History, and the National Aeronautics and Space Administration—and especially Cary Browse in the Technical Information Center of the Charles Stark Draper Laboratory, Inc.— helped me find relevant historical documents. Matthew Bunn generously gave me free range of the files of technical papers he had collected on missile guidance and reentry vehicles, and he and David Morrison kept sending me further useful clippings.

The documents alone, however, would have made little sense to me. The most important input into the research came from people involved in the development of missile guidance and inertial navigation. I began by imagining that it would be hard to get anyone to talk to me. It was not. The first place I met with cooperation far beyond my expectations was at the Charles Stark Draper Laboratory in Cambridge, Massachusetts, where I have particularly to thank its founder, the late Professor Charles Stark Draper, and also Brigadier General Robert A. Duffy (U.S. Air Force, rtd.), its then President, and Joseph O'Connor, Vice-President. The then Director of the U.S. Navy Strategic Systems Program Office, Rear Admiral Glenwood Clark, both gave me his own views on the development of U.S. submarine-launched ballistic missile guidance and navigation and allowed me to flesh these out by talking to his staff. Two of his predecessors, now retired, Vice Admiral Levering Smith and Rear Admiral Robert H. Wertheim, were also exceptionally helpful. Major General Aloysius G. Casey, Commander of the U.S. Air Force Ballistic Missile Office, and his predecessors

General Bernard Schriever and Major General John W. Hepfer led me through the corresponding matters for ICBMs.

These were only the central figures, however. Many others helped too. Some are famous, such as former Secretaries of Defense Robert McNamara and James Schlesinger and National Security Adviser McGeorge Bundy. Most are not. Barring a small number of interviews with members of the U.S. intelligence community that were conducted under conditions of anonymity, the full list of those interviewed can be found in appendix C.

They entrusted their stories to a stranger and a skeptic. For that more than anything else, but also for giving me many documents and frequently helpful written comments on chapters of the book, I bear them the greatest debt of all.

Most of these interviews were conducted in the United States. When weary from airports, freeways, and motels, it was good to have friends there who would take me in and do much else besides, such as teach me cross-country skiing. Special thanks go to Diane Paul, Joel Kallich and Sue Jennings, Everett Mendelsohn and Mary Anderson, Dan and Betty-Ann Kevles, Merriley Borrell, Joan Countryman and Ed Jackmauh, and Mary McClymont; also to two sets of old friends from Edinburgh, Ben Kobashigawa, Barb Bates, Lorin and Jun-Dai, and Jim Bearden, Liz Alpert, Janey, and Emma.

Back home in Edinburgh, Terry Inkster heroically and skilfully transcribed the mountain of tapes I had accumulated, and Margaret Tomlinson efficiently turned into usable discs the bits of the manuscript I had written before buying a word processor. Elena Cook drew the thankless task of trying to teach me Russian. She, and also my mother, Anne MacKenzie, were generous in their linguistic assistance.

My obsession with missiles and gyroscopes probably puzzled many friends. Some, especially Lynn Jamieson, knew what it was about and why it mattered. That helped, perhaps more than was obvious. Two colleagues and then friends even came to share the obsession. Graham Spinardi wrote a Ph.D. dissertation on the fleet ballistic missile program, re-interviewing participants to whom I had spoken and talking to many others besides. We have freely shared transcripts, documents, drafts, and ideas. Where this book discusses that program, as it does in chapters 3 and 5 particularly, Graham's contribution is central. Wolfgang Rüdig worked for a year on the project as

Research Fellow and developed a fascination with the German contribution to inertial guidance and navigation. The section of chapter 2 dealing with that topic could not have been written without his help.

All books, though, eventually come down to what can be a lonely encounter between the author and his or her pen, typewriter, or word processor. Caroline Bamford kept the encounter relatively peaceful, but kept it from being lonely. Thank you, Caroline.

I experimented with many of the ideas in this book in the form of seminar papers and articles, and colleagues too numerous to list gave me useful comments on these. Only one article became a chapter of the book: chapter 6 appeared, in a rather different form, in the Fall 1988 issue of the MIT Press journal, *International Security.* Other articles that have fed into this book are "Missile Accuracy—An Arms Control Opportunity," *Bulletin of the Atomic Scientists,* Vol. 42, No. 6 (June/July 1986); "Missile Accuracy: A Case Study in the Social Processes of Technological Change," in Wiebe E. Bijker, Thomas P. Hughes, and Trevor J. Pinch (eds.), *The Social Construction of Technological Systems: New Directions in the Sociology and History of Technology* (Cambridge, Mass.: MIT Press, 1987); "The Problem with 'The Facts': Nuclear Weapons Policy and the Social Negotiation of Data," in Roger Davidson and Phil White (eds.), *Information and Government* (Edinburgh: Edinburgh University Press, 1988); (with Wolfgang Rüdig and Graham Spinardi) "Social Research on Technology and the Policy Agenda: An Example from the Strategic Arms Race," in Brian Elliott (ed.), *Technology and Social Change* (Edinburgh: Edinburgh University Press, 1988); (with Graham Spinardi) "The Shaping of Nuclear Weapon System Technology: U.S. Fleet Ballistic Missile Guidance and Navigation," *Social Studies of Science,* Vol. 18 (1988); "Stellar-Inertial Guidance: A Study in the Sociology of Military Technology," in Everett Mendelsohn, M. Roe Smith, and Peter Weingart (eds.), *Science, Technology and the Military: Sociology of the Sciences Yearbook,* Vol. 14 (1988) (Dordrecht: Reidel, 1988); "From Kwajalein to Armageddon? Testing and the Social Construction of Missile Accuracy," in David Gooding, Trevor Pinch, and Simon Schaffer (eds.), *The Uses of Experiment: Studies of Experiment in the Natural Sciences* (Cambridge: Cambridge University Press, 1989); "Towards an Historical Sociology of Nuclear Weapons Technologies," in Nils Petter Gleditsch and

Olav Njølstad, eds., *Arms Races: Technological and Political Dynamics* (London: SAGE, 1990).

I would also like to thank the following people and organizations who either provided me with illustrations to use or gave me permission to reproduce others that had already been published: Scientific American, Inc., Thomas P. Hughes, Dr. Fritz Mueller, the Charles Stark Draper Laboratory, Inc., the Institute of Navigation, John M. Wuerth, University of Illinois at Urbana-Champaign, Honeywell Inc., Litton Guidance and Control Systems, the Kearfott Division of the Singer Corporation, David Lynch, the Delco Electronics Division of General Motors, the Electronics Division of the Northrop Corporation, Rear Admiral Robert Wertheim (U.S. Navy, rtd.), and Rockwell International. For help in the production of other illustrations, I am grateful to Audiovisual Services, University of Edinburgh.

1

A Historical Sociology of Nuclear Missile Guidance

Look out the window of the room in which you are now sitting. Focus on a tree or a building about a hundred yards or meters away. Imagine a circle with your room at its center and that object on its edge.

That circle defines the accuracy of the most modern U.S. strategic missiles. Fired from a silo or submarine on the other side of the earth, then arching up through space, an MX or Trident II missile is designed to deposit its nuclear warheads within little more than that circle. All this is to be achieved without human intervention beyond the order to fire. A basketball-sized sphere of instruments and a little computer harness the brute power of the missile's rocket motors as it climbs out of the earth's atmosphere. As the missile coasts up in space, this guidance system then orders a final gentle nudge to be given to the missile's deadly cargo, sending the warheads, now unguided, on their path through space toward their final plunge back through the atmosphere to their targets.

In most strategic ballistic missile guidance systems this is achieved without recourse to external references, landmarks, or radio signals. The computer memorizes launch point, target, and what it needs to know about the shape and gravitational field of the earth. Blind inside their sphere, tiny sensors that you could hold in the palm of your hand measure acceleration and rotation with exquisite sensitivity. The result is pure con-tradiction. The enormous raw violence of a nuclear explosion is delivered with the precision—taking the different scales into account—of a surgeon performing the most delicate of operations.

It is an astonishing technical achievement, yet it is one that can be viewed only with ambivalence. What is that accuracy for? Missing the center of Moscow by a hundred yards, or half a mile, makes little difference to the death and damage a nuclear warhead would cause. Great accuracy is unnecessary in the classic deterrent role for the missile as bearer of unspeakable retaliation against the cities of a nuclear aggressor. It is not needed for "assured destruction."

Very considerable accuracy is needed, however, if strikes against opposing missile silos and underground command posts are contemplated. But "counterforce," as the strategy of striking at an opponent's nuclear forces is called, raises an awkward question. If one is retaliating, will the silos not already be empty? Does counterforce, and thus missile accuracy, not imply a preparedness to strike first?

The defenders of accuracy and counterforce said that matters were not as simple as that. The Soviet Union, they speculated, might use only a portion of its intercontinental ballistic missile (ICBM) force to attack the United States, so the latter might well want to destroy silos in a retaliatory strike. Nuclear war, some went on, might not be the immediate all-out holocaust implied by "assured destruction," but might be more limited, more drawn-out, and possibly even winnable. A few added that there were even circumstances in which the United States might wish to strike first.

Whether these arguments are found to be credible or not, the key point for this book remains: missile guidance forms a technological "window" into crucial divides in nuclear thinking. Central tensions of the nuclear age, such as whether there is any sense in which a nuclear war can be won, express themselves in the little spheres and ultrasensitive sensors at the heart of missiles.

My goal in this book is to throw light on this key modern technology. I am not primarily concerned with explaining it technically. Others have done that better than I can and I do not want any reader interested in the general issues to be put off by undue technicality.[1] My aim, rather, is to understand nuclear missile guidance as a historical product and social creation.

1. The classic technical account is D. G. Hoag, "Ballistic-missile Guidance," in B. T. Feld et al., eds., *Impact of New Technologies on the Arms Race* (Cambridge, Mass.: MIT Press, 1971), 19–108. See also Kosta Tsipis, *Arsenal: Understanding Weapons in the Nuclear Age* (New York: Simon and Schuster, 1983), chapter 5 and appendix E.

The implicit consensus among the few authors who have dealt with missile guidance in other than a narrowly technical way has been that there could be little to say about this. They have almost all agreed that increasing missile accuracy is just a natural way for technology to develop. Some have gone on to argue that changing nuclear strategies are, as a consequence, determined by technology. Here are three typical statements: from a "hawk" who believes nuclear wars can be fought and won, from a "liberal" proponent of arms control, and from a leading scholar on the Marxist left.

Teams of scientists and engineers do and inevitably will discover ways of improving system performance.

On the issue of guidance accuracy, there is no way to get hold of it, it is a laboratory development, and there is no way to stop progress in that field.

The possibility of greater accuracy in targeting missiles led to the shift from the "countervalue" approach, aimed at cities and economic targets, to one aimed at specific military targets, i.e., "counterforce."[2]

This book reveals just how wrong it is to assume that missile accuracy is a natural or inevitable consequence of technical change. Nor, however, has it come about simply because governments, for good or bad reasons, have desired it. Rather, it is the product of a complex process of conflict and collaboration between a range of social actors including ambitious, energetic technologists, laboratories and corporations, and political and military leaders and the organizations they head. The invention of accuracy has fueled, and has itself been fueled by, the cold war. It has been a shaping force, but has itself been shaped.

Does it matter how missile accuracy has come about? Should we not accept missile accuracy as a fact of life, and learn to live with it as best we can, rather than digging into its origins? I

2. Colin S. Gray, *The Future of Land-Based Missile Forces*, Adelphi Paper No. 140 (London: International Institute for Strategic Studies, 1977), 4; Jack Ruina in 1969 Congressional testimony, as quoted, approvingly, ibid., 4fn.; Fred Halliday, *The Making of the Second Cold War* (London: Verso, 1983), 225. For Gray's "war-winning views," see Colin S. Gray and Keith Payne, "Victory is Possible," *Foreign Policy*, No. 39 (1980), 14–27; he does actually admit strategic doctrine as well as natural technological development as a force shaping missile accuracy (Gray, *Future*, 4–5).

think not. First of all, it is important that we understand the nature of the nuclear world. This world is not the product of technology developing autonomously, it is not brought about by benignly rational decision making, even on "our" side, and it is not the result of a coherent malign conspiracy. These views are not just intellectually wrong—though that is what this book shows—but disabling. They render us passive and pessimistic. To see the mundane social processes that form the nuclear world is to see simultaneously the possibility of intervening in them, of reshaping that world.

Understanding the invention of accuracy as a social process also carries with it more particular consequences. For example, it is often asserted in debates about nuclear weapons that they "cannot be uninvented." Although trivially true, that assertion is also profoundly misleading. A technology is not just social up to the point of invention and self-sustaining thereafter. Its conditions of possibility are always social. As we shall see, it follows from this that there may be a very real, and politically important, sense in which accuracy can be uninvented.

I began the research that led to this book in the early 1980s, when deepening antagonism between East and West, the deployment of new generations of missiles, and the revival of the peace movement had all pushed the nuclear weapons question to the forefront of the political agenda. In the more hopeful atmosphere of the start of the new decade, the attention of many has moved on to other things. Yet the nuclear arsenals remain, diminished only slightly, and, equally importantly, the social structures that generated them are still in place. Even if neither were the case, it would be important to take stock of this strange period of human history. It is imperative to understand as best we can the processes that shape the nuclear world.

In studying guidance, we are of course examining but a single aspect of the processes of the creation of nuclear weapons technology. But my aim in studying guidance has been to throw light on these processes more generally. Only further research can show how typical are these findings.[3] At the very least, guidance forms a set of threads central to the nuclear skein, because all nuclear strategies have to address the ques-

3. As discussed below, the dominant way most existing studies of technology and the arms race have been framed limits their usefulness in answering the sort of questions of interest here.

tion of what to target. In constructing a historical sociology of nuclear missile guidance, I therefore hope to have made some contribution to untangling that skein.

The Structure of the Book

This book is organized as follows. The rest of this chapter outlines a little further what it means to write a historical sociology of a technology such as missile guidance and addresses the obvious practical question of why military secrecy did not prevent it from being written. The chapter continues with a simple introduction to nuclear missile guidance, the organizations involved in its development, and the issues in nuclear strategy connected to it, my aim being to help orient the reader in the chapters that follow. The chapter ends with some brief remarks on the book's terminology and language.

Chapter 2 delves into the history of the dominant technology of missile guidance—inertial, or self-contained, guidance. It asks how this technology was invented: through sudden insight, the application of science, response to social needs, or what? Chapter 3 describes the development, in the United States, of the weapon system where this technology has found its most consequential application—the nuclear ballistic missile. How did the missile displace the bomber as the primary American strategic weapon?

In chapters 4 and 5, I analyze the processes by which the relatively inaccurate missiles of the 1950s became the ultraprecise American weapons of today. In these chapters we will see at work the process outlined above, the complex, conflictual interaction of different social groups—technological, military, and political. We will also see, however, how boundaries are created around projects, resources channeled into them, and a gradual and cumulative process of technological change institutionalized. That process may then look like a natural trajectory of technology.[4] But, as we shall see in these chapters, there is nothing natural about it.

A straightforward, though not failsafe, way of grasping the role of the social in technical change is to see how different

4. The notion of a technological trajectory, sometimes qualified by the adjective "natural," has become widespread in recent analyses of technological change. The idea is discussed further in chapter 4.

societies develop the same technology. In chapter 6, I analyze the development of missile guidance in the Soviet Union, the only country comparable to the United States in its sustained indigenous tradition of strategic guidance system development. Certainly there are similarities: the invention of accuracy has been as important an enterprise in the Soviet Union as in the United States. But how it has been achieved is subtly different, both organizationally and technically.

Chapter 7 brings into the open an issue lurking beneath the surface of the other chapters. Is there not a bedrock of "technical fact" underlying the processes of social choice and contention? We may dispute whether we want missiles to be accurate. But can we dispute how accurate a given missile actually is? In this chapter we see that we can; that the accuracy of a missile, or even of one of the tiny components of its guidance system, is a complex and contestable construction. Here, too, we shall find the clue to the possibility of uninventing accuracy.

Chapter 8 pulls together the lessons from the previous chapters. What has been learned about the process of technical change, the politics of the nuclear world, and the relationship between the two? To borrow an expression from one of the great scholars of technology, what, in these matters, is the relationship between interpreting the world and changing it?[5]

The Historical Sociology of Technology

Though the particular topic of this book is new—no extensive history of missile guidance has previously been written—its general subject matter, technical change, is not.[6] Without burdening the reader with an extensive review of the considerable body of existing literature, it is worth discussing what a "historical sociology" of technology means and why one is needed.

5. "The philosophers have only *interpreted* the world . . . the point is to *change* it." Karl Marx, eleventh thesis on Feuerbach, in, e.g., Marx, *Early Writings* (Harmondsworth, Middlesex: Penguin, 1975), 423. I have discussed Marx as a scholar of technical change in "Marx and the Machine," *Technology and Culture*, Vol. 25 (1984), 473–502.
6. Technologists involved in the field have written some histories of particular episodes in the development of missile guidance and some general accounts of the emergence of inertial navigation. These are cited in later chapters.

The reader uninterested in such questions can skip to the next main section of this chapter, where I discuss the practical question of the sources of data for such a study of a militarily sensitive technology.

Why Historical?

The "historical" is the easier part to explain. Those of us who research social processes are seldom able to set up our own experiments. We have to wait for the world to do it for us. The passage of time, and changes it brings in the factors and phenomena that interest us, are our single best resource.

That may seem obvious to the point of truism. But consider for a moment the dominant form of social research on weapons technology: the weapon-system case study. This is a detailed investigation of the decisions surrounding the development and deployment of, for example, a particular aircraft or missile. The popularity of this form was such that one could for a while have said that every U.S. missile carried a political science Ph.D. dissertation along with it! Although that popularity has now waned, no satisfactory alternative methodology has been found, and even the best modern work on technology and the arms race retains something of a case-study flavor.[7]

The weapon-system case study has clear virtues. The best of them were remarkable in the extent of their probing into the processes of decision making, and it is highly regrettable that the decline of the form means that we lack case studies of the

7. Within arms race studies in general there is also a tradition of mathematical modeling. That tradition has had little to say about technological change, and has focused instead on parameters such as the size of military budgets and the numbers of weapon systems. For a useful review, see Håkan Wiberg, "Arms Races, Formal Models, and Quantitative Tests," in Nils Petter Gleditsch and Olav Njølstad, eds., *Arms Races: Technological and Political Dynamics* (London: SAGE, 1990).

I cite as my example of the best modern work on technology and the arms race within the political science tradition, Matthew Evangelista, *Innovation and the Arms Race: How the United States and the Soviet Union Develop New Military Technologies* (Ithaca: Cornell University Press, 1988). For a discussion of this book, which breaks from the case-study approach in several ways, notably in its international comparison, but retains other aspects of the form, see Donald MacKenzie, "Technology and the Arms Race," *International Security*, Vol. 14, No. 1 (Summer 1989), pp. 161–175.

most modern missiles, particularly MX and Trident, of a quality comparable to studies of their predecessors.[8]

In even the best case studies, however, a focus on one particular weapon system inhibits examination of much of what is important concerning technology and the arms race. Because the crucial decisions concerning a particular system are usually taken in a relatively small number of years, a researcher investigating them tends to treat as constant those factors that change only slowly. The state of the art of a gradually changing technology, such as missile guidance, is precisely such a factor.[9] One consequence is that a key category of political actor, the organizations within which such technical changes take place, tends to be missed in case-study analyses.[10]

In short, the weapon-system case study freezes time. Unfreezing it, and bringing to attention phenomena that escape individual case study, is the virtue of a historical perspective. Emphatically, though, nothing restricts the latter to past events alone. The present is equally open to examination, just so long as it is studied not in isolation from the past but as a moment in continuing processes. That, I hope, is how I have examined it here.

Why Sociological?

There are two reasons why the perspective employed should be sociological as well as historical. One is relatively obvious,

8. For U.S. Navy missiles, for example, it is sad but instructive to compare D. Douglas Dalgleish and Larry Schweikart, *Trident* (Carbondale and Edwardsville, Ill.: Southern Illinois University Press, 1984) with Harvey Sapolsky, *The Polaris System Development: Bureaucratic and Programmatic Success in Government* (Cambridge, Mass.: Harvard University Press, 1972) and Ted Greenwood, *Making the MIRV: A Study of Defense Decision-Making* (Cambridge, Mass.: Ballinger, 1975).
9. The case-study method is better suited to the analysis of discontinuous, "radical," technical changes. See Evangelista, *Innovation and the Arms Race*.
10. Consider Charles Stark Draper's Massachusetts Institute of Technology (MIT) Instrumentation Laboratory. None of the missile-system case studies do more than devote a few lines to the Instrumentation Laboratory, and some do not mention it at all. In each particular case, authors presumably saw its role as unimportant. Yet when one notes that the Instrumentation Laboratory designed the guidance system for every U.S. Navy Fleet Ballistic Missile (Polaris A1, A2, and A3, Poseidon C3, Trident C4 and D5), as well as the Air Force's Thor, Titan II, MX, and (in effect) Small ICBM, one begins to wonder whether this neglect was justified.

another less so. The more obvious reason is that technological change is simultaneously economic, political, organizational, cultural, and legal change, to enumerate just some of the aspects of "the social." Technologies, especially large, complex, technical systems, are not always simple tools that can be improved without wider consequences. Changes in technology go hand-in-hand with changes, small and large, in the preconditions of their use, in the ways they are used, in who uses them, and in the reasons for their use.[11] Nor is this a question of a one-way "impact of technology upon society." For the way technology changes cannot be explained in isolation from the economic, political, and other social circumstances of that change.[12]

These considerations need not always be spelled out formally. Indeed, they simply describe the practice of good, modern history of technology, and while social science disciplines contain useful resources to help flesh out the issues involved, these resources are many and various.[13] No single approach or framework has a monopoly on insight.

When it comes to the second reason for adoption of a sociological perspective, however, a more particular inheritance becomes relevant, that of the sociology of scientific knowledge. This field has developed over the last twenty years as researchers, by no means all sociologists in disciplinary affiliation, have moved beyond the older "sociology of scientists," which consisted of the study of matters such as the career patterns of scientists and the reward system of science, its ethos and norms. At the heart of the new field are a set of empirical studies that have demonstrated social processes—both internal to science and involving the wider society—at the heart of the knowledge-generating and knowledge-assessing activities of science.[14] The existence of these processes is no indicator of false, inadequate,

11. The most useful single discussion of this remains Langdon Winner, *Autonomous Technology: Technics-out-of-Control as a Theme in Political Thought* (Cambridge, Mass.: MIT Press, 1977).

12. See Donald MacKenzie and Judy Wajcman, eds., *The Social Shaping of Technology* (Milton Keynes, Bucks, England: Open University Press, 1985).

13. John M. Staudenmaier, S.J., *Technology's Storytellers: Reweaving the Human Fabric* (Cambridge, Mass.: MIT Press, 1985).

14. An excellent bibliographic introduction, even if no longer wholly up-to-date, is Steven Shapin, "History of Science and its Sociological Reconstructions," *History of Science*, Vol. 20 (1982), 157–211.

or "ideological" science; rather, the central claim of the field is that they are to be found equally in all science, "good" and "bad."[15]

The most direct bearing of this upon technology arises because technology consists not just of artifacts, nor even of artifacts together with human activities, but also of knowledge.[16] Along with scientific, mathematical, and medical knowledge, but no other areas, our society treats technological knowledge as "hard fact."[17] It is, for example, a vital resource of technologists (as distinct from political leaders, generals, or corporate executives) that in questions of weapons design they are the arbiters of what is feasible as distinct from the "softer" issues of what is acceptable, needed, or affordable.

But if a sociology of scientific knowledge is possible, so should be a sociology of technological knowledge. Like the former, it is not a question of debunking, of exposing the causes of false or inadequate knowledge. A useful starting point, common to the study of both science and technology, is the realization that no knowledge possesses absolute warrant, whether from logic, experiment, or practice. There are always grounds for challenging any knowledge claim. But not all knowledge is challenged, nor is all challenge successful or even credible. Why some knowledge claims are challenged and some are not, and why some challenges succeed and some fail, thus become interesting empirical questions. Central to the answers are matters of the interests, goals, traditions, and experiences of the social groups (technological and other) involved; of the conventions surrounding technological testing; and of the rel-

15. For two early statements of this, see Barry Barnes, *Scientific Knowledge and Sociological Theory* (London: Routledge and Kegan Paul, 1974); and David Bloor, *Knowledge and Social Imagery* (London: Routledge and Kegan Paul, 1976).

16. Other possible connections are discussed in Trevor J. Pinch and Wiebe E. Bijker, "The Social Construction of Facts and Artefacts: Or How the Sociology of Science and the Sociology of Technology Might Benefit Each Other," *Social Studies of Science*, Vol. 14 (1984), 399–441. There is a shortened version of this paper in Wiebe E. Bijker, Thomas P. Hughes, and Trevor Pinch, eds., *The Social Construction of Technological Systems: New Directions in the History and Sociology of Technology* (Cambridge, Mass.: MIT Press, 1987), 17–50.

17. That this is culturally specific can immediately be seen by comparing the status of religious knowledge in different kinds of societies.

ative prestige and credibility of different links in the network of knowledge.[18]

Technological knowledge, in other words, is social through and through. We are not dealing here just with the pathologies of technological knowledge, such as the form of technological parochialism practitioners describe as the "NIH [not invented here] syndrome." Nor are we dealing with overt controversy alone, where technological knowledge claims openly clash, as in the debates about the safety of nuclear power. If I am correct—and I hope the chapters that follow will make this abstract argument concretely convincing—what is at issue is an ineradicable aspect of all technological knowledge.

A consequence is that a certain way of thinking common to even the best historically and sociologically informed studies of technology must here be avoided. One manifestation of that way of thinking is to contrast "technical" and "social" (e.g., political) reasons for a decision such as the choice of a particular design for an artifact. A related manifestation is to explain some choices by their inherent "technical superiority," greater "efficiency," and so on. Technical reasons for a course of action, technical superiority, and technical efficiency are all vitally important; in practice, they often seem sufficient to determine a given outcome. But it is important, as far as possible, to investigate why a given technical reason was found compelling, when, abstractly, it could have been challenged; and to ask what counts as superiority and efficiency in particular circumstances.

"As far as possible" needs to be emphasized. Opening the "black box" of technology in this way is no easy matter. From the analyst, it requires detailed understanding of the technical field in question, and it also requires data of a kind that is difficult or impossible to obtain without in-depth interviewing of those involved. Some black boxes must remain shut in

18. More complete presentations of the argument outlined here can be found in Donald MacKenzie, "The Problem with 'The Facts': Nuclear Weapons Policy and the Social Negotiation of Data," in Roger Davidson and Phil White, eds., *Information and Government* (Edinburgh: Edinburgh University Press, 1988), 232–251, and in MacKenzie, "From Kwajalein to Armageddon? Testing and the Social Construction of Missile Accuracy," in David Gooding, Trevor Pinch, and Simon Schaffer, eds., *The Uses of Experiment: Studies of Experiment in the Natural Sciences* (Cambridge: Cambridge University Press, 1989), 409–435.

the narrative that follows, some decisions must remain unexplained. Nevertheless, I hope I have gone part of the way to providing a detailed sociology of technical knowledge for my chosen area, guidance technology.

Studying a "Secret" Technology

Are all these considerations beside the point? Is my basic requirement, an adequate historical narrative, possible for a military technology as sensitive as strategic missile guidance? Perhaps surprisingly, the answer turns out to be yes, provided that those involved in its development cooperate with the research, which in this case they did with great generosity. The chief restriction that flows directly from security classification concerns the performance figures for guidance systems and their components. The missile accuracies given in appendix A, for example, could not be derived from interviews. But provided interviewees were not asked for figures, or, for example, for detailed blueprints, they were able to discuss in considerable technical depth not just missile guidance in general but the specifics of particular systems and their components. Equally important, they also were able to tell me with almost complete freedom about the processes, technical and social, that lead to the selection and design of these systems and components.

The willingness of guidance and navigation technologists to do this was, I think, premised on a sense that the writing of the history of their technical field was a worthwhile endeavor. But neither was any difficulty found in securing interviews with serving and retired officers in the ballistic missile offices of the U.S. Air Force and Navy, even those without a background in guidance. These offices play a vital role at the interface of technology (in the form of the development of the techniques of missile guidance) and politics (in the form of the Department of Defense, Congress, and the demands of missile users such as the Strategic Air Command). Somewhat greater problems arose with civilian officials in the Department of Defense, such as former Secretaries of Defense. Unlike many of the others approached for this study, for them to be asked for an interview is not an unusual experience. Straightforward refusal was again rare, but arranging a time to meet was often difficult and the length of interviews sometimes limited. Furthermore, top decision makers' recall of particular decisions among the many they were involved in is often poor. Nevertheless, speaking with former key decision makers allowed some insight into how the development of missile guidance appeared to those close to the pinnacle of formal power, while the recollections of others in less high positions in Washington greatly fleshed out my knowledge of the processes leading to particular outcomes.

Naturally, matters proved less straightforward in investigating developments in the Soviet Union. Direct interviewing of those involved, although not completely impossible, proved to be much more difficult. So nearly all of my evidence here is indirect. Most important, I was able to speak to some American "guidance systems analysts." This esoteric community takes what U.S. intelligence satellites, radars, and listening stations have detected from Soviet missile testing and turns it into knowledge of how accurate Soviet missiles are and thus of the threat they pose to America's nuclear forces. From their information I was able to construct a history of Soviet missile guidance, presented in chapter 6. Though this can be connected to what a wider literature has had to say about Soviet nuclear strategy and military research and development, it remains, inevitably, a thinner, less sociological, and less reliable history than is possible for American missile guidance.

Of course, even the results of the direct interviewing possible in the United States are inherently problematic. Memories fade. People have partial perceptions of events. They want a particular view of events to be accepted. The dynamics of the interview as a social interaction affect what is said and what is understood by what is said. So it would be naive to take what one is told for "what really happened."

Nevertheless, at the most basic level, interviews can be checked against each other. One of the advantages of doing a large number of interviews, as here, is that different participants' perceptions of the "same" events can be compared. Interviews can also be checked against documentary sources: published material, archival material, and written material provided by interviewees.

As far as possible, then, I have used written as well as oral sources. It is worth noting that the former are in principle no less problematic than interviews: a patent application, article in a technical journal, internal report, or minutes of a meeting have their purposes just as much as something said in an interview. Drawing on and collating these different sources, it is possible to build up a picture of events. It has to be done in stages, though, because part of that picture of events is a sense of the processes that give rise to particular pieces of "evidence."

The picture built up in this way is, implicitly or explicitly, theory-laden, because the interpretation of evidence is bound up with the evolving overall account. This is true even at the most basic sensory level. Influenced by my beliefs as to what someone "must" have said, I would sometimes hear different things on my tapes than would the skilled secretary who transcribed them for me. Nevertheless, an account built up in this way by no means simply reflects the investigator's preconceptions.[19] Mine, certainly, were changed in both subtle and radical ways by the experience of doing this research. The inevitably problematic relation between "reality" and the historical and sociological accounts of it does not, in any case, seem to me to be grounds for abandoning the attempt both to describe and to explain what has taken place and what is taking place.

Missile Guidance: Technology, Nuclear Strategy, and Organizations

The picture that has emerged is complex. To guide the reader, it is therefore worthwhile to provide a preliminary sketch of the technology I discuss and some of the circumstances of its development. This sketch is simply for orientation: nothing in the narrative will be lost if a reader, confident in his or her understanding of these matters, moves now to the start of chapter 2.

Of the many ways in which nuclear weapons can be delivered to their targets, three have been of particular importance: bombers, cruise missiles, and ballistic missiles. A cruise missile is rather like a pilotless airplane: it flies through the atmosphere drawing from it the oxygen its engine needs for combustion. A ballistic missile, on the other hand, carries its oxygen

19. On the particular issue of quotation from tapes: all those quoted were sent copies of the relevant chapter(s) so that they could check the accuracy of the quotation, tell me if they felt the quotation was being used out of context or in any other distorting way, and have their names removed if they wished.

along with it. Its rocket engines can operate outside the atmosphere, and long-range ballistic missiles climb into space in their trajectory from launch point to target.

The ballistic missile is my focus here, though aircraft navigation and cruise missile guidance will be touched upon.[20] The chief military advantage of the ballistic missile is its speed and the consequent extreme difficulty of defending against it. From launch point in the United States to target in the Soviet Union, or vice versa, takes thirty minutes or less, as distinct from several hours for even the fastest bomber or cruise missile. The final plunge of a ballistic missile warhead back from space through the atmosphere to its target—the period when defense of the target might seem most plausible—takes on the order of a couple of minutes, and the small warhead, normally detached from the missile that bore it, will be traveling at many times the speed of sound.

The customary description of such a missile as "ballistic" is both helpful and misleading. It points to the analogy with the artillery shell, and that link was originally strong. As we shall see, both in pre-1945 Germany and, to begin with, in the United States, the ballistic missile was the province of the Army, while the more airplane-like cruise missile was developed by the Air Force. But the analogy misleads if the ballistic missile is thought of as unguided in the sense that a conventional artillery shell, once it leaves the barrel of the gun that fired it, is unguided.

True, the actual warhead (or warheads) is unguided: after separation from the missile carrying it, its trajectory, like that of an artillery shell, is affected only by gravity and external forces such as air resistance.[21] But all ballistic missiles apart from the very crudest, short-range ones are guided. Steering commands alter the direction—and in some systems also the magnitude—of rocket thrust.

20. In the book I will use "guidance" to describe the process of directing a missile to its target; "navigation" to describe the process of providing information on position (and sometimes velocity and orientation as well) in a vehicle with a human crew.

21. Providing a warhead with a miniature guidance system has been a widely canvased possibility, but all strategic-range ballistic missile warheads are still unguided. The U.S. Pershing II intermediate-range ballistic missile, withdrawn under the Intermediate Nuclear Forces agreement, had a guided warhead.

The system issuing these steering commands varies greatly in complexity from missile to missile. A simple guidance system for a single-warhead missile might seek only to keep the missile stable on a preplanned trajectory and to shut off its rocket motor once it had reached the speed needed to bring its warhead to the target. That was essentially how the earliest operational guided ballistic missile, the Second World War German V-2, was guided. At the other extreme, a modern American missile's guidance system issues steering commands on the basis of constantly recomputed trajectories to the target. Such a missile carries several warheads destined for different targets (in the system known as MIRV, or multiple independently targetable reentry vehicle), and will maneuver in space so as to release these one after the other at locations, speeds, and in directions such that their subsequent unguided paths will take them to their targets.

How can the basic information required for guidance be obtained, whether that information consists simply of the speed and orientation of a V-2 or the more sophisticated data needed for a modern missile? More generally, how is it possible to navigate automatically, without a human navigator looking at landmarks or sighting on the stars?

One approach is to rely on external sources of information, such as radio signals or star-sightings taken by an automated star-tracker. As we shall see, strategic missiles have been and are guided by such means. But another approach has become dominant in missile guidance. Most strategic missiles are guided by it alone, and where external information is used too (in the form of star-sightings, as radio signals are not currently used in strategic ballistic missiles), it is in a supplementary role.

That second approach is what is called inertial guidance. Its dominant characteristic, and the chief military argument for it, is that it is self-contained. Such a system is "told" where it is at the point of launch. It then senses the acceleration and changes in orientation of the missile in flight and uses this information either to return the missile to the preplanned trajectory if it has departed from it, or, in more sophisticated systems, to work out its current position, velocity, and orientation.

The sensing of acceleration and changes in orientation by an inertial system is achieved without use of external references: it is in an almost literal sense a black box.[22] Acceleration can be measured in several ways; the simplest is to measure the extent of the force needed to prevent what is, in effect, a little pendulum from swinging backward as the missile accelerates forward.

Change in orientation is measured by gyroscopes, ultrasophisticated versions of the child's spinning toy. When a gyroscope is rotated around its sensitive axis, its spinning wheel reacts by changing orientation in a predictable way, and from that change in orientation the extent of the original rotation can be detected.[23] The normal way of using gyroscopes in missile guidance is to mount both them and the accelerometers on a platform that is free to rotate with respect to the missile. This platform is then kept in the same orientation whatever the twists and turns of the missile. The gyroscopes detect any tendency of the platform to rotate, and send feedback signals to little electric motors that keep it stable. Figure 1.1 shows this sort of system in outline.

In missiles of the 1940s and 1950s (and 1960s in the case of Soviet missiles), the calculations a guidance system has to perform were done by simple mechanical or electrical means. The advent of digital computers, which are small, light, and rugged enough to be carried on board a missile, opened new opportunities for the guidance system designer. Information needed for guidance could be stored in the computer's memory. Aside from the location of the launch point and target, the most important such data concern the earth's gravitational field. The latter affects the missile's trajectory but cannot—as early opponents of inertial navigation had noted—be measured directly by the guidance system. Onboard computers can also correct

22. There can be reasons for not actually painting them black. U.S. Navy aircraft inertial navigation engineers found it advantageous to paint them gold instead: "Inertial systems do not look expensive or delicate; therefore field maintenance technicians do not handle them with the care that is required. The I[nertial] N[avigation] S[ystem] cases are painted gray or black and they look like any other 'black box,' even though they may be 10 to 20 times more expensive, and much more susceptible to handling damage" (Anonymous, "INS Appearance Does Not Induce Careful Handling," Crosstalk Item Summary: Item 5, *Proceedings of the Fourteenth Joint Services Data Exchange Group for Inertial Systems* [Clearwater Beach, Florida: November 18–20, 1980], 411).

23. This is the effect known as "precession," discussed in chapter 2.

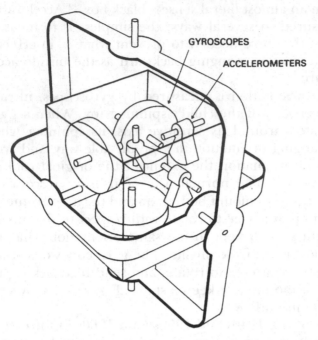

GYROSCOPES

ACCELEROMETERS

Figure 1.1
"Stable Platform" Inertial Guidance System (highly schematized)
Source: Redrawn from figure in Kosta Tsipis, "The Accuracy of
Strategic Missiles," *Scientific American*, Vol. 233, No. 1 (July 1975), 18.
Copyright © 1975 by Scientific American, Inc. All rights reserved.

for predictable errors in the output of the gyroscopes and
accelerometers.

Described in this way, missile guidance sounds like a straightforward technical matter. But this description glosses over a
great deal. It makes it sound as if a good guidance system is
all that it is needed for accuracy, when that is not so: the quality
of the reentry vehicles, for example, also is crucial.[24] It makes
missile guidance sound easy to achieve, while nothing could be
further from the truth. To achieve the accuracy needed to
destroy targets at intercontinental ranges, even with the enormous destructive power of nuclear warheads, requires gyroscopes and accelerometers of extraordinary precision. Even
today, only a small number of organizations worldwide possess
the knowledge, facilities, and skills needed. Controlling access

24. Reentry vehicle design is one of the "black boxes" that will not be opened
in this book.

to guidance technology is currently very important in the belated and ineffectual attempt to prevent the proliferation of ballistic missiles in the Third World.

Furthermore, the simplicity of the above description hides the wide range of alternative possibilities for the design of a guidance system. There is the basic decision of whether to seek to build an inertial system, a system relying on external information, or some combination of the two. If an inertial system is chosen, then, as we shall see, there is a range of different possible gyroscope and accelerometer designs. One could, for example, abandon mechanical gyroscopes altogether and detect changes in orientation by a device employing a different physical principle. Nor is it necessary to place the gyroscopes and accelerometers on a stable platform. One could, for example, simply attach them to the body of the missile and perform calculations in the computer to correct for the fact that their orientation would no longer be fixed.

These alternatives mean more than that guidance system design is a complex technical matter rather than a straightforward technical matter. Different technical pathways are seen by those involved as having different military and political meanings. Most important, how accurate one strives to make one's missiles is related intimately to the targets one envisages. Hence a small technical detail such as the design of accelerometers for U.S. Navy missiles can reflect, literally in miniature, the history of conflict over the strategic purpose of these missiles, as we shall see in chapter 5.

For the sake of simplicity, aware I am doing injustice to the many issues and nuances involved, I have sorted out different approaches to nuclear strategy into two clusters in figure 1.2.[25] Countercity targeting, with its modest demands on accuracy, is central to the strategic posture known as "assured destruction." This position, effectively official U.S. policy in the later 1960s and early 1970s, meant that the United States had to have nuclear forces sufficient to inflict unacceptable levels of destruction on the Soviet Union, even if the Soviets attacked first. A closely related idea is "finite deterrence": deploying

25. A major omission, for example, is any treatment of the use of "tactical," battlefield nuclear weapons. There is a clear account of the development of nuclear strategy in Lawrence Freedman, *The Evolution of Nuclear Strategy* (London: Macmillan, 1981).

Main Issue in Nuclear Strategy Connected to Missile Guidance

Targeting large, "soft" targets such as cities.
Key to the retaliatory strategies of "assured destruction" and "finite deterrence."

versus

Targeting small, "hard" targets such as missile silos and command posts.
Key to "counterforce" strategy, "first strike," and "preemption."

Figure 1.2

only a force of such a size and nature as is needed to deter a nuclear attack on one's territory. Advocated by the U.S. Navy in the 1950s, the idea is now particularly popular in Gorbachev's Soviet Union.

The other cluster of strategies involves the technically more demanding targeting of an opponent's nuclear forces. A "first strike" is not merely any first use of nuclear weapons, but a first use designed to destroy or disable as large a proportion of an opponent's nuclear force as possible, thereby minimizing retaliation. "Preemption" is the main circumstance where a first strike might be attempted. Certainly a very real option in the U.S. war plans of the 1950s and early 1960s, it meant attack upon receipt of intelligence warning that one was about to be attacked.

As the nuclear forces of the two sides grew larger and better protected—especially as large missile forces were deployed in submarines—the fear (or hope) of a fully successful first strike faded. For some strategists this implied acceptance that the assurance of destruction was mutual: the famous acronym MAD (mutual assured destruction) dates from this period. Others maintained that nuclear war could still be winnable, or at least could be thought winnable. They entertained a variety of notions as to how this might be so, but "counterforce" strikes were central to almost all of them.

Since issues of nuclear strategy such as these are of considerable importance to the armed services and to some of a nuclear power's political leadership, a range of military and political actors potentially have an interest in the details of guidance system design. But this interest is by no means always

---

Main "Political" Actors Directly Involved in Missile Guidance in the United States

Secretary of Defense (sometimes)

Director of Defense Research and Engineering

Congress (sometimes)

Figure 1.3

manifest. For example, no U.S. President has ever intervened directly in guidance design, though presidential decisions have certainly affected it less directly. Congress, too, has generally, but not always, considered guidance a technical matter in which it should not meddle. Some Defense Secretaries have engaged directly with missile guidance and, moving one step down the hierarchy, their technical deputies, known for most of the period under consideration as the Directors of Defense Research and Engineering, have often been deeply involved. Figure 1.3 lists these main political actors involved in guidance decisions in the United States.

The two U.S. armed services directly involved in strategic missiles are the Air Force and Navy. As we shall see in chapter 3, the Army was restricted to relatively short-range systems as the result of interservice battles in the 1950s. In the 1950s both the Air Force and the Navy established special organizations to superintend ballistic missile development: the Air Force's Western Development Division, now Ballistic Missile Office, and the Navy's Special Projects Office. These have at different times gone under different names, but for simplicity I use only these three in the text. They are shown in figure 1.4.

A similar issue arises for the military organizations as for the political actors: not all of those who might be seen as having a reason to involve themselves directly in guidance system design actually have done so. Sometimes the Ballistic Missile Office and Special Projects Office have been left to get on with guidance system design. In the Air Force, the Strategic Air Command—the ultimate "user" of Air Force strategic missiles—was, for example, effectively excluded from early ICBM design, though it now has a permanent presence in the Ballistic Missile Office through which it can express its opinions. The Special Projects Office has had considerable autonomy within the

---

The Two Main Organizations Responsible for U.S. Strategic Ballistic Missile Development

Air Force Western Development Division (now Ballistic Missile Office)
Responsible for land-based intercontinental ballistic missiles (ICBMs), such as Atlas, Titan, Minuteman, MX, and the Small ICBM.

Navy Special Projects Office
Responsible for the Fleet Ballistic Missile program of submarine-launched ballistic missiles, such as Polaris, Poseidon, and Trident.

Figure 1.4

Navy, but others in that service, especially strategic planners in the Office of the Chief of Naval Operations, have on occasion sought to influence guidance system specifications.

Most of the detailed design of guidance systems and almost all basic research on systems and components has been done outside the armed services. The major exception was work performed by a team of German guidance specialists who came to the United States after the end of the Second World War and were settled at the Army's Redstone Arsenal outside Huntsville, Alabama. As we shall see in chapter 3, two main American teams also became heavily involved in ballistic missile guidance. One was university-based. At the Massachusetts Institute of Technology (MIT), Charles Stark Draper's Instrumentation Laboratory, after successful work on anti-aircraft fire-control systems during the Second World War, took as its postwar mission the development of inertial guidance and navigation. The other American team was the Autonetics Division of the aircraft and guided missile company, North American Aviation, which is now part of Rockwell International.

The German team left ballistic missile guidance when the Army lost its battle to design and deploy long-range missiles. But both Autonetics and the Instrumentation Laboratory are still involved. Autonetics was responsible for the guidance systems of the three generations of Air Force Minuteman intercontinental ballistic missiles. It has also been the main developer of the inertial systems that navigate ballistic-missile submarines. The Instrumentation Laboratory (renamed the Charles Stark Draper Laboratory, Inc. after its early-1970s divestiture by MIT) has had design responsibility for all of the

---

The Three Main Ballistic Missile Inertial Guidance System Design Organizations in the United States

German team at Huntsville (Redstone, Jupiter, Pershing 1, Saturn)

MIT Instrumentation Laboratory/Draper Laboratory (Thor, Titan II, MX, Small ICBM, Polaris, Poseidon, Trident I, Trident II, Apollo)

Autonetics Division of North American Aviation/Rockwell International (Minuteman 1, 2, and 3; ballistic missile submarine inertial navigators)

Figure 1.5

guidance systems for the missiles fired from these submarines, as well as for the Air Force's Thor, Titan II, and MX missiles. The guidance system for what may become the next Air Force ballistic missile, the Small ICBM, is also based upon a Draper design.

If the Huntsville team's experience in Germany is added to that in the United States, these three organizations—listed together with their main missile systems in figure 1.5—each had or have thirty or more years experience in missile guidance. They employed distinctive and to a degree competing technological approaches; for example, each selected a different type of gyroscope for refinement and development. None was happy to stand still. Each advanced the technology of guidance in their chosen way and sought from the armed services the resources necessary to do so and the opportunities to make use of the innovative technology in new systems.

One of the main tasks of this book is to trace the links between issues of guidance system design, nuclear strategy, and these three types of actors: political, military, and technological. Is, for example, politics in command? Do political leaders select nuclear strategies, the military organizations procure the missiles needed to implement these strategies, and the technologists design guidance systems according to the military's specifications? Or, at the other extreme, do the technologists become seized with technical enthusiasms and persuade the military that missiles with these characteristics are needed, and does national nuclear strategy then shift to justify possession of this new arsenal? These divergent possible patterns of connection between actors correspond to quite different roles for politics and technology in the arms race, a question that has been subject to considerable if inconclusive debate.

At issue is not just the role of technologists and their organizations, but that of technical change itself. Addressing the latter's connection to politics and strategy will require us, especially in chapter 4, to compare technical change in missile guidance with that in related areas. Inertial technology is used not just to guide missiles but for a range of other purposes. The only application most readers are likely to have been close to is long-range civil aircraft navigation. Starting twenty years ago, human navigators on transoceanic flights were replaced by systems that work out the aircraft's position according to the principles outlined above, using gyroscopes and accelerometers.

Modern military aircraft are also equipped with inertial navigators, as are aircraft carriers and submarines. In particular, in the submarines that carry ballistic missiles, position, velocity, and orientation need to be known very exactly before the missiles are fired, at least if submarine-based missiles are not to be a lot less accurate than those fired from fixed points on land. Deep under the ocean, where neither radio systems nor the stars can be used, inertial navigation is particularly useful because of its self-contained nature. Space is another area in which inertial technology is widely employed. The Saturn rocket that lifted the U.S. astronauts on their mission to the moon carried an inertial system designed by the German team at Huntville, and the Apollo module was navigated by a system designed by Draper's Instrumentation Laboratory.[26] These main areas of application of inertial guidance and navigation technology are listed in figure 1.6.[27]

This breadth of applications is the cause and consequence of the development, since the 1950s, of a whole industrial sector, populated by large corporations and corporate divisions. Honeywell, Litton Industries, the Delco Electronics (formerly AC Spark Plug) Division of General Motors, and the Kearfott Division of General Precision (later part of the Singer Corporation and now of the Astronautics Corporation of America) are, in addition to Autonetics, the names that will

26. The latter was not purely inertial, but employed star-sightings as well.
27. More recently, two further applications—land navigation and the midcourse guidance of tactical missiles—have become important.

Main Applications of Inertial Guidance and Navigation

Missile Guidance

Submarine Navigation

Space

Military Aircraft

Long-range Civil Aircraft

Figure 1.6

appear most often on the pages that follow, but several other firms are involved in the United States and overseas. These firms have been important as manufacturers of strategic ballistic missile guidance technology, especially that designed by the Draper Laboratory, which has no large-scale production capabilities of its own. Occasionally, too, these firms appear as actors in the process of design, with all its political ramifications, as well as manufacture.

On Language and Terminology

How to speak and write about nuclear matters is controversial. Some argue that the language used should constantly remind the speaker and writer, listener, and reader of the human meaning of nuclear weaponry, of the fathomless death, pain, and suffering that a nuclear war would cause. Those directly involved in the "nuclear world," the strategists, planners, and technologists, use a bland and technical language, the precise function of which, its critics say, is to avoid such human reference.

There is a sense in which this book is not just a history, but also an anthropology: a journey among a tribe, which, to most of us, is unknown and alien. I hope that is one of its virtues. Accordingly, I have decided that the tone of its reference to nuclear weaponry shall be theirs.

Replicating all their linguistic habits, however, would make it incomprehensible. In particular, I have avoided acronyms as much as possible, employing them only when the desire to use direct quotation made it unavoidable, or where the acronym is at least as well known as the term abbreviated. Examples of the latter are the use of ICBM for intercontinental ballistic missile, and MIRV (multiple independently targetable reentry vehicle) to describe the system whereby a single missile can be equipped with several warheads, each capable of hitting a different target. Technical terms cannot wholly be avoided even if they can be spelled out fully. The index indicates in bold the location of the fullest account of each and should be consulted in case of difficulty. As far as possible, too, I have banished material of interest primarily to the specialist reader—the historically minded guidance engineer, the sociologist of science, and so on—into footnotes. The more general reader can happily ignore these copious notes, apart from when he or she wants to know my source for a particular claim.

A final remark is perhaps needed on the black box metaphor employed repeatedly in this book. I use it in two closely related ways. The first is to refer to a guidance or navigation system that does not require input from the outside world to operate. This, for example, is the sense in which the term was used in the first extant paper on the topic by inertial guidance pioneer Charles Stark Draper: "For the greatest practical effectiveness [a] navigational system must be able to function as a self-contained unit without dependence on information obtained by the exchange of natural or artificial radiation with the Earth or other aircraft. The ideal arrangement is a 'black box.'"[28] In the other meaning, a black box is opaque in a slightly different sense. It is a technical artifact—or, more loosely, any process or program—that is regarded as just performing its function, without any need for, or perhaps any possibility of, awareness of its internal workings on the part of users.

28. Draper and "Group B," *Fundamental Possibilities and Limitations of Navigation by Means of Inertial Space References* (February 1947), 29. This document is discussed in chapter 2.

2

Inventing a Black Box

What is it to invent a technology? The conventional wisdom of our culture has traditionally had a dominant answer. Invention was held to be a sudden flash of technical insight in the mind of an especially talented individual. This view underpinned admiration of the "genius" of the inventor. It legitimated the notion that private property in invention was possible, in the form of patents granted inventors to protect their rights to exploit their inventions. It delineated a task for the history of technology: to apportion credit to the correct individuals for the development of a technical field.

In the age of the large corporate research and development laboratory, this "individual genius" view of invention began to seem dated, though some aspects of it can still be found, for example in film portrayals of inventors. What largely replaced it was a view that routinized and depersonalized invention, making it the "application of science." Insight and creativity became the province of science. Given access to scientific knowledge, ample resources, and trained staff (geniuses were no longer needed), the results would flow. "Applied science" became synonymous with technology.

In this chapter, we shall see how inadequate both the "individual genius" and "applied science" views of technology are when confronted with the history of inertial guidance and navigation. To attribute the invention of the latter to a single individual or a single moment in time is quite impossible. And though it can now be presented as a straightforward deduction from physics, that was not how it came into being or for many years how it seemed. Indeed, one of the foremost physicists of the twentieth century ridiculed inertial navigation as an impossibility.

The most difficult point to grasp about the history of inertial navigation is precisely that its possibility was for a long time uncertain. Our modern certainty that it is possible means, to borrow a metaphor from a sociologist of scientific knowledge, Harry Collins, that inertial navigation is now like a ship in a bottle.[1] It takes a deliberate act of imagination to remember that the ship was once "just a bundle of sticks," a bundle that seemed impossible to transform into a schooner inside a bottle.

In this chapter we shall also see how varied were the "sticks" that made up the "ship in a bottle" with which we are dealing. Both elements highlighted by our conventional views of invention—insight and science—were necessary to the ensemble. But they were not sufficient, nor was the creation of inertial navigation a matter of engineering just metal, wires and equations. People had to be engineered, too—persuaded to suspend their doubts, induced to provide resources, trained and motivated to play their parts in a production process unprecedented in its demands. Successfully inventing the technology, turned out to be heterogeneous engineering, the engineering of the social as well as the physical world.[2]

Just as the materials that had to be put together were heterogeneous, so too were the people who put them together diverse. Because of the doubts about its possibility, inertial navigation lived for a long time on the border between serious technology and crankish fantasizing. Along with more sober engineers, a gentleman scientist, a polar explorer, an actor, and a German submarine commander figured in the story. But as black-box navigation became a reality, a different cast began to become involved—major state institutions and large corporations—attracted by the sense that here was a technology central to the pursuit of power.

1. H. M. Collins, "The Replication of Experiments in Physics," in Barry Barnes and David Edge, eds., *Science in Context: Readings in the Sociology of Science* (Milton Keynes, Bucks.: Open University Press, 1982), 94–116, quote on p. 94. First published as "The Seven Sexes: A Study in the Sociology of a Phenomenon, or the Replication of Experiments in Physics," *Sociology*, Vol. 9 (1975), 205–224.

2. See John Law, "Technology and Heterogeneous Engineering: the Case of the Portuguese Expansion," in Wiebe E. Bijker, Thomas P. Hughes, and Trevor Pinch, eds., *The Social Construction of Technological Systems: New Directions in the Sociology and History of Technology* (Cambridge, Mass.: MIT Press, 1987), 111–134.

Power—economic and military—has been the subtext to much of the history of navigation. The great voyages of exploration and the subsequent subjugation of the non-European world to the European maritime nations in a sense ultimately rested on the capacity to chart a course, find one's goal, and return safely.[3] State-sponsored research and development has perhaps had a longer history in navigation than in any other field.[4] As we shall see, the naval arms race before the First World War, the consequences of Germany's humiliation in 1918, and the aerial bombings of the Second World War all gave renewed impetus to the connection between navigation and power.

A black-box means of navigation that did not rely on external inputs was the ideal technology of power, since it would be invulnerable to the vagaries of the weather and the interference and deception of enemies. A "need" for inertial navigation could thus be said to have existed. But that "need" would have remained latent and without influence—like the "need" for an anti-gravity device, for example—if the sense of inertial navigation's impossibility had not been undermined. The ultimate task of the inventors of inertial navigation, in their heterogeneous engineering, was to harness their schemes for black boxes to the interests of state power. That, perhaps, is the central thread underlying the many twists and turns of our story.

Imagining a Black Box

Considerations of state power were many times removed from the episode in which an idea akin to modern inertial navigation first emerged.[5] Early in 1873, several issues of the recently

3. Ibid.; see also Carlo M. Cipolla, *Guns, Sails and Empires: Technological Innovation and the Early Phases of European Expansion, 1400–1700* (New York: Pantheon Books, 1965).

4. For example the 1714 British Act of Parliament "For Providing a Publick Reward for such Person or Persons as shall discover the Longitude at Sea." The story is told in E. G. R. Taylor, *The Haven-Finding Art: A History of Navigation from Odysseus to Captain Cook* (London: Hollis and Carter, 1971), chapter 11.

5. The work discussed here was brought to light by Claud Powell in "Inertial Navigation," *Journal of the Royal Institute of Navigation*, Vol. 30 (September 1977), 511–512.

founded British scientific weekly *Nature* contained letters discussing animal instincts. Correspondents included the eminent evolutionists Alfred Russel Wallace and Charles Darwin. In the background of the discussion was the question of whether it was "probable that instincts have any other origin than transmission by inheritance of acquisitions resulting from what we call individual experience."[6]

One of the "instincts" discussed was animal navigation, such as "the power many animals possess to find their way back over a road they have traveled blindfolded (shut up in a basket in a coach for example)."[7] Darwin suggested that there were human analogues in the way in which Siberian natives could keep a "true course towards a particular spot, whilst passing for a long distance through hummocky ice, with incessant changes of direction, and with no guide in the heavens or on the frozen sea".[8]

An Irish gentleman scientist, Joseph John Murphy,[9] wrote to the editor of *Nature* to take issue with this. "What man can do is to find the third side of a triangle after travelling the other two sides *with his eyes open*. Animals can do the same after travelling the two sides *with their eyes shut*," even when they had been asleep, he wrote. "There is nothing in man's mind similar to such a process as this. It can be made conceivable only by a mechanical analogy, if at all."[10]

Murphy's analogy was "a ball freely suspended from the roof of a railway carriage." The ball would remain stationery relative to the carriage when the latter is in uniform motion, but "every change in the velocity of the motion of the carriage, and of its direction, will give a shock of corresponding magnitude and direction to the ball." This mechanism could be connected to

6. The quotation is from the unsigned editorial, "Perception and Instinct in the Lower Animals," *Nature,* Vol. 7 (March 20, 1873), 378.

7. Alfred R. Wallace, "Inherited Feeling," *Nature,* Vol. 7 (February 20, 1873), 303.

8. Charles Darwin, "Origin of Certain Instincts," *Nature,* Vol. 7 (April 3, 1873), 418.

9. Murphy was born in Belfast in 1827, and published on a range of subjects from logic to geology and meteorology. See the entry for him in J. C. Poggendorf, *Biographisch-Literarisches Handwörterbuch zur Geschichte der Exacten Wissenschaften* Vol. 3(2) (1858–1883)(Leipzig: Barth, 1898), 951.

10. Joseph John Murphy, "Instinct: A Mechanical Analogy," *Nature,* Vol. 7 (April 24, 1873), 483.

a clock to register the times of the shocks, and a mathematical integrating device added so that the "position of the carriage, expressed in terms of distance and direction from the place from which it had set out" could simply be "read off."[11]

What Murphy did was thus to deploy some commonplace ideas of the physics of his day to explain how navigation without recourse to external references might be possible. According to that physics it was impossible in a self-contained system to detect position or uniform motion in a straight line. But accelerations or changes in orientation could be detected. The simplest case would be when the carriage starts from rest at a known position and travels along a known straight line. If one can measure its acceleration at all points in time, then, according to Galileo and Newton, one need only mathematically integrate that acceleration twice, with respect to time, in order to know the carriage's position.

Nothing in Murphy's presentation suggests any intention of constructing such a device, and, as far as I am aware, no one took up his ideas.[12] It was only in the twentieth century that the idea of navigation by the double integration of acceleration took on potential technological significance.

"Gyro Culture"

What gave it that significance was the development at the end of the nineteenth century and in the early twentieth century of a whole set of knowledge, skills, ideas and devices based around the gyroscope; what I refer to below as "gyro culture."

Although the device predated him—indeed it can be seen as a derivative of the age-old spinning top—the person who christened it in the 1850s was the French physicist Léon Foucault.[13]

11. Ibid.
12. The idea that animal navigation could be explained in this way resurfaced in the early 1960s, when real, not merely hypothetical, mechanical analogies could be cited. See John S. Barlow, "Inertial Navigation as a Basis for Animal Navigation," *Journal of Theoretical Biology*, Vol. 6 (1964), 76–117. I am grateful to Philip J. Klass for bringing this paper to my attention. Barlow cites Murphy's work (ibid., 94).
13. One early version was the "horizontal top" of the English seafarer and instrument maker, Captain John Serson, working in the 1740s. Serson intended his device—essentially a top with a flattened upper surface—to be used as an artificial horizon for taking astronomical sights. See James Short,

He conducted a series of experiments to demonstrate the rotation of the earth, the most famous of which employed a large swinging pendulum. As it swung, the plane of its swing gradually rotated, as can be seen in modern replicas, for example that in the National Museum of American History in Washington, D.C. Foucault argued that what was actually happening was that the plane of the pendulum's swing remained in the same orientation in "inertial space" (that is, with respect to the fixed stars) while the earth rotated. The consequent apparent movement allowed the earth's rotation to be "seen."

Foucault hoped that the "gyroscope" too would allow it to be seen: hence his name for the device, from the Greek *gyros* (a circle or rotation) and *skopeein* (to see). He knew from physics that a spinning body also tended to remain in fixed orientation in inertial space, unless disturbed. A rapidly spinning wheel, suspended in such a way that it has freedom to change orientation, would seem slowly to do so. But actually it would be the earth that was turning, while the "gyroscope" would keep pointing in the same direction with respect to the stars.[14]

"An Account of an Horizontal Top, Invented by Mr. Serson," *Philosophical Transactions*, Vol. 47 (1751/52), 352–353; "B.J.," "An Historical Account and Description of Mr Serson's whirling Horizontal Speculum, with its Use in Navigation," *The Gentleman's Magazine and Historical Chronical*, Vol. 24 (1754), 446–447; Anonymous, "The Speculum Described," *The Gentleman's Magazine and Historical Chronicle*, Vol. 24 (1754), 447–448; Anon., "The Manner of observing with the Speculum at Sea," *The Gentleman's Magazine and Historical Chronicle*, Vol. 24 (1754), 448–449. H. W. Sorg drew attention to Serson in his "From Serson to Draper—Two Centuries of Gyroscopic Development," *Navigation: Journal of the Institute of Navigation*, Vol. 23 (1976–77), 313–324. For such biographical details of Serson as are available, see E.G.R. Taylor, *The Mathematical Practitioners of Hanoverian England* (Cambridge: Cambridge University Press, 1966), 37, 190. A proposal similar to Serson's for a gyroscopic artificial horizon was apparently made to the Russian Academy of Sciences in 1759 by the celebrated Russian natural philosopher and physicist (also poet and dramatist) M. V. Lomonosov (1711–1765); see Sorg, "Serson to Draper," 313. According to B. M. Kedrov, "Lomonosov, Mikhail Vasilievich," in C. C. Gillispie, ed., *Dictionary of Scientific Biography*, Vol. 8 (New York: Scribner's, 1973), 470–471, "Interested in navigation especially of the northern seas, in 1759 [Lomonosov] invented a number of instruments for astronomy and navigation, including a self-recording compass, and reflected on the precise determination of a ship's route."

14. Léon Foucault, "Sur une nouvelle Démonstration expérimentale du Mouvement de la Terre, Fondée sur la Fixité du Plan de Rotation," (On a New Experimental Demonstration of the Earth's Motion, Based upon the

Foucault's actual experiment had limitations, however. He could only "spin-up" his gyroscope wheels mechanically and leave them to run. Even with a sophisticated suspension, friction meant that only eight to ten minutes of useful operation was possible, too little to allow the earth's slow rotation to manifest itself clearly.[15]

In the 1860s, however, the gyroscope was brought together with the emerging technology of electric motors, and this piece of not altogether successful science slowly became successful technology.[16] With energy supplied electrically, a gyro could spin more or less indefinitely. This made conceivable the practical use of a different effect described by Foucault. When he constrained the axis of rotation of his gyroscope to remain in the horizontal plane, he found, he said, that instead of remaining in its original orientation the axis "directed itself towards the north," just like a magnetic compass.[17] Making use of electric motors, a succession of prototype gyrocompasses was built from the 1860s onward.[18]

In practice, these devices were not rivals of the magnetic compass. The person who finally developed a practical gyrocompass was a man who seemed very unlikely to succeed in such a task. Dr. Hermann Franz Joseph Hubertus Maria Anschütz-Kaempfe (1872–1931) had a breadth of interests that would lead one to suspect him of being a dilettante: he first studied medicine and then art history, and became a polar explorer.[19]

But Anschütz-Kaempfe also had an obsession—a submarine voyage to the North Pole. He realized "that there was no scientific value merely in reaching the North Pole by sledge . . .

Fixity of the Plan de of Rotation) *Comptes Rendus Hebdomadaires des Séances de l'Académie des Sciences,* Vol. 35 (1852), 421–424.

15. Ibid., 423.

16. Sorg, "Serson to Draper," 316.

17. Léon Foucault, "Démonstration expérimentale du mouvement de la Terre. Addition aux Communications faites dans les précédentes Séances," (Experimental Demostration of the Earth's Motion. Supplement to the Papers read in Previous Meetings) *Comptes Rendus Hebdomadaires des Séances de l'Académie des Sciences,* Vol. 35 (1852), 469–470.

18. Sorg, "Serson to Draper."

19. Ibid., 130–31; see also Max Schuler, "Die geschichtliche Entwicklung des Kreiselkompasses in Deutschland," (The Historical Development of the Gyrocompass in Germany) *Zeitschrift des Vereines deutscher Ingenieure,* Vol. 104 (1962), 469–476, 593–599.

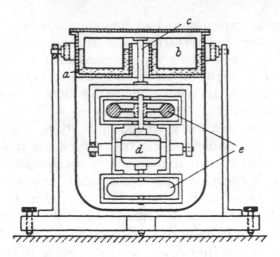

Figure 2.1
Anschütz-Kaempfe's First Gyrocompass
a = water; b = float; c = vertical support shaft; d = electric motor; e =
gyroscope.
Source: Max Schuler, "Die geschichtliche Entwicklung des Kreiselkompasses in Deutschland," *Zeitschrift des Vereines deutscher Ingenieure,* Vol. 104
(1962), 470.

one could never transport by sledge the necessary measuring
instruments. Thus he came upon the idea of diving under the
ice in a submarine. One could then blast a hole in the ice to
carry out the required measurements."[20] But as a member of
Anschütz-Kaempfe's audience at the Vienna Geographic Society put it, "That's all very fine, but how are you going to steer
your submarine?"[21] Not only were there specific problems of
the compass near the pole, but, more generally, in a submarine
a magnetic compass would be completely enclosed by the metal
hull and would also have to contend with the magnetic field
created by the submarine's motors.

Anschütz-Kaempfe turned to the gyroscope for a solution to
the problem of submarine navigation. As his work developed
(his first gyrocompass is shown in figure 2.1), his need for
resources grew, and visions of polar exploration faded as he
began the process of tying his technology to powerful interests.

20. Schuler, "Entwicklung des Kreiselkompasses," 469. It is interesting to
note that inertial navigation played an important role in the first actual
submarine voyage to the pole by the nuclear submarine USS *Nautilus.*
21. Ibid., 470.

At first he was unsuccessful. Although the German and British navies were locked in a competition that could be described as the first modern technological arms race, the German navy was not interested in his ideas.[22] "Distrusting a complete outsider," they refused to fund his work. But Anschütz-Kaempfe found a private backer, Friedrich Treitschke, who "unhesitatingly staked his whole fortune on the project," and in 1905 the two set up the firm Anschütz and Company in Kiel.[23]

Treitschke presumably reckoned that if the gyrocompass worked, the navy could not in the long run turn its back on it. All late nineteenth-century navies suffered navigational problems following the shift from wooden to iron and steel ships. The problem was understood to result from the way their metal structures interfered with the magnetic compass's operation. "Fixes" such as building the parts of the ship nearest the compass out of brass were tried but were seen as expensive and a "palliative and not a remedy."[24]

The German navy was finally persuaded to try the Anschütz-Kaempfe gyrocompass—indeed to try it on the most prestigious platform possible, the flagship *Deutschland*. The captain of the *Deutschland* reported positively on its performance in its 1908 test.[25] A new technology of power was in the making.

The trials on the *Deutschland* were no secret; they attracted wide interest, most consequentially that of the American inventor, Elmer Sperry. Despite the naval rivalry, Anschütz and Company was not prevented from selling abroad, and by 1910 the Americans had successfully tested, and the British were already buying, Anschütz gyrocompasses. Sperry threw his efforts into designing his own gyrocompass and by the autumn of 1910 had done so.[26] In April of that year he established the Sperry Gyroscope Company. The Anschütz and Sperry companies fought hard in the years 1910–1914 to dominate the world market for the new device. Anschütz and Company

22. See William H. McNeill, *The Pursuit of Power: Technology, Armed Force, and Society since A.D. 1000* (Oxford: Blackwell, 1983), chapter 8.
23. Schuler, "Entwicklung des Kreiselkompasses," 470.
24. Here I am following the analysis of Thomas Parke Hughes, *Elmer Sperry: Inventor and Engineer* (Baltimore and London: Johns Hopkins Press, 1971), chapter 5, quote on p. 130. Hughes's book is the best, indeed in a sense the only, historical account of gyro culture.
25. Hughes, *Sperry*, 131.
26. Ibid., 136.

raised a breach-of-patent action against Sperry, whom it accused of having stolen its gyrocompass work.[27] The court called as a witness a physicist who had become one of the leading technical experts in the Swiss patent office, Albert Einstein. Einstein's testimony may have helped Anschütz win their case; he even suggested at least one improvement to the gyrocompass,[28] and his contact with the gyrocompass was a resource he drew on in his study of physics.[29]

But neither Einstein nor the law could protect Anschütz's monopoly. After August 1914, the geopolitics of war replaced commercial competition and legal niceties. Sperry was left in effective command of the Allied market, though a British firm, S. G. Brown, introduced a third type of gyrocompass during the First World War.[30]

The gyrocompass was only one element—though perhaps the most central one—in a wave of practical applications of gyroscopic techniques. The gyroscope could also be used to

27. According to Anschütz-Kaempfe's cousin and mathematical collaborator Maximilian Schuler ("Entwicklung des Kreiselkompasses," 473), "When the first single gyro compass was being completed, there was announced in the, at the time, small workshop in Kiel, an elderly gentleman, Mr. E. Sperry from America who only spoke English . . . the gentlemen present took him for one of those eccentric cranks who are in the habit of making approaches to firms under the name of 'inventor.' In the course of the conversation the visitor was taken round the workshop, shown the new gyro compass and a great deal of explanation was given to him. Towards evening this gentleman left Kiel. . . . This was his first visit to Europe and when the gentlemen asked the seventy year old the purpose of his trip, he retorted laconically and tersely 'to make money.' " There is reason to doubt at least the details of Schuler's account, published half a century after the events described. At the time in question, Elmer Sperry was only in his forties, not seventy.

28. A means of centering magnetically the inner float. See K. Magnus, "Zur Geschichte der Anwendung von Kreiseln in Deutschland," (On the History of the Application of Gyroscopes in Germany) in V. D. Andreev et al., *Razvitie Mekhaniki Giroskopicheskikh i Inertsial'nykh Sistem* (The Development of the Mechanics of Gyroscopic and Inertial Systems) (Moscow: Nauka, 1973), 290–291.

29. Peter Gallison, *How Experiments End* (Chicago: University of Chicago Press, 1987), 37–39.

30. Sorg, "Serson to Draper," 318; John Perry and Sidney George Brown, "Improvements in or relating to Gyro-compasses" (U.K. Patent 124,529, March 31, 1919, filed August 3, 1916). For descriptions of the competing devices, see O. Martienssen, "Die Entwicklung des Kreiselkompasses," (The Development of the Gyrocompass) *Zeitschrift des Vereines deutscher Ingenieure*, Vol. 67 (1923), 182–187.

stabilize a vehicle—ship, automobile, monorail car, or torpedo—either directly or by connection to a control system. In the maritime sphere, where the most immediately important practical applications lay, stabilization was again a problem created by "progress." As Thomas P. Hughes puts it, "a ship under way with a broad exposure of canvas enjoyed a natural equilibrium," but an iron steamer did not.[31] Not only did this make life unpleasant for passengers, but it caused problems for accurate long-distance gunfire from warships.

These and other primarily naval applications of the gyroscope were reinforced by the growing importance of aircraft. The aircraft brought with it problems of stabilization, navigation, and, eventually, fire control. Although these problems were in a sense new, the gyro culture that was maturing initially around naval problems already possessed a repertoire of solutions that could be adapted, often with considerable ingenuity. The Sperry Gyroscope Company, in particular, moved quickly and successfully into the new field, developing a gyroscopic automatic aircraft stabilizer, and in 1916–1918 beginning work on an aircraft gyrocompass and artificial horizon.[32] Knowing the orientation of the craft was seen as both particularly important in the air and particularly difficult to achieve by visual means (e.g., when flying in cloud), and gyroscopic aircraft instruments to achieve this developed into a field of gyro culture as important as the naval applications.

Gyro culture was both international—in that its pioneers in different countries were well aware of each other's work—and national—in that it was increasingly seen as a "technology of power" too important to permit dependence on external sources, at least if a country had pretensions of playing a major world role. As late as 1930, "the only really operational gyrocompasses were German (Anschütz), American (Sperry) or British (Brown)." So the French government encouraged SAGEM (la Société d'Applications Générales d'Electricité et de Méchanique), which supplied auxiliary systems for the Anschütz and Sperry gyrocompasses used by the French navy, to develop its own gyrocompass capability.[33] SAGEM later was

31. Hughes, *Sperry*, 110.
32. Ibid., chapters 7 and 8.
33. P. Lloret, *La Navigation à travers les Âges* (Paris: SAGEM, n.d.), 26. The

to be the key resource in the guidance work necessary to develop a French nuclear missile force. In the Soviet Union, too, indigenous gyro culture—close to nonexistent, at least in the form of practical work, during the period of rebuilding in the 1920s[34]—developed fast from 1930 onward. An important role was played by the Soviet naval engineer Nikolai Nikolaevich Ostryakov (1904–1946). Before his premature death in an automobile accident, Ostryakov did much to establish gyro culture in the Soviet Union.[35]

Gyro culture had a strong theoretical strand to it. Gyroscopes behave counterintuitively. As Sperry told a 1912 audience at the U.S. Naval Academy:

If I impress a force on one end of the axis of a gyroscope it will resist this impressed force but will turn in a direction at right angles to the force impressed. This motion at right angles to the impressed forces is called "precession." It will be observed that the gyroscope does not resist any forces impressed by linear motion; nothing but angular motion causes precession, and nothing but the forces impressed by angular motion are resisted.[36]

Gyroscopic precession, shown visually in figure 2.2, found practical application in the gyrocompass and the other gyro-based technologies.[37] But it was also a source of theoretical

quotation is from the identically paginated English translation, *Navigation through the Ages,* also published by SAGEM.

34. V. D. Andreev, I. D. Blyumin, E. A. Devyanin and D. M. Klimov, "Obzor Razvitiya Teorii Giroskopicheskikh i Inertsial'nykh Navigatsionnykh Sistem (Survey of the Development of the Theory of Gyroscopic and Inertial Navigation Systems)," in Andreev et al., *Razvitie Mekhaniki,* 38.

35. A. Yu. Ishlinskii, "Inzhener Nikolai Nikolaevich Ostryakov," in Ishlinksii, *Mekhanika: Idei, Zadachi, Prilozheniya* (Mechanics: Ideas, Problems, Applications) (Moscow: Nauka, 1985), 562–571.

36. Hughes, *Sperry,* 109-110.

37. In a gyrocompass, for example, the gyroscope is suspended in such a way that it is in equilibrium only when its spin axis is parallel to the rotation axis of the earth (i.e., it is "pointing north"). If the spin axis were pointing in any other direction, the gyroscope would apparently begin to tilt as it sought to remain in fixed orientation in inertial space while the earth rotated. The suspension is so designed that gravity then creates a torque or "turning force" in the vertical plane. (A gyroscope suspended in such a way that it is sensitive to the effects of gravity or acceleration is sometimes known as "pendulous.") The gyroscope would then precess in the horizontal plane, that is start to turn toward the north. This process would continue until it "settled," pointing north. For simple but clear accounts, see Hughes, *Sperry,*

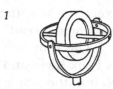

1

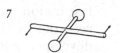

The Gyroscope.

2

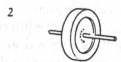

The wheel rotates rapidly.

3

Suppose the rim is split into segments.

4

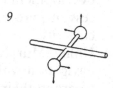

Attend to two segments.

5

The segments are connected to an axle.

6

As they turn about the axle, one moves up, the other down.

7

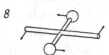

The axle is given a horizontal twist.

8

The segments then are pushed to the right and left.

9

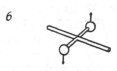

They thus have horizontal and vertical motion.

10

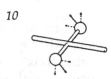

They therefore move diagonally.

11

The axle must therefore tilt.

12

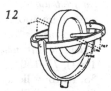

Thus, when the gyroscope is pushed, it tilts at right angles to the direction of the push.

Figure 2.2
Precession of a Gyroscope
Source: Thomas P. Hughes, *Elmer Sperry: Inventor and Engineer* (Baltimore: Johns Hopkins Press, 1971), 110–111, after G. P. Meredith, "Visual Education in the Air Age," *Aeronautics*, April 1945.

fascination—an attractive area for exercises and theoretical research in rigid body mechanics, which was already a well-developed and mature scientific field.[38] The four volumes by the well-known mathematical physicists Felix Klein and Arnold Sommerfeld, *Über die Theorie des Kreisels* (On the Theory of the Gyroscope)[39] are indicative of the extent of theoretical development by the turn of this century.

But there was much more to gyro culture than *die Theorie des Kreisels*. Building working gyroscopic systems required manual skills and "tacit knowledge" as well as theoretical understanding.[40] That practical foundation was to be central to the development of inertial navigation. Equally important, the success of gyro culture moved the gyroscope from the status of curiosity to that of a highly visible, prestigious technology. Unlike Murphy, the twentieth-century dreamers of black-box navigation without exception incorporated a gyroscope somewhere in their dreams.

Dreaming of Black Boxes—and Trying to Build One

Dreams they were, however. It is true that the gyrocompass and other gyroscopic devices such as the artificial horizon were "black-box" instruments. Because a gyrocompass told one where north was and an artificial horizon indicated the direction of the vertical, they permitted knowledge of orientation without resort to the stars or landmarks. But, though useful devices and important "technologies of power," they were far from being entire navigation systems. Nevertheless, the existence and success of gyroscopic devices was probably the key factor changing Murphy's nineteenth-century theoretical spec-

143, or the entry "Compass," in Anonymous, *How Things Work: The Universal Encyclopedia of Machines*, vol. 1 (London: Paladin, 1972), 552–553. Those who find them too simple might care to turn to Charles T. Davenport, "The Gyrocompass," in Paul H. Savet, ed., *Gyroscopes: Theory and Design, with Applications to Instrumentation, Guidance, and Control* (New York: McGraw-Hill, 1961), 77–130.

38. Its roots lie in Newton's physics, and important contributions were made, for example, by Leonhard Euler (1707–1783).

39. Leipzig: Teubner, 1897–1910.

40. See Michael Polanyi, *Personal Knowledge* (London: Routledge and Kegan Paul, 1958) and Harry M. Collins, "Tacit Knowledge and Scientific Networks," in Barnes and Edge, *Science in Context*, 44–64.

ulation into a number of twentieth-century technological knowledge claims.[41]

In the early years of the century at least five patents were awarded for black-box navigation systems. This work can be divided into two different categories. The first attempted to use the new black-box indicators of orientation as substitutes for star sightings—to pull the stars inside the black box. "Using these devices [gyroscopic indicators of orientation], knowing the point of departure of a ship, and taking into account with the help of a chronometer the earth's rotation in relation to the stars during the voyage, one can in principle work out the coordinates of the ship in the way it is done with the help of a sextant."[42]

The second strand could be seen as loosely analogous to Murphy's idea, in that it relied upon the double integration of acceleration. But the gyroscope was used to solve a problem never addressed by Murphy: how to keep the system for measuring acceleration in stable orientation. The first such patent was filed from Kiautschou, China, in 1905 by one Reinhard Wussow. His "apparatus for determining velocities and distances traveled" ("Apparat zur Bestimmung von Geschwindigkeiten und Wegelängen"[43]) used as an accelerometer a mass on top of a flexible support. The mass was restrained by horizontal springs and connected to two mechanical integrators,

41. I make the contrast with Murphy only to point out differences between the nineteenth- and twentieth-century contexts. As far as I am aware, none of the twentieth-century developers of inertial navigation knew of Murphy's letter.

42. Andreev et al., "Obzor Razvitiya," 55. This strand is exemplified by M. E. Carrie, "Gyroscopic Compass" (U.S. Patent 1,253,666, January 15, 1918, filed March 24, 1903); V. Alekseev (Russian Patent 28,451, April 30, 1916, on basis of 1911 application); F. R. Sweeny, "Geographic Position Indicator" (U.S. Patent 1,086,246, February 3, 1914, filed July 22, 1911); C. G. Abbot, "Gyroscopic Navigation Instrument" (U.S. Patent 1,501, 886, July 15, 1924, filed October 27, 1921). I take the first three of these references from Andreev et al., "Obzor Razvitiya," and have not been able to inspect the second and third of them.

43. German patent, December 11, 1906, filed September 19, 1905. I have been unable to discover any biographical details for Wussow. I owe the reference to L. I. Tkachev, *Sistemy Inertsial'noy Orientirovki* (Inertial Navigation Systems)(Moscow: Moskovskiy Energeticheskiy Institut, 1973). See pp. 30–35 of the Joint Publications Research Service translation (Arlington, Va.: JPRS, July 20, 1976).

with the whole device held level by a gyroscope. In 1930, the French space-flight enthusiast R. Esnault-Pelterie also argued for black-box navigation by double integration of acceleration and noted the possible use of a gyroscopic system to stabilize the accelerometers.[44]

It is not possible from the patent record to say whether any of these early proposals "worked," in the eyes of their inventors or anyone else; indeed in only one case can we be sure that the design did not merely remain on paper.[45] Furthermore, this early-twentieth-century work was essentially sporadic and noncumulative. Practical engineers concentrated on improving

44. R. Esnault-Pelterie, *L'Astronautique* (Paris: Lahure, 1930), 173–189. He proposed (ibid., 176–179) as a possible accelerometer a device similar to what later became known as a PIGA, pendulous integrating gyro accelerometer. He worried about the problem of keeping the accelerometers in constant orientation, writing (ibid., 187) that "the gyroscope naturally offers the means [to achieve constant orientation] but at the price of enormous complication."

45. Some parts of a prototype of Abbot's "Gyroscopic Navigation Instrument" were still extant when Frithiof V. Johnson joined the Aeronautics and Marine Department of the General Electric Company at Schenectady, New York, in 1931 (F. V. Johnson to Edwin Layton, February 25, 1966). Abbot, Secretary of the Smithsonian Institution of Washington, D.C., had during the First World War asked the General Electric Company to develop his invention as a "war measure" (D. B. Rushmore, Power and Engineering Department of General Electric to C. E. Tuller, Patents Department, January 15, 1920). This latter letter goes on to describe the device:

"The purpose of this instrument is to give both direction and geographical location. Direction is given by direct readings similar to those of the magnetic compass, but with reference to the true worth [sic; obviously 'north' is intended] instead of the magnetic worth. Readings for location must be corrected for time. The instrument consists of a hollow sphere in flotation in a suitable liquid, within a containing vessel, with defining bearings that are practically frictionless for fixing its location, and with means ['three motor driven gyroscopes with axes mutually at right angles to each other'] for maintaining a fixed orientation in space which is independent of the rotation of the earth, or changes in its position on the earth's surface. If a map of the earth's surface were drawn upon the surface of such a sphere, and if the earth itself were not in rotation, the orientation of the map on the sphere corresponding with the earth itself, then its location on the earth's surface would be read at that point on the map which is uppermost, in the vertical axis through the center of the sphere . . . With the earth in rotation, the correct latitude is still read at the top of the sphere. . . . [The] correct longitude at any time can be determined from the reading taken at the time corrected by 15 degrees per hour for the number of hours intervening between the base time at which the orientation of the sphere is correct and the time at which the reading is made. . . . The reading can be taken through a small peephole in the top of the containing reservoir."

Frithiof V. Johnson drew the attention of my colleague Wolfgang Rüdig to these documents (interview, Binghampton, N.Y., January 17, 1987). They are in the Special Collections, Freiberger Library, Case Western Reserve University, where they were deposited by Johnson.

proven instruments such as the gyrocompass, leaving black-box navigation to what appears to have been a periphery of largely isolated "dreamers."

By the 1930s, however, there were the beginnings of cumulative work. The year 1932 saw the first Soviet patent application for an inertial navigation system consisting of two accelerometers held horizontal by gyroscopes.[46] One of its authors developed another system to correct what he saw as a flaw in the original design.[47] A leading member of Soviet gyro culture, B. V. Bulgakov, then teaching at Moscow State University, also sought to improve the proposed device, and wrote an important paper on its theory. Like the original patent, his paper was not published at the time,[48] but a group of engineers led by Bulgakov went on in 1939 to build such a system, apparently with the intention that it be used for marine navigation.

46. For the history of Soviet work in this period, see Andreev et al., "Obzor Razvitiya"; A. Yu. Ishlinskii, "K Istorii Giroskopicheskikh i Inertsial'nykh Sistem (On the History of Gyroscopic and Inertial Systems)," "Istoki Prostranstvennoi Inertsial'noi Navigatsii (Origins of Inertial Navigation in Space)," and "O Giroskopakh i ikh Primeneniyakh (On Gyroscopes and their Applications)" in Ishlinskii, *Mekhanika,* 288–369, 369–375, 375–379; and Ishlinskii, *Klassicheskaya Mekhanika i Sily Inertsii* (Classical Mechanics and Forces of Inertia) (Moscow: Nauka, 1987), especially pp. 124–157.
The patent was E. B. Levental' and L. M. Kofman, "Navigatsionnyi Pribor dlya Registratsii Proidennogo Puti i Skorosti (Navigational Apparatus for the Registration of Travel Path and Speed)," Soviet Patent 184,465, on application no. 120951/40–23 of December 26, 1932. I have not seen the original application, only the version published in the *Byulleten' Izobretenii,* No. 15 (1966). I owe the reference to Andreev et al., "Obzor Razvitiya," 56–57 and 68. Levental' and Kofman identify what later became called the "Schuler period" of an inertial system (84.4 minutes).
In 1934 a group of Soviet aeronautical engineers used gyroscopes and accelerometers to track the motion of an aircraft entering and leaving spin. See V. S. Vedrov, S. A. Korovitskii, and Yu. K. Smankevich, "Issledovanie Shtopora Camoleta R-5 v Polete," (Investigation of the Spin of the Aircraft R-5 in Flight) *Trudy Tsentral'nogo Aero-Gidrodinamicheskogo Instituta,* 228 (1935), 3–49. I owe the reference to Ishlinskii, *Mekhanika,* 138. As Ishlinskii points out, this was a much easier problem than the general problem of inertial navigation, because of the short period of time for which the system had to operate.
47. Kofman's "universal orientator," described in Andreev et al., "Obzor Razvitiya," 57. According to the latter, Kofman realized that the early design would be subject to undamped oscillations.
48. B. V. Bulgakov, "Teoriya odnoi Giroskopicheskoi Sistemy Navigatsii (Theory of a Gyroscopic Navigation System)," *Mekhanika Tverdogo Tela*

According to the man who became the leading Soviet theorist of inertial navigation, Academician A. Yu. Ishlinskii, who collaborated with Bulgakov around this period, Bulgakov's system was "not perfect": "The technology was not developed enough."[49] From 1943 onward Bulgakov's pupil, L. I. Tkachev, in the Automation Department of the Moscow Power Engineering Institute, worked to develop the system.[50] Though, according to Tkachev, the Institute "had the necessary base for advanced electrical automation," there is again no record of a working system being produced.[51]

By the 1940s, however, the work done in the Soviet Union and elsewhere had been eclipsed by that carried out in Germany. That country's existing strength in gyro culture had come together with its peculiar interwar situation to provide uniquely favorable circumstances for the development of black-box navigation, and, especially, missile guidance.

The Black Box and the Third Reich

Germany's defeat in World War I increased rather than decreased the strategic significance for the German state of gyro culture. The Treaty of Versailles attempted to place strict limits on German rearmament but in doing so created a climate in which resources were channeled into technological innovation to compensate for or circumvent these limits.

The most dramatic innovation was the guided ballistic missile. During the 1920s and 1930s, a remarkable "technological social movement" emerged in many countries around amateur enthusiasm for rocketry and space travel.[52] A certain aura of science fiction, crankishness and amateurism surrounded this

Vol. 4(3)(1969), 14–23; in the English language series of *Mekhanika Tverdogo Tela* it is paginated from 12 to 19. Again, I have not see the 1938 original, only the 1969 published version. There is a brief biographical entry on Bulgakov in A. M. Prokhorov, ed., *Great Soviet Encyclopedia*, third edition, Vol. 4 (New York: Macmillan 1974), 171.

49. Interview with A. Yu. Ishlinskii, Moscow, May 23, 1988.

50. I have not been able to find a published biography of Tkachev; I learned of Tkachev's relation to Bulgakov in my interview with Ishlinskii.

51. Tkachev, *Sistemy Inertsial'noy Orientirovki*, JPRS translation, p.102. This source contains a large extract from Tkachev's unpublished 1944 thesis on "three-dimensional navigation in blind flight with pendulum-gyro systems."

52. Frank H. Winter, *Prelude to the Space Age: The Rocket Societies, 1924–1940* (Washington, D.C.: Smithsonian Institution Press, 1983)

work. Even at the end of the 1930s, when Dr. Karl W. Fieber of Siemens, a well-established and "hardheaded" firm with gyro culture interests, was given the task of helping the German rocket pioneers with guidance and control problems, he was told by his manager that the ballistic missile work was "eine Scharlatanerie (charlatanry)."[53]

The desire to escape the constraints of Versailles outweighed suspicion of crankishness in the eyes of the German army, however, and unlike its counterparts in other countries it was prepared to give serious support to the rocket pioneers.[54] So the 1930s saw the first major guided ballistic missile development in Germany, which led to the V-2, in the second World War, and a major new application area for gyro culture was opened up.

Less dramatic, but also of considerable long-term importance, was the German navy's response to the Versailles limitation of the size of its vessels to 10,000 tons. It sought to substitute quality—speed, maneuverability, and accurate gunnery—for quantity.[55] Because the consequences of this provided one important part of the gyro culture base of the missile guidance work, I shall examine them before turning to the missile program.

To help in the search for accuracy, the German navy in 1926 established a secret gyro firm. It covertly bought an Amsterdam aerial survey firm, Aerogeodetic, Inc. (Naamlooze Vennootschap Aerogeodetic), and established a Berlin branch of it as cover for the gyro work.[56] Kreiselgeräte ("gyro instruments"), as the firm became known when, following Hitler's rise to power, it no longer needed to operate in secret, played

53. K. W. Fieber, "Zur Geschichte der Deutschen Raketensteuerung (On the History of German Missile Guidance)," typescript, 1965, 11fn. Dr. Fieber kindly provided my colleague Dr. Wolfgang Rüdig with a copy of this document.
54. See Major-General Walter Dornberger, *V 2* (London: Hurst and Blackett, 1954), p. 31. Dornberger joined the Ballistic Council of the Army Weapons Department in 1930 (ibid.).
55. See Johannes G. Gievers, "Erinnerungen an Kreiselgeräte (Recollections of Kreiselgeräte)," *Jahrbuch der Deutschen Gesellschaft für Luft- und Raumfahrt*, 1971, pp. 263–291 (English translation, entitled "History of the Gyroscope," by the Defense Intelligence Agency, Washington, D.C., December, 28, 1981, translation no. LN 040–82).
56. Ibid., English translation, 2.

a central role in German inertial navigation and guidance in the following two decades.

Kreiselgeräte set to work on gyroscopic gun control systems, but its Technical Director, Johannes Maria Boykow (1879–1935), had a vision more radical than the piecemeal improvement of these systems. He was a man of rather the same stamp as Anschütz-Kaempfe, a visionary rather than a practical implementer, with an idiosyncratic career. He had been an officer in the Austrian navy, a professional actor, and a pilot in the First World War.[57] As well as directing Kreiselgeräte, Boykow remained in charge of his own firm, Meßgeräte Boykow, which was developing a gyroscopic autopilot.[58] But Boykow's real dream was black-box navigation, about which he had first published during his acting days in 1911.[59]

During the 1920s and early 1930s, Boykow worked on and patented what he called an *Übergrundkompaß* (overground compass) and the closely related *Wegemeßer* (track meter). His goal was a device that would permit determination "with respect to length and direction, the distance covered by any craft from the starting point, and thus its position at any moment," and would do this, without "observations of objects outside the craft in question," by the double integration of the output of two accelerometers on a platform held in stable orientation by gyroscopes.[60]

Unlike earlier figures, Boykow had two key resources. First was a solid base in gyro culture, in particular a skilled technical staff, notably his assistant (later Director of the Development

---

57. Magnus, "Anwendung von Kreiseln," 299–300.

58. Gievers, "Erinnerungen," English translation, 3.

59. H. [sic] Boykow, "Navigation mittels Derivators," (Navigation by means of Derivators) *Zeitschrift für Flugtechnik und Motorluftschiffahrt*, No. 11 (1911), 144–47. There is an oddity in the author's initial given here, but the author is said to be a former Naval officer (p. 144), and I follow Magnus, "Anwendung von Kreiseln," p. 306, in attributing it to J. M. Boykow.

60. The patents are J. M. Boykow, "Vorrichtung zum Anziegen der wahren Fahrt eines Fahrzeuges über Grund" (Device for Indicating the True Course of a Vehicle over the Ground) (German Patent 513,546, November 13, 1930, filed July 7, 1926) and Boykow, "Instrument for Indicating Navigational Factors" (U.S. Patent 2,109,283, February 22, 1938, filed January 10, 1934). The quotation is from the latter, p. 1. There is also a German version of it: "Einrichtung zum Messen von Wegstrecken" (German Patent 661,822, June 2, 1938, filed January 11, 1935).

Department of Kreiselgeräte), Johannes Gievers.[61] Second, Boykow managed to recruit to his dream an important patron, Captain Karl Otto Altvater, a First World War submarine commander, who by the 1920s was attached the German navy's armaments office. Altvater had been instrumental in setting up the secret gyro firm, and, as he shared Boykow's enthusiasm, it was probably he who, in the face of skepticism, kept navy funds flowing for the development of an inertial navigator.[62]

Boykow was unable to clearly communicate the detail of his ideas to others and never harnessed his resources and vision into a device that could be accepted as "working." It was probably a prototype *Übergrundkompaß* whose testing was recorded by Ernst Steinhoff:

In 1930 an attempt was made to navigate an aircraft equipped with a gyrostabilized platform and mechanically integrating accelerometers mounted on it. The flight, which departed from Berlin-Aldersdorf, was discontinued and the attempt terminated after three hours of flying time when the aircraft equipment indicated a position somewhere in Australia, while visual observations confirmed the aircraft position to be at the western border of Germany near Holland.[63]

Boykow's dream survived practical failure and lived beyond his death in 1935, in large part through the continuing activity of Captain Altvater. In 1930, Altvater joined Siemens to head its Aeronautical Division. Although originally an electrical firm, Siemens had had gyro culture interests since the end of the

61. The major source of primary material on Gievers is his file at the U.K. Admiralty Compass Observatory in Slough. I am grateful to Mr. M. W. Willcocks of the Observatory for a copy of this material. See also G. Klein and B. Stieler, "Contributions of the late Dr. Johannes Gievers to Inertial Technology—Some Aspects on [sic] the History of Inertial Navigation," *Ortnung und Navigation*, Vol. 3 (1979), 436–443.

62. G. Klein, interview by Wolfgang Rüdig; Johannes Gievers, letter to J. N. Thiry, February 25, 1977, a copy of which Mr. Thiry kindly provided to Wolfgang Rüdig.

63. Ernst A. Steinhoff, "Development of the German A-4 Guidance and Control System 1939–1945: A Memoir," in R. Cargill Hall, ed., *Essays on the History of Rocketry and Astronautics: Proceedings of the Third through Fifth History Symposia of the International Academy of Astronautics* (Washington, D.C.: NASA Science and Technology Information Office, 1977; NASA Conference Publication 2014), Vol. 2, 203–215, quote on p. 209. Steinhoff does not say what system was being tested, but other passages of his paper suggest it must have been Boykow's.

nineteenth century and was becoming a major supplier of aircraft instruments. Even before Boykow's death, his firm Meß-geräte Boykow was in financial difficulties, and in 1931 Siemens took it over.

Altvater persuaded Siemens, despite much skepticism within the firm, to buy, at considerable expense, Boykow's patents on his inertial navigation system.[64] "Turn this pipedream into a serious project," Siemens staff were instructed.[65] Without the dreamer, however, the vision languished. Johannes Thiry, who worked for Siemens from 1920 to 1945, recalls seeing a prototype sitting neglected in a Siemens laboratory around 1933.[66]

Within Siemens it was revived by another proponent of black-box navigation, Dr. Siegfried Reisch. As early as the 1920s, Reisch apppears to have been one of those who dreamed of "absolute navigation," and when he joined Altvater's Aeronautical Division of Siemens in 1940, his enthusiasm revived Altvater's interest.[67] Despite the pressures of "total war," and despite continuing deep skepticism within Siemens, Altvater was able to create a niche for Reisch to pursue his research. A special laboratory was set up for him in a villa in Vienna. By 1944 Reisch was ready to start trying to build a black-box navigator, "but this work was cut short when the tide turned against Germany and he was called on to do other work."[68] Reisch fell into the hands of the British, and though he was set to work at the Royal Aircraft Establishment at Farnborough, he did not enjoy the confidence of the technical specialists there and seems to have produced nothing concrete.[69]

64. Klein interview.
65. Gievers, "Erinnerungen," English translation, 4.
66. Johannes N. Thiry, interview with Wolfgang Rüdig, Traverse City, Michigan, July 24, 1988.
67. Gert Zoege von Manteuffel, "Über den Ursprung der Trägheitsortung" (On the Origin of Inertial Navigation), *Jahrbuch der Deutsche Wissenschaftliche Gesellschaft für Luftfahrt*, 1959, 225–230, quote on p. 225; Dr. K. W. Fieber, interview with Wolfgang Rüdig.
68. H. Hellman, "The Development of Inertial Navigation," *Navigation: Journal of the Institute of Navigation*, Vol. 9, No. 2 (Summer 1962), 90.
69. See S. Reisch, "Reports on Inertia [sic] Navigation Nr. 6" and "Reports on Inertia Navigation Nr. 9" (Royal Aircraft Establishment Technical Memos Nos. IAP 198 and 199, May 1947), now in the file AVIA 54/1869 at the U.K. Public Record Office. I am grateful to S. G. Smith, Superintendent RN1 Division, Royal Aircraft Establishment, who reviewed this file for declassifi-

A similar fate awaited the attempt within Kreiselgeräte to take Boykow's *Übergrundkompaß* further. When Siemens took over Meßgeräte Boykow, Boykow's assistant and then successor at Kreiselgeräte, Johannes Gievers, was asked to evaluate the *Übergrundkompaß*. He concluded that a workable device needed components much more accurate than those available in the early 1930s.[70] But one of his main tasks at Kreiselgeräte was the development of such components, and he kept the *Übergrundkompaß* "on the back burner," producing his own designs for it in 1940 and 1942.[71] However, the more immediate demands of the war—notably the need for a small marine gyrocompass for use in miniature submarines and a gyrocompass for use on the Russian front[72]—took precedence in the application of gyroscopes and accelerometers.

After Germany's defeat Gievers worked for some time with the Soviets, and then with the British. Both seem to have made use of his work on inertial sensors.[73] Gievers hoped that the British would also provide him with resources to develop an inertial navigation system, which he estimated was only three years from completion. He was given a laboratory at the Naval Laboratories in Teddington and ten former colleagues to help. Again, though, the British specialists, in this case at the Admiralty Compass Observatory, were skeptical,[74] and "so soon after the war conditions in England were not very good. Along with this, the whole team was suffering from homesickness."[75] In 1950 the British allowed Gievers and his team to return to Germany, where they dispersed, Gievers leaving soon for the United States, where he would eventually be appointed Chief Scientist of the Missile Division of the Chrysler Corporation.

cation, for drawing my attention to it, and to Wolfgang Rüdig for arranging for a copy of the relevant material in it to be made.

70. Gievers, "Erinnerungen," English translation, 4.

71. Klein and Stieler, "Gievers."

72. Commander W. E. May, W. G. Heatly and H. C. Wassell, Draft "Report of Visit to B.N.G.M., Minden, June 12th, 13th, 1946," June 21, 1946, in Admiralty Compass Observatory File on Gievers.

73. See chapter 6 below for the Soviet use. E. Hoy to B. Stieler, September 11, 1979, in Admiralty Compass Observatory file on Gievers.

74. W.F. Rawlinson et al., "Report on Discussions at A.C.O. with Dr. Gievers regarding his proposed Gyro-Compass and Stable Element," August 24, 1946, in Admiralty Compass Observatory file on Gievers.

75. Gievers, "Erinnerungen," English translation, p. 28.

Guiding the V-2

The work of Reisch and Gievers on inertial navigation was performed in secret and did not become public for many years after the war.[76] Though the black-box navigation ideas of the early dreamers and of Boykow had entered the patent record, they seem to have been either forgotten or dismissed. By far the main route by which the work in Germany in the 1920s, 1930s and 1940s influenced later developments was in its incorporation into the V-2 ballistic missile, a system that achieved fame and visibility beyond that of almost any military technology of the period except the atomic bomb.[77]

Though the initiation of the ballistic missile project can be traced to the effects of the Treaty of Versailles, the project's initial momentum was not sufficient to carry it to a successful conclusion in the face of its growing demands for resources, Allied attacks, and the particular circumstances of wartime Germany. Hitler's support was not automatic, especially after he had dreamed "that no A-4 (V-2) will ever reach England."

76. The first published discussions were von Manteuffel's 1959 paper on Reisch, "Ursprung der Trägheitsortung" and Gievers's 1971 "Erinnerungen." Dr. E. M. Fischel, who had assessed the work of Reisch for the Luftwaffe, came to the United States after the war with the V-2 rocket technologists. According to Hellman ("Inertial Navigation," p. 90), theoretical work was done by this group at Fort Bliss (where it was located before the move to Redstone Arsenal), and in 1946–1947 both government agencies and private companies, including North American Aviation, were briefed about it.

77. The V-1's guidance system was a combination of existing technologies similar to one designed during the previous world war by Elmer Sperry and his son Lawrence in their attempts to develop a flying bomb they called the "aerial torpedo." See Hughes, *Sperry*, 243–274, especially 263–264 on the "aerial torpedo's" guidance system. Because both the "aerial torpedo" and the V-1 were conceived of essentially as pilotless aircraft, it must have seemed natural to adapt existing aircraft technology to the task of cruise missile guidance. Both were stabilized gyroscopically, while altitude was regulated by a barometer. Both were designed to fly straight and level toward the target (in the V-1 assisted by a magnetic compass), until in the case of the aerial torpedo, "a revolution counter connected to the main propellor shaft of the engine initiated the torpedo's dive after the predetermined distance (number of revolutions) had been flown" (ibid., 264). The V-1's dive was initiated by a similar propeller device, though in its case it was auxilliary, the missile being powered by a pulsejet. On V-1 guidance, see Kenneth P. Werrell, *The Evolution of the Cruise Missile* (Maxwell Air Force Base, Alabama: Air University Press, 1985), p. 43.

He could not be persuaded to visit the missile center that was established near the little fishing village of Peenemünde on the island of Usedom. Instead, successful missile flights were filmed, a rousing commentary was provided by the rocket team's leader, Wernher von Braun, and the resultant movies were shown to Hitler.[78] The Allied bombing campaign meant shortages of all kinds of components and placed missile development and production facilities in great danger. In a network of tunnels under Kohnstein Mountain in the Harz Mountains, an underground production facility, known as the Mittelwerk, was created, which at its peak employed over 8,000 people. Many came from the nearby concentration camp, Dora, and lived, worked, and died in atrocious conditions.[79]

In the 1930s these were problems for the future, though, and in any case the technical specialists who worked with von Braun focused their attention on the narrower but crucial matter of getting a missile to fly successfully. Central problems were control, in the sense of achieving stable flight, and guidance, in the sense of directing the rocket to its target. Early efforts by the German team, by early Soviet workers, and by American rocket pioneer Robert Goddard addressed the first alone. Thus the two A-2 test rockets fired from Borkum Island in the North Sea in 1934 were stabilized "by brute force" using a large gyro wheel in the center of the rocket.[80]

78. Dornberger, *V-2*, 93 and 100–103.

79. Frederick I. Ordway III and Mitchell R. Sharpe, *The Rocket Team: From the V-2 to the Saturn Rocket* (Cambridge, Mass.: MIT Press, 1982), chapter 5. For an account of Dora by an inmate, and a condemnation of its omission from much of the historiography of the V-2, see Jean Michel, *Dora* (New York: Holt, Rinehart and Winston, 1980).

80. The A-1 design never flew. The intention had been to stabilize by "brute force" like the A-2, but with the wheel in the nose, not the center. See Ordway and Sharpe, *The Rocket Team*, 23. For A-2 guidance, and for these developments more generally, see F.K. Mueller, *A History of Inertial Guidance* (Redstone Arsenal, Alabama: Army Ballistic Missile Agency, n.d. but c. 1960); Mueller's history has now been reprinted in the *Journal of the British Interplanetary Society* , Vol. 38 (1985), 180–92. Goddard used gyroscopic stabilization, but of a more sophisticated kind involving feedback from a gyro to jet-vanes: see ibid., p. 4, and R. H. Goddard, "Mechanism for Directing Flight" (U.S. Patent 1,879,187, September 27, 1932, filed February 7, 1931). He did not, however, go on to develop a guidance system analogous to that used for the V-2. The 1930s Soviet experimental rockets R-05 and R-06 also contained a gyroscopic stabilization system designed by P.I. Ivanov (Winter, *Prelude*, 69 and 122).

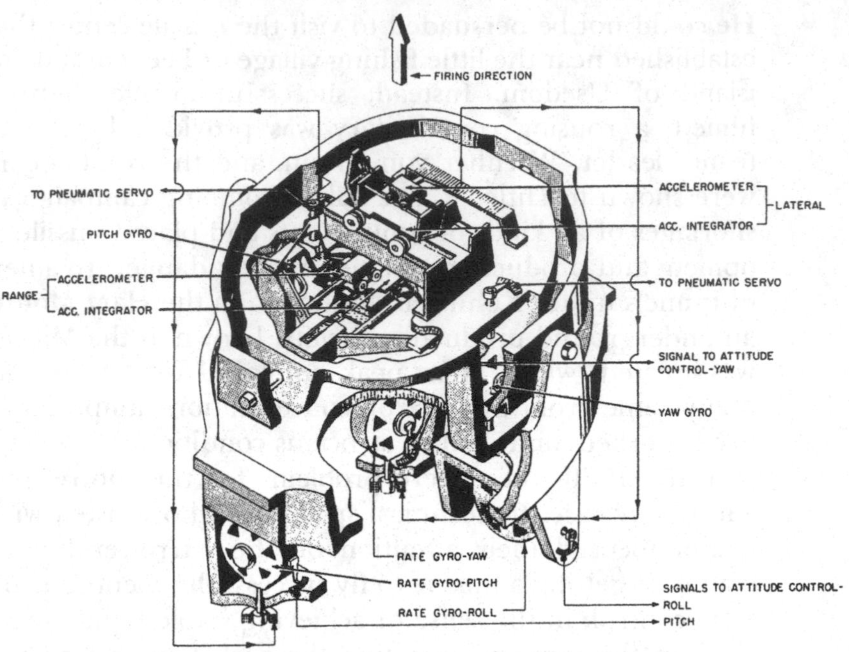

Figure 2.3
Kreiselgeräte Stable Platform for Control of A-3 Rocket
Source: Fritz K. Mueller, *A History of Inertial Guidance* (Redstone Arsenal, Ala.: Army Ballistic Missile Agency, n.d.), 5. For the meaning of "pitch," "roll," and "yaw," see figure 2.4.

But it was already becoming clear by then that the project was going to lead not just to a rocket but to a missile, and so guidance as well as control had to be addressed. In 1934 von Braun approached Kreiselgeräte for technical assistance. Not surprisingly, Boykow suggested that the rockets be guided by the black-box navigation system he had been developing, and Fritz Mueller, who had worked at Kreiselgeräte since the end of 1930, was seconded to assist the rocket team.[81]

By 1937 Boykow's proposed system, shown in figure 2.3, was ready for incorporation in tests of the A-3 rocket. Three gyroscopes provided input into the control system that attempted to stabilize the rocket and, through a pneumatic feedback mechanism, held a platform in fixed orientation. Two acceler-

81. Interview with Dr. Fritz K. Mueller, Huntsville, Alabama, March 23, 1985.

ometers were mounted on the platform. They consisted of little "wagons" restrained by springs and worked by measurement of the movement of the wagons under the influence of acceleration. Their outputs, integrated by a simple analog integrator, were fed into the control system to correct deviations from the desired vertical launch direction.

In hindsight, this could be seen as an inertial guidance system, with a stable platform, accelerometers, and gyroscopes, as described in chapter 1. But at the time it was scarcely that. The goal of its use in the A-3 was simply stable vertical flight, rather than any more ambitious task of guidance, but even that could not be achieved. The five A-3 tests were failures.[82] The failure could be blamed not on the platform but on the inadequacy of the control forces generated by vanes placed in the path of the rocket exhaust. But, in the words of Kreiselgeräte's Mueller, there was also a feeling that though the "principle of [Boykow's platform] was sound," it was "too complex to be practical with the instruments which could be built at that time."[83]

Von Braun's team threw themselves into the search for alternatives, developing their own in-house guidance expertise and drawing in other parts of German gyro culture to supplement the work of Kreiselgeräte. To achieve controlled, stable flight of their operational missile, the V-2,[84] they drew on the experience of Siemens in aircraft instrumentation and control. The seconded Siemens engineer, Karl Fieber, adapted aircraft gyroscopes in the successful construction of a control system known as *Vertikantsteuerung*, which is shown in figure 2.4.[85]

82. Mueller, *History,* 4.
83. Ibid.; letter to author from Dr. Fritz K. Mueller, December 29, 1988.
84. A different, but also successful line of development was followed for the further test rocket, the A-5. "Kreiselgeräte developed the Sg 52 and Sg 64 stable platforms and control systems without accelerometers. Askania developed a simplified guidance and control system, and wind tunnel tests helped to improve the effectiveness of the jet-vane control. The testing of the A-5 accordingly turned out to be successful" (Mueller letter).
85. On the role of Siemens, see Stefan Karner, "Die Steuerung der V2: Zum Anteil der Firma Siemens an der Entwicklung der ersten selbgesteuerten Grossrakete (The Guidance of the V2: On the Role of the Siemens Company in the Development of the first large Guided Missile)," *Technikgeschichte*, Vol. 46 (1979), 45–66. In *Vertikantsteuerung* the missile would be launched vertically, with any tendency to roll or yaw controlled by a gyro that those involved referred to as the *Vertikant*. A second control gyro was preprogrammed in such a way as to cause the missile then to pitch gradually downward until it

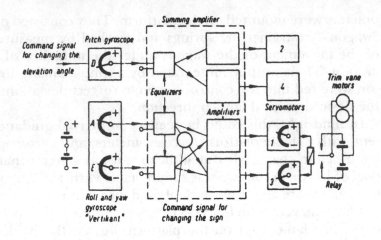

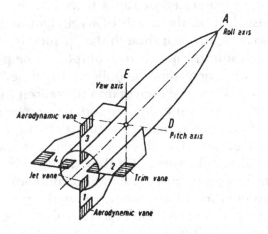

Figure 2.4
Control System of the V-2
Source: Otto Müller, "The Control System of the V-2," in Th. Benecke
and A. W. Quick, eds., *History of German Guided Missiles Development* (Brunswick: Appelhans, 1957), 103.

Two alternative technical approaches to guiding a stabilized missile were pursued. One, largely the in-house effort of the rocket team at Peenemünde, abandoned the black-box principle of Boykow's device and sought to guide missiles by radio. The other, located primarily in external gyro culture, held to the idea of guidance without use of external inputs but sought to achieve this in a way much simpler than the Kreiselgeräte system. Both approaches attempted to do the same thing: correct deviations of the missile from the direction of the target and thus stop its warhead from missing the target by falling either to its left or to its right, and shut off the rocket motor once the missile reached a speed such that it would neither fall short of the target nor overshoot it.

The radio method employed two transmitting antennas placed some distance behind the launch site, symmetrically on either side of the line running through the launch point on to the target. The signals received by the missile from them enabled detection and correction of deviation from that line. The Doppler effect—the shift in frequency when source and receiver are in motion relative to one another—was used to detect missile velocity.[86]

This radio guidance worked; indeed it was believed to be the most accurate of the range of available techniques. But, despite being an in-house development, it was handicapped by not being a black box. "There was always a fear that the Allies would jam the control signal. It was also feared that the additional vehicles required for radio equipment would present a more appealing target for Allied air attack. Finally, there was the logistical burden of maintaining additional equipment and vehicles."[87]

reached an angle of about 45 degrees. Deviations from the desired orientation detected by either gyroscope would produce electric signals, which were processed through a simple analog computation and amplification network to move vanes which caused the missile to return to the correct orientation. See Otto Müller, "The Control System of the V-2," in Th. Benecke and A. W. Quick, eds., *History of German Guided Missiles Development* (Brunswick: Appelhans, 1957), 80–101, especially 80–83.

86. Walter Haeussermann, "Developments in the Field of Automatic Guidance and Control of Rockets," *Journal of Guidance and Control*, Vol. 4 (1981), pp. 225–239, pp. 226–228.

87. Gregory P. Kennedy, *Vengeance Weapon 2: The V-2 Guided Missile* (Washington, D.C.: Smithsonian Institution Press, 1983), 71.

Ninety percent of the V-2s used operationally against British targets were guided by a black-box system in which the two gyroscopes of the *Vertikantsteuerung* system (now known as the LEV-3) were used to stabilize the missile and correct deviations from the plane of flight toward the target. An integrating accelerometer measured missile velocity and initiated rocket motor shut-off.[88]

The version most commonly used is shown in figure 2.5. It consisted of the two control gyroscopes, and a gyroscopic accelerometer designed by Mueller of Kreiselgeräte. The latter is shown in figure 2.6.[89] Because the gyro is made acceleration-sensitive or "pendulous," and because it mechanically performs the mathematical operation of integration, this accelerometer is what would now be called a PIGA, or pendulous integrating gyro accelerometer. Devices of this general kind were to be— and still are—of great significance in missile guidance, though other types of accelerometers were also developed for the V-2.[90]

88. Ibid.

89. The gyroscope suspension was designed so that acceleration of the missile would create a torque or turning force on the gyroscope. The gyroscope would then precess at right angles to the acceleration-induced torque: that is, its spin axis would rotate around the direction of flight. Since the rate of rotation would be proportional to the acceleration, the angle turned through would be proportional to the time integral of acceleration—i.e., to velocity. When it had turned to the angle corresponding to the precalculated cut-off velocity, it would trigger a signal that shut down the rocket motor. The reader will notice that this simplified account says nothing about the effect of gravity on the process of measuring acceleration. This point will be returned to later.

90. They were of the type we would now call a "restrained pendulum." No gyro was involved. Instead a mass was suspended in an electromagnetic field. As the mass sought to move relative to its surroundings under the influence of acceleration (strictly, it sought to remain in the same place while its surroundings moved), this movement was detected and the restraining electromagnetic field increased. The size of the feedback current needed to do this was then an indicator of the acceleration being measured. Two different "restrained pendula" accelerometers were developed, one for range control (by Professors Th. Buchhold and C. Wagner), the other for lateral control (by Dr. H. Schlitt). Like Mueller's PIGA, the Buchhold-Wagner accelerometer was designed to measure acceleration along the flight path. Its most distinctive feature, which was not replicated in later accelerometers, was that the necessary integration of acceleration was achieved by passing the feedback current through an electrolytic cell. "The cell accepts a charge or discharge proportional to the current. Before the flight this cell had to be charged by a current in the opposite direction, corresponding to the required cutoff

Figure 2.5
LEV-3 Guidance System for the V-2
Photograph courtesy Dr. Fritz K. Mueller.

This simple but effective guidance system worked. Though the V-2 was no precision weapon, it had as its primary target the sprawling conurbation of Greater London, and it could hit this often enough to cause significant loss of life. With its trajectory arching up almost into space, its extraordinarily rapid descent toward its target, and its black-box guidance, it seemed impossible to defend against, and this, rather than its negligible immediate effect on the course of the war, estab-

velocity. In flight, the cell was discharged by the current constraining the accelerometer pendulum; it indicated its complete discharge with a small voltage jump on its electrodes. This signal initiated propulsion cutoff through a vacuum tube amplifier" (Haeussermann, "Automatic Guidance and Control," p. 228). Schlitt's accelerometer, by comparison, was for "lateral control," that is detection of deviations from the plane of flight. On the various accelerometers, see ibid.; also Mueller, *History,* 16–20; Kennedy, *V-2,* 71–74; and Thomas M. Moore, "German Missile Accelerometers," *Electrical Engineering,* November 1949, 996–999.

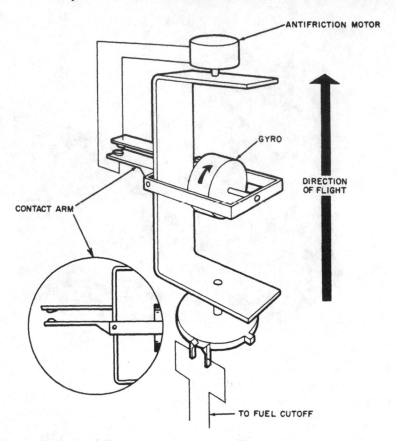

ANTIFRICTION MOTOR

GYRO

DIRECTION
OF FLIGHT

CONTACT ARM

TO FUEL CUTOFF

Figure 2.6
Pendulous Integrating Gyro Accelerometer (PIGA) for the V-2
Source: Fritz K. Mueller, *A History of Inertial Guidance* (Redstone Arsenal,
Ala.: Army Ballistic Missile agency, n.d.), 19.

lished the ballistic missile as a crucial factor in the postwar world.

But even this "success" did not fully establish inertial guidance and navigation. The system of control gyroscopes and an integrating accelerometer could be seen as merely an *ad hoc* device to keep a particular vehicle roughly on a preplanned trajectory, rather than a black box which would allow any vehicle to roam the world and always know where it was without recourse to external input. In that sense, by 1945 inertial navigation remained a dream.

Kreiselgeräte, it is true, developed in the early 1940s a missile guidance system more in tune with Boykow's original vision, at least in its mechanical design.[91] Known as the SG-66, this had originally been a failed competitor of the LEV-3.[92] It consisted of gyroscopes and a single accelerometer to initiate rocket motor cut-off.[93] But instead of being fixed rigidly to the missile body as in the LEV-3, in the SG-66 the gyroscopes and accelerometer were supported on a platform stabilized by the gyros (see figure 2.7), in the manner of a modern inertial guidance system.

By early 1945, the SG-66 was performing satisfactorily in test flights.[94] But before it could be used operationally, the Allied armies had entered Germany and were pressing toward Berlin.[95] The victors already had an eye on a potential future conflict among themselves and were hungry for German high

---

91. It was still intended for the same limited guidance task as the LEV-3, however, as witnessed by the presence on most versions of it of only one accelerometer.

92. Interview with Walter Haeussermann, Huntsville, Alabama, March 22, 1985. The firm chiefly responsible for the LEV-3 was Siemens. Dr. Haeussermann recalls that a third competitor for the inertial guidance of the V-2 was Anschütz.

93. Interview with Mueller. Otto Müller, "Control System," 97–98, records a single test flight in which the SG-66 carried an accelerometer for lateral control as well as the accelerometer for range control.

94. Elektromechanische Werke GmbH., Werk Halle, Nordhausen, an OKH/Wa Prüf (BuM) 10 (March 15, 1945). I am grateful to Dr. Fritz Mueller for a copy of this document, which contains the test results for the SG-66 test flights.

95. Interview with Haeussermann. Otto Müller, "Control System," 97–98, records a single test flight in which the SG-66 carried an accelerometer for lateral control as well as the accelerometer for range control.

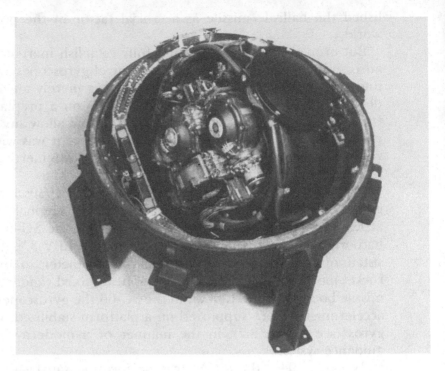

Figure 2.7
SG-66 Guidance System for the V-2
Photograph courtesy Dr. Fritz K. Mueller.

technology. The rocket team and German gyro culture made whatever accommodations they could. The last act in the prehistory of black-box navigation was to be performed not in Germany, but in the United States.

The Black Box in America

Nothing automatically guaranteed that those who gained access to the human and material resources of German technology would go on to use them successfully to develop black-box guidance and navigation. Great Britain is a case in point. As we saw, the British gained the services of two of the leading German specialists, Gievers and Reisch, but no successful inertial navigation project developed. By 1951, possibly in response to the American work described below, a project to develop a fighter aircraft inertial navigator had started, but it never came to fruition.[96] Although British work on inertial guidance and navigation continued in the 1950s, its relative weakness was a major factor in the growing technological dependence of the British "independent deterrent" on the United States. Eventually, in the 1960s, the current pattern emerged whereby the navigation and guidance systems for British strategic weapons are bought from the U.S. as black boxes, while British firms seek to maintain a presence in the inertial guidance and navigation market for nonstrategic weapons.[97]

From today's point of view this could be presented as failure to advance down an obvious technical path. That hindsight judgment is possible, however, only because we now know that inertial navigation works. What happened before 1945 can now been seen as partial realizations of today's established technology. In 1945, however, it would still have been perfectly arguable that they had been failures because they had been attempts at an impossible task, and that limited special-purpose devices such as the LEV-3 were all that was ever going to be possible. What we now see as the path to follow could equally have appeared as a blind alley.

The central aspect of the work done in the immediate postwar years in the United States was the irreversible establishment of the sense that inertial navigation was possible. These years finally saw the ship in place in the bottle, as it were, and were in that sense crucial, despite all the refinements that were yet to come. Shortly, I will outline the two main obstacles that had to be overcome in proving inertial navigation possible. But before I do so, I need briefly to describe the main technical groups involved.

One consisted of the guidance specialists from the V-2 work who came to the United States with Wernher von Braun or shortly thereafter. They worked for the U.S. Army, first at Fort Bliss, Texas, and then at the Redstone Arsenal in Huntsville, Alabama. They continued the work begun in Germany on ballistic missile guidance, concentrating on inertial guidance

96. The surviving documentation of it is in the file AVIA 54/1869 at the Public Record Office.
97. Two British firms are currently active: Ferranti and British Aerospace. The latter incorporates what used to be the U.K. division of Sperry Gyroscope. Elliott Brothers, now part of GEC, were important in the early years of inertial navigation in Britain, but are less so now.

rather than radio guidance and developing stable platforms of the SG-66 sort.[98] They increased the sophistication of these, moving from the SG-66's single accelerometer to systems with two and then ultimately three accelerometers, one for each direction in space. They chose Mueller's PIGA as their preferred accelerometer design,[99] and, in at least some systems, they used a distinctive "knuckle-joint" or "inside-out" form of stable platform (compare figure 2.8, with the now more orthodox design shown schematically in figure 1.1). In the German design the gyros and accelerometers were clustered round a central universal joint, shown in figure 2.9, which permitted at least limited freedom of movement in all three dimensions.

In the Gyro and Stabilizer Laboratory at Redstone Arsenal, a group headed successively by two former Kreiselgeräte staff, Fritz Mueller and Heinrich Rothe, refined an approach to inertial sensor design that had been experimented with at Kreiselgeräte but had not been applied to the V-2. In this approach, bearing friction, already identified as a major barrier to increased accuracy, was reduced by eliminating mechanical contact. Instead, the gyroscope was supported by a film of gas pressurized by a gas bottle or pump. This design is discussed further in chapter 6, because of its significance for Soviet missile guidance.

Their work was successful. For example, they achieved much greater accuracy with the mid-1950s U.S. Army Redstone mis-

98. For a description of the guidance work the German group did for the U.S. Army, see Richard H. Parvin, *Inertial Navigation* (Princeton, N.J.: Van Nostrand, 1962), 5–9.; the source of information for this section of Parvin's book is acknowledged (ibid., p. xi) to be Dr. Walter Haeussermann of the German group. See also F. K. Mueller, "Considerations on Inertial Guidance for Missiles," *Navigation: Journal of the Institute for Navigation,* Vol. 6 (1959), 240–251. Some aspects of the wider context of the Redstone work can be found described in Wernher von Braun, "The Redstone, Jupiter, and Juno," in Eugene M. Emme, ed., *The History of Rocket Technology: Essays on Research, Development and Utility* (Detroit: Wayne State University Press, 1964), 107–121.

99. Other V-2 accelerometer work also informed U.S. developmerts. Dr. H. Schlitt, designer of one of the pendulous mass accelerometers fo. the V-2, ended up working for Bell Aerospace, and was influential in establishing the line of accelerometer development at Bell that continues to this day. The accelerometers in the gravity gradiometer designed for the Trident program, for example, can be traced back to Schlitt's design (interview with Ernest Metzger, Buffalo, March 14, 1985).

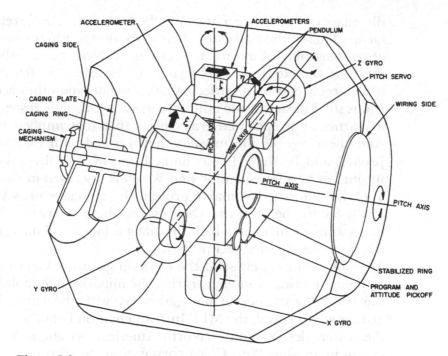

Figure 2.8
"Inside-Out" Stable Platform
Source: Fritz K. Mueller, *A History of Inertial Guidance* (Redstone Arsenal, Ala.: Army Ballistic Missile Agency, n.d), 25.

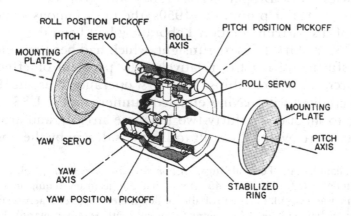

Figure 2.9
Universal Joint Internal Gimbal System for Inside-Out Stable Platform
Source: Fritz K. Mueller, *A History of Inertial Guidance* (Redstone Arsenal, Ala.: Army Ballistic Missile Agency, n.d), 26.

sile than they had with the V-2.[100] But several interrelated factors limited the impact of their work on the wider question of the feasibility of inertial navigation. It developed slowly: by the time the Redstone was flying successfully, the feasibility issue largely had been settled by other means and other actors. It was still ballistic missile guidance work, with all the restriction to particular preplanned trajectories that that implied then. And the range of the missiles they worked with (Redstone, Jupiter, and Pershing I) was limited, not by any physical constraint but by a bureaucratic one, which is discussed in chapter 3. Only with their last major project, the guidance of NASA's 1960s Saturn booster, did they escape these restrictions, but that was in a climate in which inertial guidance and navigation was a well-established technology.

So for all the expertise of the German group, acknowledged by their American contemporaries, the initiative in establishing the feasibility of inertial navigation lay with the latter. Two groups were central: the MIT Instrumentation Laboratory and the Autonetics Division of North American Aviation. A third group in another West Coast corporation, Northrop, was also important, though the company staked its future in the area on a form of navigation where the black box remained at least a little open.[101]

Charles Stark Draper (1901–1987) was professor of aeronautics at MIT. Up until the 1950s, the most famous achievement of his Instrumentation Laboratory had been the Second World War Mark 14 gyro gunsight, which had been extremely successful in solving the relatively new problem of effective anti-aircraft fire from ships. In its first operational use in 1942, it was credited with having enabled gunners on the USS *South Dakota* to shoot down thirty-two Japanese aircraft, and around 100,000 were eventually built to Draper's design by the Sperry

100. William Lucas, "Political Bugs," *Astronautics and Aeronautics*, Vol. 10, No. 10 (October 1972), 44–52, p. 45, states that the accuracy requirement for Redstone was a circular error probable of 300 meters; Walter Haeussermann recalls it as having been 1 kilometer (interview with Haeussermann). Either figure represents a substantial improvement on the V-2.

101. Work was also done in the immediate postwar years at Hughes Aircraft Company, but this stopped after some initial development of components. Walter Wrigley, "The History of Inertial Navigation," *The Journal of Navigation*, Vol. 30 (1977), 61–68.

Gyroscope Company.[102] Draper had started to build a team with considerable gyro culture expertise and was eager to see it take on new, challenging problems. He had been thinking for some time about the problem of black-box navigation, the most challenging problem of all, and in 1945 he secured an Air Force contract that enabled his team to get to work on it in the form of a system for automatic bombing. It was the beginning of a line of work for which the Instrumentation Laboratory's staff was to grow from around a hundred to over two thousand.[103]

The groups at North American and Northrop were part of aircraft companies, but were established not primarily to navigate aircraft but to develop guidance systems for intercontinental cruise missiles. As we shall see in chapter 3, in the early postwar United States, cruise missile programs were more favored than ballistic missile programs. Northrop was responsible for a subsonic cruise missile known as Snark, and North American for the supersonic Navaho. Each was supposed to have a range of up to 5,000 miles, and the daunting guidance tasks this vast range posed seem to have been the primary reason for the establishment of major guidance efforts in the two companies.[104]

102. Charles Stark Draper, "On Course to Modern Guidance," *Astronautics and Aeronautics*, Vol. 18, No. 2 (February 1980), 60, 61.
103. Dating the start of Draper's interest in inertial navigation is difficult. It clearly existed by 1945, but there is no documentary evidence of how long before that it emerged. In interview with him (Cambridge, Mass., October 2, 1984), Draper dated it to the late 1930s. In 1941, Walter F. Wrigley completed a thesis under Draper's supervision on *An Investigation of the Methods Available for Indicating the Direction of the Vertical from Moving Bases* (D.Sc. thesis: Massachusetts Institute of Technology, 1941). Though, as we shall see, the problem of the vertical is central to inertial navigation, there is no indication in the thesis of interest in the latter. Professor Wrigley does not recall it being a subject of conversation between him and Draper at the period when he was working on his thesis (interview with Walter F. Wrigley, Cambridge, Mass., November 8, 1984).

In the early postwar years the Air Force was not yet formally separate from the Army, but was the Army Air Forces. For the significance of this, see chapter 3. I shall, however, use the term "Air Force" for this period unless in the particular context it is important to emphasize the distinction.

For the Laboratory's expansion, see the Charles Stark Draper Laboratory, Inc., *1980 Annual Report: the First Fifty Years* (Cambridge, Mass.: Draper Laboratory, 1980).
104. The best review of cruise missile development in this period is Kenneth

Unlike the Huntsville group, the groups at MIT, North American, and Northrop had no tight upper limits of range placed upon them. Indeed, a large part of the problems they faced arose because their systems would have to work for the hours of bomber or cruise missile flight from the United States to the Soviet Union rather than the minutes of ballistic missile boost phase. At least some of their systems would in practice have limited the vehicle carrying them to a preset trajectory, such as a great-circle flight path. But their work was largely evaluated as if unrestricted, globe-roaming black-box navigation was their goal, as it was in an ultimate sense.

Their task was not to establish the desirability of this goal. No one in the U.S. Air Force needed convincing that navigation was a key problem. Navigational failure had dogged World War II bombing campaigns, often forcing an unpleasant choice between indiscriminate nighttime terror bombing and highly dangerous daytime sorties for attempts at precision bombing. The radio navigational systems developed during the war were much better than nothing, but attempts to jam them were commonplace. So a black-box system was patently to be preferred. It was its possibility, rather, that was the issue.

Proponents of inertial navigation had to combat two counterclaims: that the idea was impossible in principle, and that it was impossible in practice, at least for the foreseeable future. "The problem of the vertical" and "the problem of performance" were the two central issues. Neither was new, but both came to a head in the immediate postwar years. Both could be described as technical problems; both, however, required a solution that went beyond "the technical" as narrowly conceived.

The Problem of the Vertical

Like many of their predecessors, the MIT and Autonetics groups proposed to navigate by the double integration of acceleration. But to a physicist in tune with Einstein's general theory of relativity, this could raise an apparently profound difficulty that seemed to imply that the search for a black-box navigator

P. Werrell, *The Evolution of the Cruise Missile* (Maxwell Air Force Base, Alabama: Air University Press, 1985), chapter 4.

was futile. One had not been constructed because one could not be constructed.

Einstein had argued that inside a black box, imagined as "a spacious chest resembling a room with an observer inside who is equipped with apparatus," the effects of linear acceleration of the box were indistinguishable from the effects of a gravitational field.[105] In the form of the law of the equivalence of inertial and gravitational mass, this argument had become a centerpiece of the general theory. Its negative consequences for black-box navigation were spelled out in at least one physics text: "By no mechanical experiment, indeed, can an apparent gravitational field thus produced by acceleration of a frame of reference be distinguished from a true field due to gravitational attraction. This fact constitutes a practical difficulty in the blind navigation of airplanes, since it makes impossible the construction of a device to indicate the true vertical unaffected by accelerations of the airplane when in curved flight."[106]

Suppose, as in many of the inertial navigation proposals, one were trying to navigate using two accelerometers held horizontal, that is at right angles to the local direction of gravity, or "local vertical." How could one ascertain the direction of the vertical? A plumb line would no longer indicate the true vertical when the vehicle carrying it accelerated. And if one did not know the direction of the vertical, one could not know whether or not one's accelerometers were horizontal, so their readings could not be interpreted. A "vicious circle" seemed to exist, making impossible simultaneous knowledge of acceleration and orientation. Inertial navigation seemed to rest on being able to measure independently physical phenomena whose effects within a black box were inseparable.

This argument became a serious risk to U.S. inertial navigation projects in the 1940s, when the eminent physicist George Gamow began to circulate a paper ridiculing them, entitled "Vertical, Vertical, Who's Got the Vertical?"[107] Gamow

105. Albert Einstein, *Relativity: The Special and General Theory* (New York: Crown, 1961; first published in 1916), chapter 20.
106. F. K. Richtmyer and E. H. Kennard, *Introduction to Modern Physics*, third edition (New York: McGraw-Hill, 1942), 151.
107. Several of my interviewees remembered it, quite independently, but I have not been able to find the original document. It is not in the Gamow papers in the Library of Congress, Washington D.C.. I have therefore had to reconstruct its contents from what interviewees told me of it. My suspicion

was not merely a prestigious scientist: he was a member of the Guidance and Control Panel of the Air Force Scientific Advisory Board.[108] The threat, recalls one pioneer of inertial guidance, was that a sense would develop that "the military was wasting money on something that fundamentally couldn't be done."[109]

Gamow was not alone in raising the objection that within a black box physical phenomena crucial to navigation could not be disentangled. That in essence was what the specialists at the U.K. Admiralty Compass Observatory had said when asked to evaluate Gievers's proposed inertial navigator, and Gievers had been unable to formulate a convincing reply to his interrogators.[110] The British physicist P. M. S. Blackett also reportedly objected to inertial navigation schemes on these grounds.[111]

There was, however, an analogy to be found at the beginning of the development of gyro culture, and it was this analogy that the proponents of inertial navigation seized upon as the basis for their defense. While Anschütz-Kaempfe's gyrocompass was still in laboratory development, Oskar Martiennsen, a physicist and electrical engineer who worked at the University of Kiel, calculated theoretically that the device would be seriously disturbed by acceleration of the vessel carrying it.[112] For

is that since the inertial navigation projects were classified, the paper was never cleared for public release.

108. Thomas A. Sturm, *The USAF Scientific Advisory Board: Its First Twenty Years, 1944–1964* (Washington, D.C.: U.S. Air Force Historical Division Liaison Office, February 1, 1967), 148.

109. Wrigley interview. Professor Wrigley recollects that Gamow's argument got to the ears of members of Congress, but I have found no confirmation of this.

110. Rawlison et al., "Discussions with Dr. Gievers," 3:

"The compass and stable vertical are intended for use on board ships or aircraft and must therefore work, not only when the ship is at rest, but also under way, when it must eliminate all errors due to speed over the earth in any direction, errors due to rolling and pitching and errors due to acceleration in any direction. Dr. Gievers was pressed to provide either physical explanations or theoretical justification for his claims and assumptions that the compass would operate satisfactorily under sea-going conditions. He stated that he fully understood the questions being put to him, but that whilst convinced himself that his ideas were sound he could not at present put down convincing explanations, though he was confident that, given time and suitable conditions such explanations could be provided."

111. Jerry Ravetz, personal communication.

112. O. Martienssen, "Die Verwendbarkeit des Rotationskompasses als Ersatz des magnetischen Kompasses (The Suitability of the Rotation Compass as a

the gyrocompass work he thus created what Edward Constant calls a "presumptive anomaly," under which "assumptions derived from science indicate either that under some future conditions the conventional system will fail (or function badly) or that a radically different system will do a much better job"; except that in this case the gyrocompass was the radical alternative rather than the established system.[113]

Anschütz-Kaempfe's cousin and loyal supporter Max Schuler, a specialist in applied mathematics and mechanics, was not inclined to conclude that the gyrocompass project had to be abandoned. He pored over Martiennsen's calculations, seeking a way to minimize acceleration-induced errors in the gyrocompass. The solution he constructed, simple in outline, but complex in its practical implications and intriguing theoretically,[114] became central to the "technical" response by the proponents of inertial navigation to the "problem of the vertical."

It is more easily grasped in its more general version, made public by Schuler in 1923, than in its specific application to the gyrocompass.[115] Imagine trying to determine the direction of the vertical by suspending a weight from a string in a vehicle traveling over the surface of the earth. Such a method would indeed fail immediately when the vehicle accelerated, unless— and this was Schuler's key point—if the string were so long that the weight was at the center of the earth. If that were so, the weight could remain undisturbed, and the string would continue to indicate the direction of the vertical, whatever the movements of the vehicle.

An "earth-radius pendulum" was of course not feasible in any literal sense, Schuler admitted. But the same effect—freedom from the disturbing effects of acceleration—could be produced by designing a gyroscopic system such that it had the

substitute for the Magnetic Compass)," *Physikalische Zeitschrift*, Vol. 7 (1906), 535–543; see Schuler, "Entwicklung des Kreiselkompasses," part 1, 471.

113. Edward W. Constant II, *The Origins of the Turbojet Revolution* (Baltimore: Johns Hopkins University Press, 1980), 15.

114. See, for example, Frank Coffman Bell, "Schuler's Principle and Inertial Navigation," *Annals of the New York Academy of Sciences*, Vol. 147 (1969), 493–514.

115. M. Schuler, "Die Störung von Pendel- und Kreiselapparaten durch die Beschleuniges des Fahrzeuges (The Disturbance of Pendulum and Gyroscopic Apparatus by the Acceleration of a Vehicle)," *Physikalische Zeitschrift*, Vol. 24 (1923), 344–350.

same period of oscillation that an earth-radius pendulum would have: 84 minutes.[116]

By 1908, Schuler and Anschütz-Kaempfe appear to have succeeded in building a gyrocompass with an 84-minute period, and it became a feature of other gyrocompass designs as well.[117] Schuler did not go on to use his analysis as the basis for a more general black-box navigator: he ruled out a black-box aircraft navigator based on the integration of acceleration as "ein ganz unmögliches Beginnen," an entirely impossible undertaking.[118] But those who sought to construct such a system turned increasingly to systems that could be seen as analogous to the earth-radius pendulum.[119]

A simple way of thinking about how this was achieved is to imagine a system in which two accelerometers, mutually at right angles, are placed on a platform stabilized by gyroscopes. The platform is horizontal at the point where the vehicle starts. As the vehicle carrying the platform moves, the output from the accelerometers is integrated twice with respect to time to yield the vehicle's displacement from its starting point. Under the undisturbed action of the gyroscopes, the platform would move away from the horizontal as the vehicle moves around the earth, because the gyroscope spin axes will tend to remain in fixed orientation with respect to the stars. But since the earth can be taken, to a first approximation, to be a sphere, it is easy to calculate from the displacement of the vehicle the correction necessary to return the platform to the horizontal. The necessary scaling factors can be built into a simple feedback loop that continuously causes the platform to be tilted so as to keep it horizontal. Such a system could be seen as working as if kept horizontal by an earth-radius pendulum (see figure 2.10). Putting matters in a different way, one could say that the cunningly designed feedback mechanism was artificially separating gravity from the accelerations one wished to measure.[120]

116. Ibid., 345–346.
117. Schuler, "Entwicklung des Kreiselkompasses," 471; Schuler, "Pendel- und Kreiselapparaten," 348.
118. Schuler, "Pendel- und Kreiselapparaten," 349.
119. This, for example, seems to have been key to Reisch's work and the foundation of his claim to be the inventor of inertial navigation. See Hellman, "Inertial Navigation," 89.
120. See Wrigley, "History," 61.

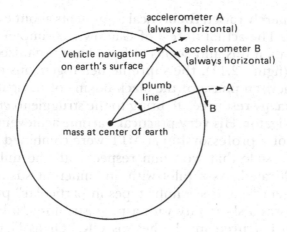

Figure 2.10
Earth-Radius Pendulum

Such a system would be described by some as "Schuler tuned." We need not enter here into the debate as to whether or not such an attribution is justified.[121] The point is that those working in the field in the United States in the late 1940s felt confident that they had an answer to the "problem of the vertical" raised by Gamow.[122] They might have fared ill how-

121. See, for example, Bell, "Schuler's Principle." Whether modern inertial navigators are actually based on Schuler's principle, and indeed what Schuler's principle really is, are central issues in dispute over priority in the invention of inertial navigation. The phrase "Schuler tuning," was introduced by Walter Wrigley, "Schuler Tuning Characteristics in Navigational Instruments," *Navigation*, Vol. 2 (1950), 282–290, but the general idea was by then at least moderately well known to those working in the field. Boykow's 1934 proposal for an "Instrument for Indicating Navigational Factors," for example, contained a "device for compensating the apparent earth rotation," required "simply because in the absence of such corrections the gyros would tend to remain with their impulse shafts in a horizontal plane of the starting point . . . the corrections are required in order to maintain the platform always in a position at right angles to the earth radius passing through the craft at any given moment" (p. 14). Kofman and Levental's "Navigatsionnyi Pribor" of 1932 made explicit use of the 84–minute period. We have only the 1966 published version of their proposal to go on, but the 1949 article by L. I. Tkachev, "O 84-Minutnom Periode dlya Sistem so Cvyazannymi i Svobodnymi Giroskopami (On the 84-Minute Period for Systems with Coupled and Free Gyroscopes)," *Prikladnaya Matematike i Mekhanika*, Vol. 13 (1949), 217–218, is clear evidence of the currency in the Soviet Union of Schuler's work, and it connects Schuler to black-box navigation.

ever, had they merely raised theoretical arguments about earth-radius pendula. The actual defeat of Gamow was simpler and more dramatic than this and seems to have been organized by MIT's Draper (figure 2.11), the supreme heterogeneous engineer of the American phase of the black-boxing of navigation.

Draper had many resources to bring to the struggle to create a black-box navigator. His very practical wartime achievements and the status of a professorship at MIT were combined with a down-to-earth style that won him respect with the military contacts he cultivated. As a pilot with an "inherent suspicion of civilians in general and scientific types in particular" put it, "Doc [Draper] was a short guy with a pugilistic nose and cauliflower ears so I figured maybe he was OK." Or as General Bernard Schriever, the key figure in the Air Force's ballistic missile program (see chapter 3), described him: "Not like an Oppenheimer . . . very aloof and hard to talk to. Stark [Draper] is just the opposite, outgoing, extrovert."[123]

Like Gamow, Draper was a member of the Scientific Advisory Board of the Air Force. He used this position to force a confrontation, arranging a major classified conference to review progress in the field of automatic navigation—to my knowledge the first such meeting held. All the Air Force-funded groups working on black-box navigation and Gamow were invited. Perhaps realizing that the meeting was "stacked" against him, Gamow did not appear. His absence was taken as acceptance that he was wrong. As one participant in the meeting recalls, "the fact that Gamow did not attend the meeting," together with papers from the various groups that showed that progress was being made, "allowed the military to continue their support for such activities."[124]

122. See, though Gamow's name is not mentioned, J. J. Gilvarry, S. H. Browne and I. K. Williams, "Theory of Blind Navigation by Dynamical Measurements," *Journal of Applied Physics*, Vol. 21 (1950), 753–761.
123. Chip Collins, in "Looking back at Draper," *Mass[achusetts] High Tech*, August 17–30, 1987 (I am quoting from an unpaginated reprint kindly sent to me by Peter Palmer); interview with General Bernard Schriever (U.S. Air Force, rtd.), Washington, D.C., March 25, 1985. General Schriever's reference is of course to the leader of the atom bomb project, J. Robert Oppenheimer.
124. Interview with Wrigley. The now declassified papers from one such conference, in February 1949, are in the Draper files, Oral History Collection,

Figure 2.11
Professor Charles Stark Draper
Photograph courtesy Charles Stark Draper Laboratory, Inc.

The problem of the vertical did not go away. But after the defeat of Gamow I know of no attempt to raise it again as an objection in principle to inertial navigation. Instead, it became a "technical difficulty," just as it had been for the V-2 guidance specialists. For them it took the form, in the words of a later writer, that an accelerometer "cannot measure the component of acceleration due to the force of gravity. This part of the motion must be determined analytically from the known magnitude and direction of gravity as a function of position."[125] So

MIT Archives and Special Collections, MC 134, as Scientific Advisory Board, Office of the Chief of Staff, United States Air Force, *Seminar on Automatic Celestial and Inertial Long Range Guidance Systems* (no date or place of publication given). It may however be that the meeting recalled by Professor Wrigley was an earlier one.
125. D. G. Hoag, "Ballistic-Missile Guidance," in B. T. Feld et al. (eds.), *Impact of New Technologies on the Arms Race* (Cambridge, Mass.: MIT Press, 1971), 24.

the "trigger" for rocket motor cut-off had to be set according to a prelaunch calculation that took this into account. This could be done only approximately.[126]

The matter was troublesome: it was reckoned to be on account of this that inertial guidance of the V-2 appeared less accurate than radio guidance.[127] And that troublesomeness was not to disappear with the development of inertial technology. Rather, it changed form. "Schuler tuning" could be said to rest on an implicit model of the earth's gravitational field. With the development of digital guidance computers and growing demands for accuracy, that model became explicit, embedded not just in an electromechanical feedback loop but in data in computer memory. How good that data was, and the extent to which possible deficiencies in it would be the decisive constraint on a black-box navigation system, were to be recurrent issues.

But that was part of a different debate. The decisive shift of "the problem of the vertical" in the late 1940s to the status of technical difficulty meant that the remaining arguments against inertial navigation were practical ones rather than objections in principle.

The Problem of Performance

The practical objections were nevertheless formidable. In particular, even if black-box navigation was sound in principle, could sufficiently accurate gyroscopes and accelerometers be constructed to make it possible for a black-box navigator to work in practice? Disposing of this difficulty proved much harder than disposing of Gamow.

Again the problem was not new: it was well known, for example, to the Kreiselgeräte group. It was highlighted in the 1940s when the Air Force quantified for the Instrumentation Laboratory and the groups at Northrop and Autonetics exactly what it would mean for an inertial navigator to "work." Their

126. Otto Müller, "Control System," 85–86. As Müller puts it, the output of an integrating accelerometer with its sensitive axis along the direction of rocket thrust is not the rocket velocity but the sum of the rocket velocity and the time integral of g cos ϕ, where ϕ is the angle of inclination of the trajectory and g the acceleration due to gravity. The approximation arises because the correction had to be preset, but only an average, not an exact, time to cut-off was known in advance.
127. Ibid., 86.

navigation and guidance systems had to have an average accuracy of 5,000 feet (1.5 kilometers or, roughly, a mile) at the target.[128] Asuming that the vehicle being navigated or guided was a bomber or cruise missile—possibly traveling at subsonic speed—and that the flight path was the United States to the Soviet Union, the time of flight to the target could be as much as ten hours. So a black-box guidance or navigation system had to have an accumulated average error of no more than about a mile after ten hours.

There was of course nothing ultimately sacrosanct about this requirement: the slightly later British inertial navigator project had an Air Staff accuracy requirement of five miles after 20 minutes.[129] (I suggest the probable reason for the U.S. Air Force requirement in chapter 3.) But the Draper, Autonetics, and Northrop groups were in no position, on their own, simply to ignore the requirement, even though they patently felt it to be onerous.

A classified theoretical analysis by Draper and colleagues at the Instrumentation Laboratory, dating from February 1947, spelled out what an average accuracy of a mile after ten hours of flight meant in terms of instrument performance. They found the key issue to be the performance of the gyroscope rather than the accelerometer and calculated that to achieve an accuracy of a nautical mile after just one hour of flight required gyroscopes a hundred times more accurate than the state of the art.[130] And since the Air Force was envisaging a flight of up to ten hours, the problem was even worse than that.

128. S.E. Weaver, "The MX-775 Midcourse Guidance System," in Scientific Advisory Board, *Seminar,* Vol. 1, 37, and J. R. Moore, "North American Inertial Guidance System," in Scientific Advisory Board, *Seminar,* Vol. 1, 53.
129. Anonymous, "Methods of Lining up Azimuth Gyro," undated, in file AVIA 54/1869, Public Record Office. This project was, of course, designed for application to a fighter not a bomber. But this brought with it a different difficulty: an extremely stringent constraint on the weight and volume of the system.
130. Draper and "Group B," *Fundamental Possibilities and Limitations of Navigation by Means of Inertial Space References* (February 1947). No publication details are given on the copy of this document which I have inspected, which is in the library of the Charles Stark Draper Laboratory Inc., and is paginated from 29 to 56. "Group B" consisted of W. Wrigley, J. F. Hutzenlaub, A. G. Bogosian, H. B. Brainerd, E. B. Dane, E. V. Hardway, Jr., J. J. Jarosh and M. Petersen. They chose the slightly looser criterion of one nautical mile accuracy: a nautical mile (1.85 kilometers; a statute mile is about 1.6 kilo-

The Instrumentation Laboratory, Autonetics, and Northrop adopted various mixes of two different strategies for trying to solve the problem. The first was to circumvent it. The function of the gyroscopes was to keep the stable platform of the inertial system in exactly known orientation; the difficulty of doing this for literally hours on end was what made gyroscope performance seem a more crucial constraint than accelerometer performance.[131]

By opening the black box a little and making use of some external input, the difficulty of the task could be greatly reduced. The relevant input was star-sightings. Any drift of the gyroscopes, and consequent misorientation of the stable platform, could be detected and corrected by periodic—or maybe even continuous—automatic star-sighting. The problem of gyroscope performance could thus apparently be circumvented.

This solution—the construction of a "stellar-inertial" navigation system—had, perhaps, the added attraction of being, superficially at least, more familiar to Air Force officers, who were by then well accustomed to the use of star-sightings in navigation. Thus, although Draper was dubious of this breach in the black-box approach, noting that "stellar observations . . . are subject to inteference by weather, aurorae, meteors and countermeasures (star shells)," his successful 1945 proposal to the Air Force was entitled "Tentative Specifications for a Stellar Bombing System," and the stellar-inertial system developed by the Instrumentation Laboratory was christened FEBE, after Phoebus, the sun god.[132] (A descendant of the FEBE system is shown in figure 2.12.)

meters) is the distance on the earth's surface corresponding to a minute of arc—a sixtieth of a degree. One could not hope for this level of accuracy, they reasoned, unless one had gyroscopes available that would "drift" by less than this amount in the requisite time. But "the lowest available drift rate in good existing equipment is approximately 30 milliradians [1.7 degrees] per hour" (ibid., 50). In the V-2 work, it is worth noting, gyro drift rate was "somewhat less than 10 degrees per hour" (Parvin, *Inertial Navigation*, 6).

131. This was true only for bombers and cruise missiles, with their long flight times and relatively modest accelerations. For ballistic missiles, with very short powered flights and high accelerations, accelerometer performance was seen as a more crucial issue than gyroscope performance.

132. Draper and "Group B," *Possibilities and Limitations*, 36. I have not seen the proposal document. Its title is given in J. Scott Ferguson, *Historical*

Figure 2.12
FEBE's "Descendant": MIT/AC Spark Plug Stellar-Inertial Bombing
System
Stellar-inertial bombing system inertial measurement unit being installed in
U.S. Air Force B-50 in June 1954. Note the star-tracker on top of the unit,
in protective cover. Photograph courtesy Jack Becker, Delco Systems Operations, General Motors Corporation.

Autonetics too had only a "marriage of convenience" with
stellar-inertial guidance, though the stellar work was a more
distinct trend there than at the Instrumentation Laboratory.[133]
But Northrop embraced the technology permanently; it is now,
and has been for some time, unique among U.S. corporations
in this field in actively marketing stellar-inertial navigation for
bomber and cruise missile use.[134] During the 1940s and 1950s,

Collection Projects (Cambridge, Mass.: Charles Stark Draper Laboratory,
August 3, 1979, C-5249, unpaginated).
133. See J. M. Slater, *Twenty Years of Inertial Navigation at North American
Aviation* (Anaheim, Calif.: Autonetics Division, North American Aviation,
1966).
134. Anonymous, *Reaching for the Stars: NAS-26 Precision Astroinertial Navigation System* (Hawthorne, Calif.: Northrop Corporation, Electronics Division,
n.d.).

the main guidance project at Northrop was the corporation's Snark intercontinental-range cruise missile. Snark was guided by a stellar-inertial system, which the company held to be capable of an average error of 1.4 nautical miles, or 8,500 feet, not too far away from the goal of 5,000 feet. But the project as a whole fell foul of the unstable technological politics of missilery in the U.S. Air Force: in 1950, for example, the accuracy requirement was suddenly tightened from 5,000 feet to 1,500 feet. A series of disastrous tests led to the jibe, repeated by interviewees even today, that the waters off the test station at Cape Canaveral were "Snark-infested." Finally, by the time Snark was deployed, consensus had swung against the cruise missile and in favor of the ballistic missile. A small force of Snarks was deployed at Presque Isle in northern Maine in May 1959 but was decommissioned only two years later by the incoming Kennedy Administration. Stellar-inertial navigation (as distinct from stellar-inertial guidance of ballistic missiles) thereafter moved in the United States into the shadowy world of the "black programs"—programs protected by more than the normal barriers of military secrecy.[135]

The second solution to the problem of inadequate gyroscope performance was of course to seek directly to improve it to the necessary level. This was the primary path taken at both the Instrumentation Laboratory and Autonetics; the impetus in this direction was less strong at Northrop because of the commitment there to the stellar-inertial solution.[136]

As Draper's analysis had shown, it was a daunting path. The reluctance of the traditional firms such as Sperry to commit themselves to it was one major reason why North American

135. Werrell, *Cruise Missile*, 85, 96–97; see chapter 3 for the probable reasons for accuracy requirements. See Anonymous, "SR-71 to use Skybolt Guidance Sensor," *Aviation Week and Space Technology*, November 23, 1964, 22. This article reports use of the stellar-inertial system developed by Northrop for the canceled Skybolt air-to-ground ballistic missile as a navigator for the SR-71 "Black Bird" strategic reconnaissance aircraft, successor to the famous U-2.

136. I do not, of course, want to imply that these were the only sites in the United States at which improved gyroscope performance was sought. Thus use of the externally-pressurized air-bearing enabled the Huntsville team to improve gyroscope performance from the V-2's ten degrees per hour to "a few tenths of a degree per hour without compensation" (Parvin, *Inertial Navigation*, 6).

Aviation (Autonetics's parent) and Northrop, both aircraft firms rather than gyro culture ones, developed in house guidance efforts to support their cruise missile projects. Gyro culture was a crucial resource for them, nevertheless. Autonetics recruited a "gyro genius," John Slater, from Sperry. Autonetics's charismatic manager and leader, John Moore, recalls recruiting engineers in this period: "I would go out and take pictures of roadside stands where . . . cantaloupes were maybe 2 cents . . . we'd always go back East if we could find a snow shower—that's when we would go and hit the colleges!"[137]

The Instrumentation Laboratory's approach to improving gyroscope accuracy was simple in form, even if complex and ramified in details. It bore the clear stamp of Draper's dominant, driving style. Early in his work on black-box navigation, certainly by 1947–1948, Draper selected his "paradigm" gyroscope design.[138] The Instrumentation Laboratory then devoted enormous efforts over many years to achieving the two to three orders of magnitude improvement in performance that they had calculated was necessary to make inertial navigation work.

The resulting pattern of technological change is well described by Thomas P. Hughes's notion of the "reverse salient."[139] The analogy is military. A reverse salient is something that holds up technical progress or the growth of a technological system, just as enemy forces may hold out in one particular spot even though in other areas they have been pushed back. System builders typically focus inventive effort, much like generals focus their forces, on the elimination of such reverse salients; they identify critical problems whose solution will eliminate them.

---

137. Interview with John Slater, Fullerton, Calif., February 22, 1985, and with John R. Moore, Los Angeles, February 25, 1985.
138. I am drawing the term from the classic, T. S. Kuhn, *The Structure of Scientific Revolutions* (Chicago: University of Chicago Press, second edition, 1970). "Paradigm" is often used in a loose, unhelpful way in discussions of technology. In the more specific sense in which Kuhn uses the word—"the concrete puzzle-solutions which, employed as models or examples, can replace explicit rules as a basis for the solution of the remaining puzzles of normal science" (ibid., 175)—it does, I think, point us by analogy to important phenomena in technological change.
139. See Hughes, *Networks of Power: Electrification in Western Society, 1880–1930* (Baltimore: Johns Hopkins University Press, 1983).

The term reverse salient is more apt than static, mechanical metaphors such as "bottleneck" because it captures the flux, dynamism, and confusion of the process of technical change. Not only can change bring into being reverse salients where previously components functioned satisfactorily—we saw an example of this in the shift to iron and steel ships making the existing compass a reverse salient—but it may not always be clear where progress is being held up, nor what should be done about it. Even with agreement on goals, without which the notion of reverse salience becomes inapplicable because "progress" cannot be defined unambiguously, the nature of the obstacles to the achievement of these goals and the best means of removing them may be the subject of deep disagreement. There were instances of this in gyroscope development at the Instrumentation Laboratory, but in general, the authority of Draper held sway. The goal was increased gyroscope accuracy. The chosen means was the single-degree-of-freedom, floated, integrating gyroscope.

The idea of floating a gyro to reduce friction was not new, having been suggested in the 1880s and experimented with at Kreiselgeräte in the 1930s. The work there was not successful: "we learned that the road to floating gyros is a difficult and thorny path."[140] But toward the end of the war when Draper was shown a gyroscopic accelerometer from a V-2 that had failed to explode, his reaction was, "I said, 'well, hell, I could fix that,' and so I floated the thing. "[141] Flotation was already characteristic of Draper's gyro culture work, and this presumably was the source of his confidence.[142] He was not alone. Another group, led by Frithiof V. Johnson at General Electric,

141. It was suggested (without further elaboration) in William Thomson [Lord Kelvin], "On a Gyrostatic Working Model of the Magnetic Compass," *British Association Reports,* 1884, 627 (this paper is also to be found in *Nature,* Vol. 30 [1884], 524–26); I owe the reference to Martienssen, "Entwicklung des Kreiselkompasses," 182. The quotation is from Gievers, "Erinnerungen," English translation, 21–22.
141. Interview with Charles Stark Draper, Cambridge, Mass., October 12, 1984.
142. The Mk 14 gunsight's gyros were floated, and in the 1930s Draper had sought to improve the contemporary aircraft turn indicator by floating its gyro, only to find a lack of interest on the part of the Sperry Gyroscope Company. See Draper, "On Course," 59, and the description of the Mk 14 sight in Draper, "Origins of Inertial Navigation," *Journal of Guidance and Control,* Vol. 4 (1981), 449–463.

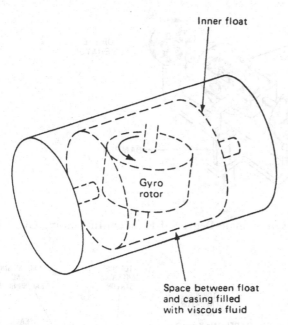

Inner float

Gyro
rotor

Space between float
and casing filled
with viscous fluid

Figure 2.13
Single-Degree-of-Freedom Floated Integrating Gyroscope (highly
schematized)

independently began development of a floated gyro for use in
early postwar experimental missile programs.[143]

The device most central to American missile guidance, even
in the epoch of MX and Trident II, was being born. In very
simple schematic, a single-degree-of-freedom floated gyro is
shown in figure 2.13. Figure 2.14 shows a little more detail,
though it omits the gyro case and "can" containing the gyro
wheel or rotor. And figure 2.15, though still schematized,
shows a 1950s design from Draper's Instrumentation Laboratory. Note the three axes in space labeled in figure 2.15. They
will be important in what follows. Figure 2.16 is a cutaway view
and figure 2.17 an assembly drawing of the actual instrument.

The essence of what the device does is simple. If it is rotated
around its input axis, the gyro reacts, like all gyros do, by

143. Frithiof V. Johnson and Harold H.P. Lemmerman, "Gyroscope" (U.S.
Patent 3,060, 752, October 30, 1962, filed September 7, 1949). The reason
for the long gap was that the patent application was placed under an order
of secrecy (letter to author from F. V. Johnson, November 7, 1986). Many
General Electric floated gyros were built, but for more general "gyro culture"
uses such as fire and flight control (ibid.) rather than inertial navigation.

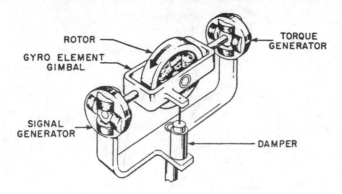

Figure 2.14
Essential Elements of a Single-Degree-of-Freedom Integrating Gyroscope

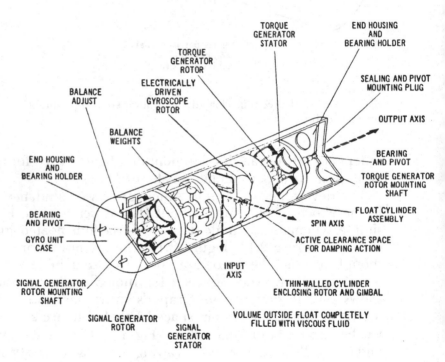

Figure 2.15
Generalized Diagram of a 1950s Instrumentation Laboratory Gyroscope
(10 IG floated integrating gyroscope)
Source: Lester R. Grohe and Hugh H. McArdle, *The 10-Series of Floated
Instruments* (Cambridge, Mass.: MIT Instrumentation Laboratory, 1955,
R-66), 2.

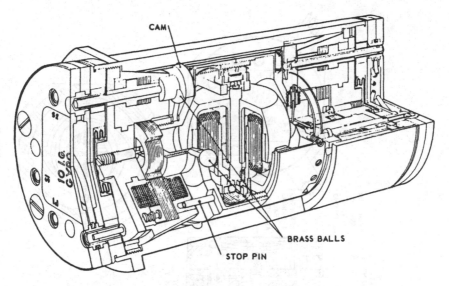

Figure 2.16
Cutaway View of 10 IG Floated Integrating Gyroscope
Source: Lester R. Grohe and Hugh H. McArdle, *The 10-Series of Floated Instruments* (Cambridge, Mass.: MIT Instrumentation Laboratory, 1955, R-66), 5.

precessing at right angles to this, around its output axis. It does not precess entirely freely. It is restrained by a damping mechanism. Shown schematically in figure 2.14, damping is provided in actual devices by the viscosity of the flotation fluid described in the next paragraph. This damping has a most useful mathematical effect. It causes the device's output to be proportional not to the rate of turning around the input axis but to its time integral, the angle turned through. Because of this the device is called an "integrating" gyro. It measures only rotations about the one input axis shown in figure 2.15: hence the term "single degree of freedom" is used to describe it.

Flotation is achieved by enclosing in a "can" the rotor and gimbal structure supporting it (a gimbal is a means of supporting something while giving it the freedom to rotate: the structure surrounding the gyro rotor in figure 2.14 is an example). The space between the can and the instrument's outer case is filled with fluid. The fluid is heated so that, as closely as possible, neutral buoyancy is achieved. This reduces to a minimum the load and friction in the bearings on the output

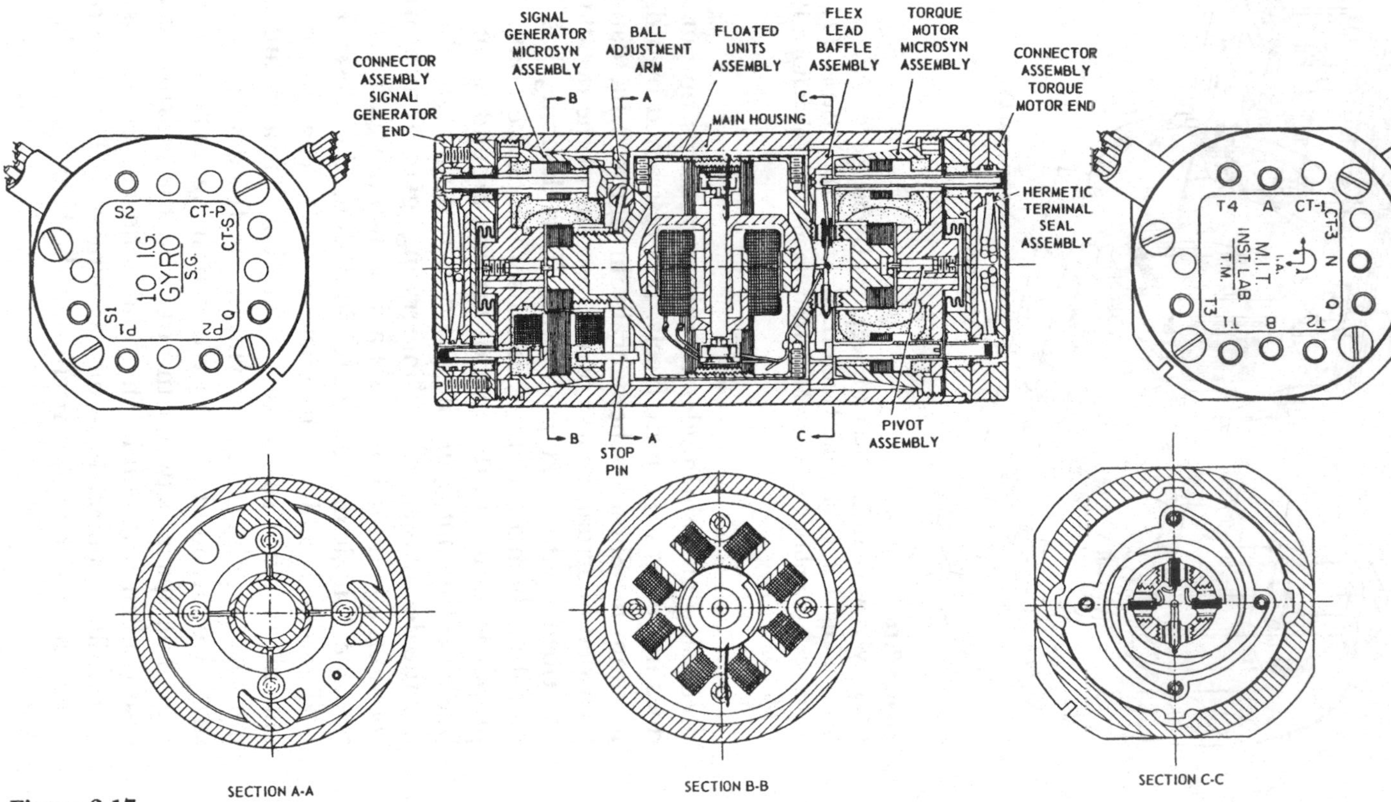

Figure 2.17

Assembly Drawing of 10 IG Floated Integrating Gyroscope

Source: Lester R. Grohe and Hugh H. McArdle. *The 10-Series of Floated Instruments* (Cambridge, Mass.: MIT Instrumentation

axis, which were originally of the pivot and jewel type used in watches.

Precession of the gyro around the output axis in response to rotation about the input axis causes an electrical output from a signal generator, shown in figures 2.14 and 2.15. This electrical output is proportional to the angle through which the device has been rotated around the input axis, for the reasons indicated above. In a stable platform navigation or guidance system such as that shown in figure 1.1, this output signal then operates electric motors that move the platform to cancel out the initial rotation, thus keeping the platform in desired orientation through feedback. A change in orientation can be achieved by feeding a signal to the gyro's torque generator (figures 2.14 and 2.15), which can forcibly precess the gyro.

All of these general features of the floated gyroscope remained a fixed point in the Instrumentation Laboratory's work. But no detail escaped attention in the search for the sources of inaccuracy and the means to remove them. This was a major technological project, ultimately stretching over decades and involving hundreds of Instrumentation Laboratory staff, other departments of MIT (such as metallurgy), and many outside firms, including by 1968 at least some of the thirty set up as "Route 128" spin-offs from the Instrumentation Laboratory.[144]

It was a project of heterogeneous engineering in at least two senses. First, the obstacles to the achievement of accuracy included human behavior, as of course did the resources available to achieve it. One major obstacle was dirt. Particles of foreign matter in the flotation fluid were quickly seen as making the achievement of high accuracy impossible. The most scrupulous cleanliness was therefore required of those undertaking the skilled and painstaking task of assembling gyros. Women assemblers, for example, were not to wear make-up to work, and if a worker returned sunburned from a vacation, he or she was asked not to go into the clean room.[145] Less obvious

144. E. B. Roberts, "A Basic Study of Innovators: How to Keep and Capitalize on their Talents," *Research Management*, Vol. 11 (1968), 249–266. "Route 128" is the road encircling Boston, along which many firms established themselves.

145. The latter points are from my interview with Peter Palmer, Cambridge, Mass., October 10, 1984, but the concern about dirt was pervasive among those involved with gyro production at this time.

forms of behavior also had to be taken into account, a point Draper understood well and in which he gradually educated his initially uncomprehending engineers:

We had this test lab running round the clock. Doc Draper would come in at the oddest hours—on his way home from a trip at 2 or 3 in the morning. . . . Any time you walked into the test lab there's Doc Draper talking with a technician or talking to the janitor. So, in my ignorance, one day I said to him, . . . "How do you find time to do this?" And he said, "Pete, those guys have information on what's running, what's going on in your test lab." What could you find out from the janitor? You'd find out from the janitor that every once in a while he'd actually do something kind of ridiculous down there, like he'd hit something with his broom, would throw something out of calibration and didn't want to tell anybody.[146]

So the internal social world of the Instrumentation Laboratory—and then ultimately of the contractors who build the gyroscopes, since the Laboratory itself built only prototypes—had to be engineered. Inertial instrument production could not be routinized. Though the Laboratory made enormous efforts to lay down exactly the sequence of operations to follow in machining, assembling, and testing a gyroscope and extremely detailed drawings were produced, the knowledge of "how to do it" could not be transferred without face-to-face contact and hands-on experience.[147] Accordingly, it became standard practice for selected staff from the contractors to become "residents" at the Instrumentation Laboratory, for substantial periods of times. In 1965, for example, there were 287 such company residents. Laboratory staff were also seconded to the contractors during the crucial early stages of production.[148]

The external social world had to be engineered so that support was generated and sustained for the large and expensive activity in which the Instrumentation Laboratory was engaged. This was the second sense in which the project was one of heterogeneous engineering. Here Draper's key resource was

146. Ibid.
147. See, for example, Lester R. Grohe et al., *Handbook on the Assembly and Testing of the M.I.T. 20IG Integrating Gyro Unit* (Cambridge, Mass.: MIT Instrumentation Laboratory, December 1956, Report R-101).
148. Instrumentation Laboratory, *Annual Report* (Cambridge, Mass.: MIT Instrumentation Laboratory, April 1965, unpaginated).

people. He had extraordinarily extensive contacts in the armed services, many of whom had been seconded for periods to the Laboratory. Though this practice predated 1945, Draper formalized it around that time by establishing a "Weapons Systems Section" of the MIT Aeronautical Engineering Department that offered a range of courses in the basics of gyro culture and applications such as fire control. By 1955 5 Air Force officers had received a doctorate through the Laboratory, while 57 Air Force officers and 105 Navy officers had received masters degrees.[149]

Time and time again the contacts established in this way were crucial to the Instrumentation Laboratory's work. To take but one example, the original Air Force contract that started the inertial navigation development was negotiated through an Air Force colonel, Leighton I. Davis, who had done a masters degree under Draper in 1940–1941. During the war Davis had arranged for Draper to work on applying his gyro gunsight to aircraft, and in 1945 they were flying back together from inspecting test results on gunsights when they heard on the plane's radio the news of Japan's surrender and the end of World War II. Davis told Draper that funds would be freed up because many short-term projects would be canceled. To Draper, "the situation seemed to offer an unusual opportunity," and he immediately and successfully proposed to Davis that he be funded to begin work on a self-contained navigation system.[150]

The successes of the "internal" and "external" forms of heterogeneous engineering were of course wholly intertwined. The resources achieved by the second made the first possible, while the results achieved by the first sustained the second. Draper developed a reputation as a person who could "deliver." General Bernard Schriever recalls that in the early years of the ballistic missile program the aspect about which "the technical experts" were most negative was "the ability to come anywhere close to the target": "The one exception was . . . Stark

149. Aeronautical Engineering Department, Weapons Systems Section, *Report Prepared for Dean Soderberg*, September 1955 (MIT Archives and Special Collections, AC 12, Dean of Engineering, Box 1).
150. U.S. Air Force Oral History Program, interview with Lieutenant-General Leighton I. Davis (Office of Air Force History, Bolling Air Force Base, Washington D.C., K239.0512–668); Draper, "On Course."

Draper. . . . Stark is a close friend of mine and I used to say, 'Oh Christ, Stark, you know damned well you can't do that,' but I don't know of a single prediction that Stark made that hasn't been fulfilled."[151] To assure key sponsors such as Schriever that his predictions were on the way to being fulfilled, Draper used to carry with him on his trips to Washington and military bases what became known as "Doc's dollar bills," wallet-sized plots of the latest gyro test performances, with the scale removed to avoid breaching security.[152]

The situation of the group at Autonetics was different. Not only was Autonetics part of a major corporation—North American Aviation (later Rockwell International)—but, because it manufactured inertial components and systems as well as designed them, it grew to be much larger than the Instrumentation Laboratory. At its peak in 1964, at the height of its work on the Minuteman program (see chapter 3), Autonetics employed 36,000 people.[153] So the heterogeneous engineering required there was of the kind more familiar in a business environment.[154] A small instance is the selection of the name for the new division of North American Aviation. Autonetics manager John Moore insisted it should begin with an "A" so as to come at the start of directories and lists, and one of the company's public relations staff came up with the appropriately modern sounding combination of "automatic" and "cybernetics".[155]

This was no routine corporate environment, however.[156] The early atmosphere at Autonetics was one in which "persons would put in 20-hour days, and 80-hour weeks, not from Company insistence but from a messianic sort of urge." The idiosyncracies involved in the delicate art of gyro building were,

151. Schriever interview.
152. Palmer interview.
153. Moore interview.
154. John M. Wuerth of Autonetics described to me in some detail the organizational problems that had to be solved in building a major successful technological program of this kind (interview, Fullerton, Calif., February 23, 1985).
155. Moore interview.
156. For the later reflections of Autonetics's manager on matters such as these, see John R. Moore, "Unique Aspects of High Technology Enterprise Management," *IEEE Transactions on Engineering Management*, Vol. 23, No. 1 (February 1976), 10–20.

in the early years at least, tolerated, even cherished. Thus there was a "certain M.D. in Los Angeles who got tired of treating sick people, gave up a flourishing practice, and took a job as gyro technician in which line of work he attained exceptional skill. . . . He had a palsy and, in assembly, his hands would move in an undamped high-frequency oscillation until, at the last bounce, the parts would fit together perfectly."[157]

The division was fortunate enough to have missile projects to support it throughout its early years, notably the guidance system for the Navaho missile. But here too the flux of Air Force missile policy was disconcerting. In 1950, for example, the range and accuracy requirements for Autonetics missile work were increased, and as a result Autonetics's first inertial system had to be removed from testing and converted to stellar-inertial.[158] Navaho also suffered even more immediately than Snark from the ballistic "missile revolution." It was abruptly cancelled in July 1957. The jibe attached to it, after tests that "disappointed all," was "Never go, Navaho."[159] But by 1957 the inertial guidance and navigation work at Autonetics was sufficiently established and the eye for new niches for it alert enough for this to be only a temporary setback.

In their approach to inertial instrument design, Autonetics was much more eclectic than the Instrumentation Laboratory; no single paradigmatic device dominated. Early attention focused on gas-supported sensors analogous to those being developed by the Germans at Fort Bliss, but a wider range of devices was also investigated and developed.[160] This may have been in part because Autonetics lacked a technical dictator

157. Slater, *Twenty Years*, 1.
158. Ibid., 17.
159. Werrell, *Cruise Missile*, 98–99.
160. Contemporary descriptions of the inertial sensors developed in the 1940s for Navaho are to be found in J. M. Slater, "The North American Gyro-Stabilized Platform," in Scientific Advisory Board, *Seminar*, Vol. 1, 139–145, and S. F. Eyestone, "Accelerometers for the MX-770 Guidance System," ibid., Vol. 1, 146–151. Some important early Autonetics patents are John J. Gilvarry and David F. Rutland, "Accelerometer" (U.S. Patent 2,641,458, June 9, 1953, filed March 5, 1949); John M. Slater, Robert M. Benson, and Darwin L. Freebairn, "Gyroscope" (U.S. Patent 2,649,808, August 25 1953, filed January 24, 1949); John M. Wuerth, "Accelerometer and Integrator" (U.S. Patent 2,882,034, April 14, 1959, filed November 1, 1948); Shirley F. Eyestone and Wesley E. Dickinson, "Accelerometer" (U.S. Patent 2,995,935, August 15, 1961, filed November 1, 1948).

(John Moore arrived in 1948, when the work was already well under way.)[161] It may also have been because of the rather different demands placed by the effort to gain and protect a share in several different types of navigation and guidance markets. One example of this, Autonetics's development of the electrostatically suspended gyroscope, is discussed later.

The Black Box in Place, Almost

By around 1950, the black-box navigation work at the Instrumentation Laboratory and Autonetics had begun to bear fruit. Who first successfully "flew" an inertial navigator turns out, of course, to be a definitional matter. A case for Autonetics would rest on the XN1 inertial navigator, the first test flight of which was on May 3, 1950.[162] The Instrumentation Laboratory's case rests on FEBE, a stellar-inertial system. During FEBE's flight tests, which began on May 5, 1949, cloud cover interrupted stellar sightings "for periods of some tens of minutes," so the system had to operate in pure inertial mode. Yet the "loss of sun tracking information did not disturb results."[163]

These early tests were "successes." But the term is a relative one. FEBE, for example, was credited with a mean error of 5 nautical miles after flights ranging from 500 to over 1,700 nautical miles: not bad, one might say, but nowhere near the Air Force's specification.[164] Further, these early systems were scarcely black boxes in the sense of fully autonomous operation. For example, in the XN6 system (the final version of the inertial navigator developed by Autonetics for Navaho), an onboard digital computer was used. But "establishing and maintaining the first airborne digital navigational computer, especially in the early vacuum-tube form," was a "heroic feat." Test flight "mean-time-between-failure at about two hours

161. Slater, *Twenty Years*, 16.
162. Ibid.
163. Draper, "Origins," 457.
164. Four of the ten test flights were excluded in calculating this mean error. In the first, cloud cover interfered with the photographic system used to determine where the aircraft "actually was" as distinct from where FEBE "believed it to be." Two flights were excluded because of "equipment failures," and the fourth "because the terminal error departed abnormally from the mean value" (ibid., 456). I have not seen test results for the Autonetics XN1.

could only be realized with tender loving care by one devoted and unusual genius named Fred Johnson."[165]

In retrospect, of course, these seem like "teething troubles." Black-boxing was well under way, supported in the United States not just by the pioneering groups at Huntsville, Autonetics, and the Instrumentation Laboratory but by a wider "inertial industry" as both established gyro culture firms such as Sperry and new enterprises such as Litton saw an important potential market.[166]

Creating a Technology

If asked who invented inertial navigation, most American navigation and guidance engineers would probably reply unhesitatingly: Charles Stark Draper. But what did Draper invent, in the traditional sense of invention? One could perhaps claim for him the single-degree-of-freedom floated integrating gyroscope, though that claim would be disputed. In any case, this concerns one particular component technology, which, though important, was not unique.

And yet the naming of Draper would not be altogether wrong, though it does remain both ethnocentric and unfair to the other groups working simultaneously in the United States. For we have seen that the processes of creating the new technology of inertial navigation cannot be captured by the traditional notion of invention, with its single crucial "eureka moment." Making black-box navigation real was not the same as just thinking of it, or even thinking of it and then developing what had been thought of, and it could be said that Draper did more than any other individual to make it real. Organizing Gamow's defeat, knowing why it was important to speak to the janitor, keeping sponsors satisfied with "Doc's dollar bills"—

165. Slater, *Twenty Years*, 23 fn.
166. Black-boxing is never a technical absolute. To keep inertial navigators black boxes requires, even today, an extensive network of spares and repairs. Companies cite the necessity to establish this as a major disincentive to entry into the civil air inertial navigation market (occupied in the West by only three American companies—the AC Delco Division of General Motors, Honeywell, and Litton). In the civil sphere this network has to be created and sustained by the companies themselves, while in the military the armed services take responsibility for it. Civil air navigation is discussed in chapter 4.

these sorts of things are not generally regarded as technical work, yet all were part of the heterogeneous engineering that created inertial navigation.

Ultimately, though, the desire to allocate credit is futile. Because the creation of a new technology is a complex matter, it is not surprising to find that many people doing work of quite different kinds are involved, and asking who contributed the most is like asking which leg is most important in supporting a three-legged stool. And because the creation of a new technology is not a matter of individual genius, it is not surprising to find similar, though not identical, work being pursued in parallel.[167]

We have also seen why the invention of inertial navigation was not the mere application of science. Perhaps the most straightforward case of deduction from existing scientific theory, the relativity-derived argument about the vertical, was an obstacle to the development of inertial navigation rather than its cause. It is certainly true that science was an essential among the "sticks" that had to be brought together to form the ship in the bottle: it is hard to imagine inertial navigation being created without the physics of Galileo and Newton. Yet it was not created until nearly three centuries after them. The emergence of gyro culture, with its widely diffused mix of ideas, theory, practical skills, and devices, was probably the most important intervening factor.

But even that on its own was not enough. It is difficult to be sure that, with the level of resources being devoted to the enterprise in the Third Reich, or in the 1930s and 1940s in the Soviet Union, a black-box navigation system would ever have come into being. The United States was rich beyond European imagination, producing some 40 percent of the world's wealth in 1950.[168] Only there was it plausible that a technology still tainted with the suspicion of impossibility would receive support sufficient for not one, but several, substantial development efforts.

167. See William F. Ogburn and Dorothy Thomas, "Are Inventions Inevitable?" *Political Science Quarterly,* Vol. 34 (1922), 83–98, but note also the comments by Edward W. Constant, "On the Diversity and Co-Evolution of Technological Multiples: Steam Turbines and Pelton Water Wheels," *Social Studies of Science,* Vol. 8 (1978), 183–210.
168. Fred Halliday, *The Making of the Second Cold War* (London: Verso, 1983), 180.

That is not to deny that as late as the late 1940s things could still have gone badly wrong had the necessary processes of persuasion, credibility building, and ally recruiting been neglected. Nor is it to imply that money on its own was enough. For example, creating a production process that would yield inertial-quality gyroscopes was, as I have noted, a sociotechnical task of enormous difficulty and delicacy. Money was needed, but equally important, skills had to be recruited, fostered, and organized; attitudes to cleanliness had to be changed; and enormous amounts had to be learned.

Was inertial navigation invented because it was needed by society, or by powerful groups within society? To suggest this is the standard counter to the ideas that invention results from genius or the application of science. Yet it is equally misleading. In one sense, black-box navigation systems had always been "needed." Navigation had always been vulnerable to the inability to see landmarks, to clouds obscuring the stars, to electric storms interfering with radio signals, and so on. Yet in an equally valid sense it was never needed before it existed. If it had never been proved possible, the need for it would be as ambiguous as the "need" for an antigravity device.

One might say that needs are created simultaneously with the means of fulfilling them. Once the belief developed that a black-box navigator was possible, it was obvious that it was needed; before it was felt to be possible, it was meaningless to speak of a need for it.

Establishing the possibility of inertial navigation was thus the key task of its inventors. They were certainly helped by the immediate connection that could be created between black-box navigation and state power. No state with an eye to possible war and as rich as the United States was likely to deny resources to those who promised to create the means for autonomous navigation and guidance of its weapons, so long as the promise was credible. But the crucial link in the chain of connections, the credibility of the promise, was guaranteed by neither the structures of geopolitics nor by any autonomous logic of technology. Forging that link was the key role of the inventors of black-box navigation.

In the United States that link was firmly in place by the first half of the 1950s. Just as that happened, however, a sudden major change took place, concerning not the black box itself but the vehicle it was intended to navigate or guide. Instead of being a bomber, cruise missile, or short-range ballistic missile it was to be a ballistic missile of intercontinental range equipped with neither a conventional nor even an atomic warhead but with a hydrogen bomb—a V-2 grown monstrous in range and destructive power. Henceforth, for all the importance of the uses of inertial technology in aircraft, spacecraft, and other vehicles, its connection to this defining weapon of the nuclear age was most fraught with significance.

3

Engineering a Revolution

By the early 1960s, many of the main features of the current U.S. strategic nuclear arsenal were established. It had taken the form of a "triad" of intercontinental ballistic missiles (ICBMs), submarine-launched ballistic missiles, and bombers. The first two were under the control of the Air Force; the last under that of the Navy. The Army's nuclear missile role was restricted to shorter-range systems. Though much has changed since then—the revival of the cruise missile, for example—the pattern established during the 1950s and early 1960s structured much of what was to follow. Organizational roles were stabilized. Physical dimensions, such as the diameters of missile silos and launch-tubes, were set that continued to exert an influence into the 1970s and beyond. By the time of the assassination of President Kennedy in November 1963, even the size of the strategic ballistic missile arsenal had been set at more or less its current level of around 1,000 ICBMs and 600 submarine-launched missiles.[1]

That this would be the general shape of the American strategic nuclear arsenal was far from obvious in the early postwar years. Then it appeared that it would consist primarily of bombers supplemented by long-range cruise missiles, both under the control of the Air Force. The ballistic missile was seen appropriately as a short-range, Army "tactical" weapon. The shift from this perspective to one in which the ballistic missile was seen as the key modern strategic weapon is what I call the "missile revolution." Its causes, the role of considera-

1. The arsenal did not actually reach this level until 1967, but the key decisions had been taken by late 1963. See Desmond Ball, *Politics and Force Levels: The Strategic Missile Program of the Kennedy Administration* (Berkeley: University of California Press, 1980).

tions of accuracy and guidance in it, and its relations to the changing division of responsibilities between the Air Force, Army, and Navy are the topics of this chapter.

The central theme that will emerge is the inseparability of technology, nuclear strategy, and organization.[2] The missile revolution did not occur because a President or Secretary of Defense decided that the United States needed strategic ballistic missiles. It was engineered largely (though not exclusively) from below, and the missile's proponents had to reshape not just technology but also organizational structures and eventually national strategy. They were assisted by developments in other technologies, particularly the hydrogen bomb, and by a growing sense of a missile race with the Soviet Union. In part, the latter reflected external events, most dramatically the launch by the Soviet Union of the first artificial earth satellite, Sputnik, on October 4, 1957. But in part it was the result of the ballistic missile's proponents own efforts in changing, well before 1957, U.S. "insider" perceptions of the path of technical development being followed in the USSR.

These efforts were needed because in the 1940s and early 1950s the organizational environment of the U.S. armed services was not encouraging of the serious development of ballistic missiles of strategic ranges. But by the winter of 1953–1954 the tools—technological, political, and strategic—for that engineering had become available and were skillfully brought together by the ballistic missile's proponents.

The events of that winter did not end the missile revolution. The weapons they led to were fragile, delicate, and expensive, for all their destructive power. The numbers in which they were intended to be deployed were small by the standards of the early 1960s. In a second wave of the missile revolution they were replaced by missiles that were deployed in much larger numbers (table 3.1). The two key missiles—Polaris and Minuteman—were selected by the Kennedy Administration of 1961–1963 to be the foundation of the modern American stra-

2. I claim no novelty in seeing the interconnection of technology and organization in the missile revolution. That was the central theme of Robert L. Perry's classic study, *The Ballistic Missile Decisions* (Santa Monica, Calif.: RAND, 1967, P-3686), and of several studies, cited below, of particular missile systems. Only in recent years, however, has the history of U.S. nuclear war planning become clear enough for the role of the third variable, nuclear strategy, to be incorporated fully into the equation.

Table 3.1
The Main Missiles Discussed in this Chapter

Year of first deployment	Name	Type	Service	Guidance system
Land-based missiles:				
"First Generation"—Based above ground apart from Titan I, liquid fuel:				
1959	Atlas	ICBM	Air Force	Radio; replaced by inertial in later versions
1959	Thor	IRBM	Air Force	Inertial
1960	Jupiter	IRBM	Army	Inertial
1961/62	Titan I	ICBM	Air Force	Radio
"Second Generation"—Based in underground silos, solid or storable liquid fuel:				
1962	Minuteman	ICBM	Air Force	Inertial
1963	Titan II	ICBM	Air Force	Inertial
Submarine-based missile:				
1960	Polaris	SLBM	Navy	Inertial

ICBM = intercontinental ballistic missile
IRBM = intermediate range ballistic missile
SLBM = submarine launched ballistic missile

tegic missile force. Their emergence is examined in the latter half of the chapter.

As the missile revolution proceeded, the issue quickly became no longer whether strategic ballistic missiles were going to exist at all, but which organizations should develop and control them and how they should be designed: for example, whether their guidance should be black box, and what level of accuracy should be specified. Slowly—and the reader may be surprised at how slowly—the issue of what strategic ballistic missiles were for came to the fore, at least within the circle privy to questions of nuclear strategy and war planning. Gradually, different answers to that question emerged, tied to different technological priorities and located in different organizational settings.

These different answers—connections of technology, strategy, and organization—proved almost as enduring as the general form of the nuclear arsenal. In chapters 4 and 5 I examine their evolution up to the present, including the two major changes that have taken place. In this chapter, however, I describe how they came into being in the first place and how, in that process, American missiles and their guidance systems were shaped.

The Ballistic Missile as Orphan

By at least the late 1940s, if not earlier, it was widely agreed within the U.S. political elite that the United States and its allies faced an ideological, political, and possibly military challenge from the Soviet Union. With the collapse of efforts to achieve international control over the atomic bomb and the rejection of the strategy of meeting the "Soviet threat" with massive conventional force, it was clear that atomic weapons were going to be central to U.S. defense policy.[3] But it was very far from clear that ballistic missiles were going to become the primary means for the delivery of such weapons.

The grounds for skepticism were apparent. The atomic bomb was a strategic weapon: the idea that it could be used in a tactical, battlefield role emerged only slowly and at the beginning was heretical.[4] But to use the bomb strategically meant its delivery over long ranges, thousands of miles, unless missiles could be based on the soil of American allies close to the Soviet Union. Nuclear weapons were still heavy: could one build a rocket powerful enough to loft one thousands of miles? A ballistic trajectory implied leaving the earth's atmosphere and

3. See Alice Kimball Smith, *A Peril and a Hope: The Scientists' Movement in America, 1945–47* (Cambridge, Mass.: MIT Press, 1971); Gregg Herken, *The Winning Weapon: The Atomic Bomb in the Cold War, 1945–1950* (New York: Knopf, 1980); Lynn Eden, "Capitalist Conflict and the State: the Making of United States Military Policy in 1948," in Charles Bright and Susan Harding, eds., *Statemaking and Social Movements: Essays in History and Theory* (Ann Arbor: University of Michigan Press, 1984), 233–261. There is a massive literature on the origins of the cold war, in which matters such as the relative roles of the United States and the Soviet Union, the relative weight of political and economic considerations, and so on, have been discussed at great length: see J. L. Black, *Origins, Evolution and Nature of the Cold War: An Annotated Bibliographic Guide* (Santa Barbara, Calif.: ABC-Clio, 1986). Here, however, I am not concerned with these overall questions.
4. See Matthew Evangelista, *Innovation and the Arms Race: How the United States and the Soviet Union Develop New Military Technologies* (Ithaca, N.Y.: Cornell University Press, 1988). The tactical atomic weapon was pushed as an alternative to the "super" (hydrogen bomb) by J. Robert Oppenheimer, among others.

then reentering it at great speed. Could the payload survive reentry?

Perhaps the most crucial source of skepticism, however, was the assumption that such a weapon would be hopelessly inaccurate.[5] The experience of the V-2 was not necessarily encouraging. Its original accuracy specification, expressed in the traditional vernacular of artillery, was "2–3 mils," that is an average error no more that 2 to 3 units, either laterally or longitudinally, for each 1,000 units of range.[6] This would have translated into an average error of around half a mile at the V-2's roughly 200-mile range. It seems probable that this specification was chosen simply on the grounds of some hoped-for improvement relative to the artillery accuracy standards of the day.[7]

Nothing approaching the specified accuracy was achieved in practice, however. Operationally, the V-2's average error was more like four miles. This was not sufficiently bad to prevent it being able to hit within the area of a large metropolis such as London. The problem came, however, when one extrapolated this error to intercontinental ranges. Thus Army Air Forces General Henry H. Arnold, a greater enthusiast about missiles than many of his colleagues, noted that "the same degree of control on a 3,000 mile shot would lead to an average error of 60 miles—in other words, only one in 600 rockets would hit a city the size of Washington."[8]

Vannevar Bush, a figure of great prestige because of his central role in mobilizing American science and technology for the war effort, was the most important scientific skeptic. He told the Senate Special Committee on Atomic Energy in December 1945: "I say technically I don't think anybody in the world knows how to do such a thing [make an accurate,

5. Perry, *Ballistic Missile Decisions*, 6–7, writes that "it was generally assumed that once the guidance problem had been solved something could be done about the others."

6. Major-General Walter Dornberger, *V2* (London: Hurst and Blackett, 1954), 56.

7. These were around 4 to 5 mils (ibid.).

8. H. H. Arnold, "Air Force in the Atomic Age," in Dexter Masters and Katharine Way, eds., *One World or None* (New York: McGraw-Hill, 1946), 26–32, quote on p. 30.

nuclear-armed intercontinental ballistic missile] and I feel confident it will not be done for a long period of time to come."[9]

As it finally gained organizational independence from the Army, the U.S. Air Force—it became a separate service in September 1947—approached the world of atomic strategy and weaponry from what it took to be the firm base of experience and placed no reliance on technical progress moving the ICBM from futuristic speculation to reality. Strategically, the atomic bomb was seen essentially as a means of making air offensives of the kind directed against Germany and Japan during the Second World War more effective. Doubts had been raised as to whether these had actually destroyed morale and disabled industry to the extent predicted. But even before Hiroshima and Nagasaki "strategic bombing" had been fashioned into a terrible weapon. On the night of March 9–10, 1945, the Twenty-First Air Force, commanded by General Curtis LeMay is reckoned to have destroyed 40 percent of Tokyo in a single incendiary raid, leaving 125,000 casualties. To Air Force leaders, the atom bomb seemed finally to settle the question of the effectiveness of such raids.[10]

Crucially, experience also seemed to the Air Force to demonstrate the bomber's claim to be the prime delivery vehicle for the new weapon. The bomber was tested and reliable. Its capacity to get through even to heavily defended targets had been shown, and the new technology of jet engines would make it more difficult to intercept. Accurate bomber navigation was known to be a problem by Air Force leaders such as Generals Arnold and LeMay, but it was a problem to which technical effort was devoted.[11]

Air Force leaders of that generation, it is worth recalling, saw the airplane evolve in their lifetimes from the first fragile biplanes to the jet bomber. There should be little surprise that they were confident that this evolution would continue and that it was the path of technological development upon which

9. Bush, quoted in Perry, *Ballistic Missile Decisions*, 6.
10. Peter Calvocoressi and Guy Wint, *Total War* (Harmondsworth, Middlesex, England: Penguin, 1972), 853. For a graphic description of the horrors of the raid on Tokyo, see Michael S. Sherry, *The Rise of American Air Power: The Creation of Armageddon* (New Haven, Conn.: Yale University Press, 1987), 273–282.
11. See Monte Duane Wright, *Most Probable Position: A History of Aerial Navigation to 1941* (Lawrence: University Press of Kansas, 1972), chapter 7.

they wished to stake the bulk of the resources at their command.

However it looks in hindsight, a preference for bombers and a skepticism about missiles was thus at the time by no means irrational. The bomber, too, was organizationally a crucial technology for the U.S. Air Force. The distinctiveness of the strategic bombing role had been the Army Air Forces' best claim to achieve the status of a separate service, and it resonated well with many aspects of the wider culture. More than any other nation, the United States had a love affair with the airplane.[12] Rich in aircraft technology and production capacity but unwilling for a variety of reasons to field massive conscript armies, the country had been protected by geography from fear of retaliatory bombing.[13]

The strategic bombing role was thus central in the Air Force's rapid rise to become the dominant U.S. service: by the mid-to-late 1950's the Air Force's share of the budget was around 47 percent, compared to 29 percent for the Navy and Marines and 24 percent for the Army.[14] It was not surprising, furthermore, that an organization dominated by pilots, as the Air Force was (and to a substantial extent still is), should be reluctant to see its central strategic role filled by anything other than a manned system. Not only were manned aircraft, and particularly bombers, interwoven with the career structure of the Air Force, but "the people who had grown with manned bombers before and during World War II and who mostly stayed with them through the next decade developed an abiding affection for them, an affection based in some degree on what aircraft meant as a way of life, a symbol, a means of performing their military assignment."[15]

12. Joseph J. Corn, *The Winged Gospel: America's Romance with Aviation* (New York: Oxford University Press, 1983).
13. A crucial episode was the defeat in Congress in 1948 of the Truman Administration's plan for universal military training. Congress voted instead to spend substantially more on aircraft procurement. For a stimulating analysis, see Eden, "U.S. Military Policy in 1948."
14. George A. Reed, *U.S. Defense Policy, U.S. Air Force Doctrine and Strategic Nuclear Weapon Systems, 1958–1964: The Case of the Minuteman ICBM* (Ph.D. dissertation, Duke University, 1986), 31. I am grateful to Lynn Eden for drawing my attention to this valuable piece of work.
15. Perry, *Ballistic Missile Decisions*, 26.

One might even speculate that the term "manned" is no accidentally sexist usage. For the U.S. Air Force, at least, the commonly drawn parallel between the missile and the phallus is quite misleading; "real men" fly planes and have the "right stuff" to take the risks that entails,[16] rather than sitting on or under the ground waiting to push a button.

Although the bomber received much greater priority than the missile, the Air Force knew it could not wisely ignore the latter completely. But even within the sphere of missiles both judgments of technical feasibility and organizational preferences (the two were not wholly separable) led attention away from the ballistic missile. Central to both was the resemblance of the cruise missile to the airplane. A cruise missile could be seen as little other than a pilotless plane—surely not too difficult a technical step to take. The ballistic missile was often seen as further along the trajectory of technical difficulty. This judgment as to the trajectory of technical change was later to be reversed and was questioned even at the time, but the assumption was that cruise missiles could be in service before ballistic missiles. Nor was the ballistic missile the trajectory's end, and thus a final goal in itself worth heavy investment. Technical evolution was instead seen as more likely to culminate in the manned spacecraft as a weapon.[17]

16. See Tom Wolfe, *The Right Stuff* (New York: Bantam, 1980). Note that a wholly psychological explanation does not work. Fighter pilots are commonly held to possess the "right stuff" in greater degree than bomber pilots, and yet in the period we are discussing it was the latter who dominated the Air Force.

17. "Prior to 1952 or 1953, the Air Force favored the winged cruise missile over the wingless ballistic missile despite quantitative studies indicating that the former would be less accurate and dependable, as well as more costly than the latter. The primary reasons for this situation seemed to be emotional and cultural resistance." Kenneth P. Werrell, *The Evolution of the Cruise Missile* (Maxwell Air Force Base, Alabama: Air University Press, 1985), 104. Technological trajectories are further discussed in chapter 4. For the greater perceived ease of transition from the aircraft to the cruise missile, see Robert L. Perry's appendix (119–121) to his "Commentary" on I. B. Holley, Jr., "The Evolution of Operations Research and its Impact on the Military Establishment: The Air Force Experience," in Monte D. Wright and Lawrence J. Paszek, eds., *Science, Technology and Warfare: Proceedings of the Third Military History Symposium, United States Air Force Academy, 8–9 May 1969* (Washington, D.C.: Office of Air Force History, Headquarters USAF, and United States

Organizationally, the analogy of the cruise missile to the airplane made it matter more to the Air Force than the ballistic missile. In the division of labor agreed to during World War II, cruise missiles were assigned to the Army Air Forces, while ballistic missiles—probably because of the analogy to artillery—became the responsibility of the Army Ordnance Corps.[18] If the Air Force could gain control over cruise missiles, it would "eliminate . . . the infringement of other services into the aerodynamic field," as a contemporary memo put it.[19]

So although they were not pursued at maximum speed, cruise missile programs—including the Snark and Navaho programs, whose guidance aspects were discussed in chapter 2,—tended to be more highly prioritized by the Air Force than ballistic missile programs. "During fiscal years 1951 through 1954, the Atlas [ballistic missile] program received $26.2 million, while the Snark and Navaho got a total of $450 million."[20]

During the 1940s and early 1950s, then, "the Air Force's stance towards ballistic missiles can best be characterized as neglect and indifference."[21] The major early Air Force ballistic missile program was the "MX-774" proposed by Convair, the Consolidated Vultee Aircraft Corporation.[22] The project began in April 1946. It was canceled in May 1947 because of skepticism about whether it would lead anywhere, at least in the short or medium term. Convair was allowed to use remaining funds to test-launch prototype missiles it had built. All three test flights failed when the rocket engines cut out prematurely. An attempt to gain further Air Force funding in 1949 came to

Air Force Academy, 1969). For a statement of the view that the ICBM was a "bridge" between manned aircraft and manned spacecraft, see Curtis E. LeMay, "Let's Get on with the Job," in Ernest G. Schwiebert, *A History of the U.S. Air Force Ballistic Missiles* (New York: Praeger, 1965), 17–18.

18. Perry, *Ballistic Missile Decisions,* 4fn.

19. Memo from General Crawford to Commanding General, Army Air Forces, March 26, 1946, quoted in Edmund Beard, *Developing the ICBM: A Study in Bureaucratic Politics* (New York: Columbia University Press, 1976), 32.

20. Werrell, *Cruise Missile,* 104.

21. Ibid., 8

22. Consolidated Vultee Aircraft Corporation, *MX-774 Ground to Ground Missile* (San Diego, Calif.: Consolidated Vultee Aircraft Corporation, 1949; Summary Report ZR-6002–002).

nothing. The Convair ballistic missile, which was eventually transformed into the Atlas ICBM, was taken up again by the Air Force only in 1951, and then only with quite restricted funding.[23]

Ballistic missile work was pursued much more consistently by the U.S. Army and its team of German specialists.[24] In Army circles the ballistic missile met much less hostility and much less skepticism. It is hard to avoid the speculation that the ballistic missile's analogy to the artillery shell eased its acceptance, and, because relatively short-range weapons were perfectly admissible, an evolutionary approach seemed at first to be possible, gradually working from V-2 technology toward longer ranges.

But in fact a barrier existed to the Army building missiles of greater range—the power of the Air Force and the Air Force's conception of the appropriate division of responsibility between itself and the Army. The Army's job was land combat. Artillery and short-range missiles were appropriate adjuncts to that role, but anything with longer range strayed beyond it. In 1956, indeed, this restriction was formally codified by a memorandum from Secretary of Defense Charles E. Wilson: the Army was not to deploy missiles with a range greater than 200 miles.[25]

The very fact of its formal codification, however, indicates that by then the situation had changed.[26] With remarkable rapidity—the change can be dated to the closing months of 1953 and early 1954—the ballistic missile had begun to move from the status of orphan to that of prize to be competed for. Those developments constituted the first wave of the missile revolution.

23. Beard, *Developing the ICBM*, 50, 55, 63, 132.
24. The Army also supported ballistic missile work at the Jet Propulsion Laboratory at the California Institute of Technology.
25. Michael H. Armacost, *The Politics of Weapons Innovation: The Thor-Jupiter Controversy* (New York: Columbia University Press, 1969), 119.
26. Armacost writes (ibid., 82): "Guided missiles had not even been mentioned in the Key West and Newport Agreements [on the division of responsibilities between the Services]. The Joint Chiefs of Staff had never specifically restricted the range of Army surface-to-surface missiles, though it was presumed that they would develop and deploy only tactical ballistic rockets."

Engineering the Revolution

It is conventional to cite three causes of that sudden change: the invention of the hydrogen bomb, many times more powerful than the original atomic bomb; the start of the Eisenhower presidency in 1953; and the growing realization that the Soviet Union was pursuing a major long-range ballistic missile program.[27] All of these factors were certainly important, but none should be seen as a simple cause unproblematically generating the missile revolution as its effect. Rather, each was a resource to be used by the ballistic missile's proponents.

The significance of the first of these factors, the invention of the hydrogen bomb, is that it promised warheads with much greater destructive power for a given size—a higher yield-to-weight ratio. It thus reduced the weight a ballistic missile would have to lift, the accuracy it needed to destroy a particular target, or both.

It is certainly correct to focus attention on the role of the hydrogen bomb in the missile revolution. Its significance can, for example, be seen in the single most crucial document of that revolution, a secret report, now declassified, by Bruno Augenstein of the Air Force "think tank," RAND. The report's advocacy of a revised and greatly accelerated ICBM program was influential, and changing warhead technology was central to its argument. "Recent great advances in nuclear technology now permit us to package high yields in small, low-weight warheads," wrote Augenstein, and he went on to spell out the ways in which this made building an ICBM a less demanding task.[28]

It is, however, important to notice the status of Augenstein's claim about the yield-to-weight ratio. It was a prediction, not a statement of fact. Thus he quoted a weapons scientist, who referred first to a small, relatively low-yield atomic bomb and then to the possibility of much higher-yield, but not much heavier, hydrogen bombs:

27. Some authors mention only the first, or first and third. I have chosen the fuller list from Herbert York, *Race to Oblivion: A Participant's View of the Arms Race* (New York, Simon and Schuster, 1970), 83.
28. B. W. Augenstein, *A Revised Development Program for Ballistic Missiles of Inter-Continental Range,* Special Memorandum No. 21 (Santa Monica, Calif.: U.S. Air Force Project RAND, February 8, 1954), quote on p. 5.

Informal information available at this center indicates a yield of [security deletion]. Such a version of the Mk-7 [atomic bomb] could be made available within one or two years. It is believed that yields in excess of this may be anticipated in the future in the 1500 lbs weight class.

However, I would like to mention another question of interest, relating to bombs in the yield class from 1 to 2 megatons, which it should be possible to produce in much smaller weights than was considered possible in the past. It is expected that we will get these weapons to weigh 3000 lbs. and probably even somewhat less.[29]

The prediction of greatly increased yield-to-weight ratios was to prove correct. But it could still have been doubted in the winter of 1953–1954. True, a hydrogen bomb had been exploded for the first time on October 31, 1952, in the "Mike" test on Eniwetok Atoll, and no one with access to test results was left in any doubt as to the weapon's power. "The shot island Elugelab is missing," Gordon Dean of the Atomic Energy Commission told President Truman the following day, "and where it was there is now an under-water crater of some 1,500 yards in diameter."[30]

"Mike" was a test of a "wet" hydrogen bomb, fueled by cryogenically cooled liquid deuterium, and it was very far from a small or light device: it was twenty feet high, six feet in diameter, and weighed, with its deuterium fuel, over sixty tons.[31] Predictions about greatly enhanced yield-to-weight ratios were based, instead, on the "dry" (lithium deuteride fueled) hydrogen bomb then under development. No "dry" bomb was exploded until the "Bravo" or "Shrimp" test of February 28, 1954,[32] after many key conclusions had been reached. Augenstein's RAND report, for example, is dated February 8.

29. Ibid., 6. Augenstein's source has been deleted from the declassified version of the document. The Mk-7 was an atomic, not hydrogen, bomb, developed for possible tactical use. Mk-7 yields were variable, up to about 70 kilotons, and the weapon weighed 1,700 lbs. See Chuck Hansen, *U.S. Nuclear Weapons: The Secret History* (Arlington, Texas: Aerofax, 1988), 133–138.

30. Memorandum for the President, November 1, 1952, quoted ibid., 95fn. 190.

31. Ibid., 56. See also Thomas B. Cochran et al., *Nuclear Weapons Databook, Volume II: U.S. Nuclear Warhead Production* (Cambridge, Mass.: Ballinger, 1987), 16, 152, and the "Epilogue" to Richard Rhodes, *The Making of the Atomic Bomb* (New York: Simon and Schuster, 1988). According to the time zone used, the date of the "Mike" test is sometimes given as November 1.

32. Cochran et al., *Databook*, Vol. 2, 154

And even after "Bravo," which was still a very large device, it could be doubted that the requisite small sizes could be achieved. Some nuclear tests were still what participants termed "fizzles," producing only tiny yields. These included all three 1953 and 1954 tests by the new nuclear weapons laboratory at Livermore, set up by hydrogen bomb advocate Edward Teller to further its development.[33]

So it was not wholly irrational for Air Force Headquarters to continue to refuse to "recognize certain of our basic assumptions which rested on the thermonuclear [hydrogen bomb] breakthrough," as ICBM advocate Simon Ramo put it.[34] Or, viewing the question from the other side, there was no compulsion to wait for such a "breakthrough" before commiting oneself to full-scale ballistic missile development. The hydrogen bomb existed conceptually since the early 1940s, and even if one doubted—as many did—that it could be made a reality, there was fast progress in the miniaturization of the unquestionably real atomic bomb well before 1953–1954. Indeed, it would also have been perfectly possible simply to set aside the question of warhead miniaturization and design very large missiles. That seems to have been the path taken in the Soviet Union; and these differences in historical origin still have their traces in the generally larger size of Soviet than American ICBMs, an issue that dogged arms control negotiations in the 1970s and 1980s.

In sum, the coming into existence of the hydrogen bomb did not compel acceptance of the ICBM in 1953–1954, nor can its earlier absence explain on its own the previous lack of American enthusiasm for the ballistic missile. Matters have to be put both more loosely and more actively than that. The ceaseless efforts—technological, political, and organizational— by hydrogen bomb advocates, especially Edward Teller, and their promises of greatly increased yield-to-weight ratios, were a vital resource for the proponents of the ICBM in their lob-

33. Ibid., 27, 153–154.
34. Quoted by Beard, *Developing the ICBM*, 155fn. A 1958 White House memorandum noted that "thermonuclear warheads of megaton yields and acceptably small weight . . . were predicted *but doubted* four years ago:" G. B. Kistiakowsky to James Killian, February 13, 1958; Eisenhower Library, Office of the Special Assistant for Science and Technology, Box 12, Missiles (Jan.–Mar. 1958)(2), emphasis added. I am grateful to Graham Spinardi for a copy of this document.

bying and alliance-building work. That Teller did not himself deliver immediately on his promises[35] scarcely mattered. Others did, and the belief became established that increasing yield-to-weight ratios were a predictable reality upon which missile designers could base their plans.

The second, less commonly cited cause of the missile revolution, the advent of the Eisenhower Administration, may well be more important than the chronology of hydrogen bomb development in explaining why the winter of 1953–1954 was so crucial. After taking office early in 1953, the Eisenhower Administration reviewed defense policy while seeking to bring the protracted and bloody conventional conflict in Korea to a definitive end. The "new look" at defense policy—though like many such reviews, less new than its proponents would claim—shifted U.S. policy even further toward reliance on nuclear weapons to "contain" the Soviet Union. The doctrine of "massive retaliation," as it became known, was most famously stated by Eisenhower's Secretary of State, John Foster Dulles, in a speech to the Council on Foreign Relations on January 12, 1954. Local aggression might no longer be met solely with local, conventional defense as it had been in Korea, but in Dulles's words, "by means and at places of our own choosing."[36]

The threat was implicit but widely understood—to retaliate against Soviet aggression, even conventional aggression, with a massive nuclear attack. As the key internal document of the "new look," National Security Council Paper NSC-162/6, "Basic National Security Policy," put it: "In the event of hostilities, the United States will consider nuclear weapons to be as available for use as other munitions."[37]

Exactly what Soviet actions might trigger U.S. nuclear use was never wholly clarified, either in public or internally to the administration, and the administration also knew that by 1953–

35. The then Director of Livermore writes: "I must note here with some personal chagrin that, while it had been the Livermore Laboratory that was brash enough to promise that a one-megaton warhead could be made in a small enough physical package, it was the Los Alamos Scientific Laboratory that was mature enough at the time to actually provide one." York, *Race to Oblivion*, 43.

36. Michael Mandelbaum, *The Nuclear Question: The United States and Nuclear Weapons, 1946–1976* (Cambridge: Cambridge University Press, 1979), 46, quote on p. 51.

37. Freedman, *Evolution of Nuclear Strategy*, 76, quote on p. 82.

1954 the Soviet Union was beginning to achieve the capability for a massive nuclear counterblow of its own.[38] Nor, in practice, were plans and capabilities for conventional military action abandoned. Nevertheless, the "new look" provided an even more favorable context than before for nuclear weapons programs.

But there was nothing in this level of "grand policy" that implied that missile programs specifically should be begun or accelerated. Massive retaliation was understood as retaliation by the Strategic Air Command's bombers, and the Eisenhower Administration's concern to hold down the defense budget—a major factor in its shift in emphasis from conventional forces to "more bang for the buck" nuclear weaponry—did not bode well for expensive new projects. The President himself was a skeptic. Even in 1956, after the first wave of the missile revolution, he told the Joint Chiefs of Staff that "he did not think too much of the ballistic missiles as military weapons" and had agreed to accelerate the missile programs only because of their "psychological importance."[39]

The more immediate impact of the new administration on the fortunes of the missile was thus less at the level of "grand policy" than in changing who occupied key middle-rank positions in government. The first Republican president for two decades brought about the first complete change in such personnel since the 1930s.[40] This provided the opportunity for a remarkable episode of heterogeneous engineering that constitutes the core of the missile revolution.

The story has been well told elsewhere and needs only a brief summary.[41] The administration figure central to it was Trevor Gardner, Special Assistant for Research and Development to the Secretary of the Air Force. Gardner had been convinced by Convair technical personnel, "who spent a considerable amount of time traveling about the country, and especially lobbying at the Pentagon," of the feasibility and value of the ICBM. He took advantage of a review of missile pro-

38. See ibid., chapter 6, and David Alan Rosenberg, "The Origins of Overkill: Nuclear Weapons and American Strategy, 1945–1960," *International Security*, Vol. 7, No. 4 (Spring 1983), 3–71.
39. Quoted by Rosenberg, "Origins of Overkill," 45.
40. York, *Race to Oblivion*, 83.
41. See, for example, ibid. and Beard, *Developing the ICBM*.

grams ordered by the new Secretary of Defense, Charles Wilson—whose intention was to cut spending on them—to reshape them and build momentum behind the missiles.[42]

Instead of placing the review in the hands of the Air Force Scientific Advisory Board, as the Air Force wished, Gardner created an ad hoc committee chaired by mathematician John von Neumann. Von Neumann was independent of particular service interests, and his reputation for genius would lend authority to the committee's conclusion. Crucially, though, von Neumann was involved in the hydrogen bomb project and was an old friend and ally of the project's inspiration, Edward Teller. Like Teller, von Neumann believed—and argued to others—that light but enormously destructive hydrogen bomb warheads would shortly become a reality.[43]

Gardner chose the rest of the committee equally wisely, not simply making sure that it would conclude in favor of the ICBM but that it would possess the clout to help Gardner undermine opposition to this conclusion. He unabashedly told a 1963 interviewer that the committee's members were selected with a view to the influence they would have in generating support for the ICBM program.[44] To help their deliberations, Augenstein's RAND report laying out the case for an accelerated ICBM program was made available to them in advance of publication.[45] Gardner also arranged for technical advice to the committee to be given by Simon Ramo and Dean Wooldridge. They had just left Howard Hughes's Hughes Aircraft Company, where they had run an electronics laboratory, to set up

42. Beard, *Developing the ICBM*, 155, 156.
43. Von Neumann and Teller first met in their native Hungary in 1925: Stanley A. Blumberg and Gwinn Owens, *Energy and Conflict: The Life and Times of Edward Teller* (New York: Putnam, 1976), 130. Von Neumann shared Teller's fierce anticommunism, and his experience of the Manhattan project. He had organized the crucial computations assessing the hydrogen bomb's feasibility. In October 1953 von Neumann had reported to Air Force Chief of Staff Nathan F. Twining that "high-yield weapons of one to two megatons weighing 3,000 pounds or less could be expected well before the end of the decade." John T. Greenwood, "The Air Force Ballistic Missile and Space Programs (1954–1974)," *Aerospace Historian*, December 1974, 190–205, quote on p. 191.
44. Armacost, *Politics of Weapons*, 57. For the composition of the committee, see Beard, *Developing the ICBM*, 157–158.
45. Perry, *Ballistic Missile Decisions*, 12–13; Beard, *Developing the ICBM*, 156–158.

their own company, Ramo-Wooldridge. Ramo-Wooldridge was to win the role of providing technical support to the ICBM program, a task that helped it on its way to becoming the corporate giant TRW (Thompson-Ramo-Wooldridge).

In February 1954, the Strategic Missiles Evaluation Committee, as it was called, endorsed a revamped and accelerated ICBM program. Organizational as well as technical changes were suggested, with an agency to be set up to manage the program and circumvent Air Force opposition to the ICBM. With Secretary of the Air Force Harold E. Talbott strongly in support, the committee's report could not be ignored.[46]

The turning point had come. Air Force leaders saw the threat that the new agency, and thus the new weapon, might escape their control altogether. They hastened to ensure its establishment within the Air Research and Development Command. Gardner ensured that it was headed there by an ICBM enthusiast, General Bernard A. Schriever. The Western Development Division, as it was unrevealingly titled, was set up in the summer of 1954, with the Ramo-Wooldridge Corporation as support.[47] Though several changes of name followed—to Ballistic Systems Division, Space and Missile Systems Office, and currently Ballistic Missile Office—the core of the organizational structure of U.S. ICBM development had been created.

The third commonly cited cause of the missile revolution, the Soviet ballistic missile program, was certainly important in generating a sense of urgency, and became even more so after the launch of Sputnik in 1957. Yet here too nothing was unambiguous. Intelligence evidence of Soviet ballistic missile development had existed for years. As early as 1948, "a British communications intelligence team, posing as archaeologists in Iran, evidently monitored the V-2 tests from Kapustin Yar," and a high-level defector, Colonel Georgi Tokaty-Tokaev, brought firsthand details of the Soviet program.[48]

46. Ibid., chapter 6.

47. Ibid., 170–171.

48. John Prados, *The Soviet Estimate: U.S. Intelligence Analysis and Soviet Strategic Forces* (Princeton, N.J.: Princeton University Press, 1986), 57. Tokaty-Tokaev was later to publish his experiences. See G. A. Tokaev, *Comrade X* (London: Harvill, 1956), especially chapter 27.

Yet no alarmist conclusions were drawn. In 1952 most of the German missile engineers who had been taken to the Soviet Union (see chapter 6) were allowed to return to the West and were interrogated by U.S. intelligence. Though their stories are now taken as indicating high-priority ballistic missile development in the Soviet Union, at the time it was concluded that "the Soviets were following the same cautious, step-level approach through air-breathing [cruise] missiles that the United States was."[49]

This picture of the Soviet work was changed as part of the "missile revolution," not as an independent cause of it. New intelligence estimates were generated during the review by the von Neumann committee, and Trevor Gardner summarized them: "The lump impression gained from these estimates is that the Soviets are significantly ahead of us in the strategic missile field."[50]

The change does not appear to have been brought about by new kinds of evidence about the Soviet ICBM program. Such evidence did eventually become available, but only well after the missile revolution. In June 1957, an American U-2 aircraft, on a surreptitious high-altitude reconnaissance flight over the Soviet Union, took a series of photographs that analysts interpreted as showing an ICBM on the test pad at the Soviet missile and space center at Tyuratam near the Aral Sea. In August of that year radio signals intercepted by U.S. intelligence in Turkey were interpreted as telemetry from a Soviet ICBM test flight.[51] It is interesting that the increasingly alarmist analyses fueled by such data were in their turn to be reversed. The "missile gap" that appeared to be proven by the Soviets succeeding in test flying an ICBM before the Americans was subsequently dismissed as myth. In June 1961, the Central Intelligence Agency estimated that the Soviet Union had deployed 125 to 150 ICBMs, with the Air Force estimating 300. But by September of that year the official estimate had been reduced to thirty-five, and by the end of the decade

49. Beard, *Developing the ICBM*, 163
50. Memorandum to Assistant Secretary of Defense (Research and Development), February 16, 1954, quoted in Beard, *Developing the ICBM*, 164.
51. William E. Burrows, *Deep Black: Space Espionage and National Security* (New York: Random House, 1986), 95.

former defense officials were suggesting that the correct figure may have been four.[52]

Myth or not, though, the "missile gap" gave added urgency to American ballistic missile programs. As resources swung toward them, it was suddenly the other technologies for delivering nuclear weapons that were under threat. The jet-propelled B-52, then beginning its entry into service, was to be the last U.S. long-range strategic bomber for nearly thirty years. Cruise missiles began to be seen as a more, rather than less, difficult technical problem than ballistic missiles, and they too went into a long eclipse. Not only the Air Force, but the military aircraft industry, had to adapt. Some suppliers—Lockheed, Convair, Martin, Boeing, and North American Aviation—succeeded in gaining a significant proportion of ballistic missile business, while Douglas, in 1956 the leading U.S. military aircraft supplier, found its military business more than halved as it failed to gain substantial missile work.[53]

Guidance, Accuracy, and the Black Box

A technological revolution had taken place.[54] Having examined its overall contours, let me now turn to the role of considerations of accuracy and of the black-box nature of inertial guidance.

As we have seen, in the early cruise missile work the Air Force imposed a demanding accuracy requirement of 5,000 feet (a little less than a statute mile—about 1.5 kilometers). That this was an Air Force imposition, rather than one freely adopted by the technologists, can perhaps be seen in one participant's 1949 complaint that the Air Force "will not even allow us a nautical mile" (6,076 feet).[55] When the MX-1593 (Atlas)

52. Ball, *Politics and Force Levels*, 55, 95; John Prados, *The Soviet Estimate*, 118. Navy intelligence, Ball notes (p. 95), gave a June 1961 estimate of 10.
53. G. R. Simonson, "Missiles and Creative Destruction in the American Aircraft Industry, 1956–1961," *Business History Review*, Vol. 38 (1965), 302–314.
54. Q.v. Edward W. Constant II, *The Origins of the Turbojet Revolution* (Baltimore: Johns Hopkins University Press, 1980).
55. S. E. Weaver, "The MX-775 Midcourse Guidance System," in Scientific Advisory Board, Office of the Chief of Staff, United States Air Force, *Seminar on Automatic Celestial and Inertial Long Range Guidance Systems*, February 1, 2,

contract was specified in January 1951, the requirement for the circular error probable (the radius of the circle around the target within which half of the warheads should land) had become more than three times more stringent at 1,500 feet, even though none of the cruise missile programs had achieved the 5,000-foot goal.[56]

The justification for this stringent requirement was probably—though I have no definite evidence to this effect—that 1,500 feet was roughly the circular error believed achievable in "blind" bombing by the high-flying bombers of the early 1950s: the missile had to be as accurate as the bomber to be acceptable.[57] But implicit in the requirement was that the missile would not be acceptable, for even the ICBM's greatest enthusiast would have regarded a 1,500-foot circular error probable as beyond the bounds of the state of the art: it was two decades before U.S. ICBMs achieved that accuracy (see appendix A).

So a key task of the ICBM's proponents was to relax considerably the accuracy requirement placed upon it, a task in which they were greatly aided by the belief that it would soon be possible to pack much increased destructive power into a small warhead. A review committee two years before the von Neumann group began the process by restoring the circular error probable requirement to one mile.[58] The crucial RAND Corporation report advocated a much more generous two to three nautical miles.[59] When an enlarged von Neumann committee,

3 1949 (Oral History [Draper] Box 2, MC 134, MIT Archives and Special Collections), Vol. 1, 37–46.

56. Beard, *Developing the ICBM*, 143 n. 34.

57. In March 1954, a little later than the period we are discussing, a Strategic Air Command briefing stated that "The current CEPs for all bomber crews using simulated radar bombing from 25,000 feet vs. industrial targets is about 1,400 feet. For visual bombing this drops to 600 feet. Tests were run on . . . Select and Lead crews only to see how much better they were than the average. The measurements of 202 simulated drops from 25,000 feet gave an average CEP of 1,390 feet for radar bombing and 352 feet for visual." Captain William B. Moore, U.S. Navy, 'Memorandum Op-36C/jm,' 18 March 1954, reprinted in David Alan Rosenberg, "'A Smoking Radiating Ruin at the End of Two Hours': Documents on American Plans for Nuclear War with the Soviet Union, 1954–55," *International Security*, Vol. 6, No. 3 (Winter 1981-82), 3–38, 18–28 (quote on p. 23).

58. Beard, *Developing the ICBM*, 143.

59. Augenstein, *A Revised Development Program*, 38.

set up to help steer the Atlas project, cursorily reconsidered the matter, it relaxed the requirement even further, to five miles.[60]

Even so, the demands on guidance were far from trifling. With the great range of an ICBM, a quantum leap in accuracy beyond that achieved in the V-2 was needed. Given what they saw as the urgency of the program, and the substantial technical skepticism about it, ICBM proponents knew they had to make the right choices in guidance technology.

By 1953–1954 the general outlines of a black-box ballistic missile guidance system were clear to the proponents of inertial navigation. It would consist of three accelerometers, once for each direction in space, mounted on a stable platform held in constant orientation relative to the fixed stars by gyroscopes. Because a ballistic missile's trajectory took it high above the surface of the earth, the advantages of keeping the platform horizontal did not seem as great for ballistic missile use as they did to at least some of the proponents of aircraft and cruise missile inertial navigation. The platform would be oriented prior to firing. Its own inertial sensors would allow it to find the vertical (the process called erection), while it would be aligned in azimuth (i.e., in the horizontal plane) by optical reference to an external landmark. Any in-flight change in orientation of the platform would be sensed by the gyroscopes, and a feedback system would cause servomotors to rotate the platform to counteract the change in orientation. In the most general terms, then, the "paradigm" ballistic missile inertial guidance system described in chapter 1 existed in conceptual terms by the time of the first wave of the "missile revolution."

Several problems existed, however, in turning it from concept to reality, at least for strategic ranges. Only simple analog computing devices were available: the digital computer was far from being light, rugged, or reliable enough for ballistic missile use. Yet these devices had to do something quite complicated. They had to make use of the output of the inertial sensors to guide the missile to a position and velocity such that rocket motor cut-off could be initiated, or the warhead separated from the missile, and the subsequent unguided trajectory

---

60. "Draper, always a technical optimist, said he could foresee much better CEPs than five miles, and we simply took that figure as a conservative estimate." York, *Race to Oblivion*, 89.

would take the warhead to the target. The solution to this problem had to bear in mind the need, discussed in chapter 2, to "program in" the earth's gravity field. Considerable mathematical ingenuity was devoted to addressing this problem, and different solutions to it are described below and in chapter 6.

Another problem was the familiar one of the performance of the inertial sensors. Gyroscope performance was seen as less of the issue here, because the stable platform had to be held stable only for the few minutes of ballistic missile boost phase, not the hours of flight of a bomber or cruise missile.[61] (This changed later when the missile guidance system itself, rather than an external reference, began to be used for prelaunch alignment.) Instead accelerometer performance became the perceived "reverse salient." Very large accelerations had to be measured very accurately. So the Instrumentation Laboratory, for example, moved around the time of the missile revolution from the fairly simple "pendulum" accelerometers it had developed for aircraft navigation to an updated version of the pendulous integrating gyro accelerometer (PIGA).[62]

A third problem was weight. Intercontinental range placed a very high premium on lightness of the guidance system, since every extra pound in the latter meant a smaller and less destructive warhead. But early inertial navigation systems were large and heavy. In part this arose from the perceived need to combat friction in the bearings of the inertial sensors by building big gyro wheels. A wheel weighing six pounds would not be uncommon in an early system, and a system using PIGAs and single-degree-of-freedom gyros would have six such wheels.[63] Gradually a different philosophy began to evolve,

61. Thus the German guidance specialists were able to obtain perfectly respectable missile accuracies in the 1950s (albeit at ranges shorter than that of an ICBM) with gyroscope drift rates of a few tenths of a degree per hour, rather than the hundredth of a degree per hour or thereabouts held necessary for aircraft or cruise missile inertial navigation. See chapter 2.

62. Charles Stark Draper, "The Evolution of Aerospace Guidance Technology at the Massachusetts Institute of Technology, 1935–1951: A Memoir," in R. Cargill-Hall, ed., *Essays on the History of Rocketry and Astronautics: Proceedings of the Third through Sixth History Symposia of the International Academy of Astronautics*, Vol. 2 (Washington, D.C.: NASA, 1977), 240–242.

63. John Slater of Autonetics wrote in 1949: "The precision gyro problem comes down to minimizing the quantity M/H, where M is uncertainty torque about the input axis and H is angular momentum. We must make H as large as is practicable and M as small as possible." J. M. Slater, "The North Amer-

especially at the Instrumentation Laboratory, which emphasized the benefits in terms of accuracy of small size in inertial sensors—acceleration and temperature changes, for example, might have less detrimental effect on a small than on a large gyro. But in 1953–1954 the process of miniaturizing inertial guidance or navigation systems was still in a very early phase.

So, despite the progress made toward black-box navigation, the candidacy of a black-box missile guidance system for an ICBM was no fait accompli. Inertial guidance had a powerful rival in radio guidance. The RAND report outlined the case for the latter, its author being, he recalls, "almost delegated" to advocate radio guidance:

Two general classes of guidance can be considered—a self-contained purely inertial system, with the guidance components on the missile, and one of several types of radio guidance systems, in which the important components are on the ground. The latter type is preferable for several reasons: Radio guidance techniques have been demonstrated on the Corporal missile, which is now essentially operational. There is no assurance that inertial components of the required accuracy, reliability and producibility can be provided at an early date. Radio guidance also has the operational advantages that since the complex equipment is on the ground, no miniaturization is needed, parallel operations can be performed for greater reliability, environmental control is possible, and the system is re-usable for many missiles.[64]

ican Gyro-Stabilized Platform," in Scientific Advisory Board, *Seminar*, Vol. 1, 140–141. Large angular momentum could of course also be obtained by increasing the speed at which the gyro rotor spun, but "increasing wheel speed to increase angular momentum so that error torques will cause less gyro drift can require so much wheel motor power and associated thermal problems that the center of gravity shift caused by temperature gradient can rise faster than the angular momentum so as to result in poorer performance" (D. G. Hoag, "Ballistic-missile Guidance," in B. T. Feld et al., eds., *Impact of New Technologies on the Arms Race* [Cambridge, Mass.: MIT Press, 1971], 82–83).

64. Interview with Hyman Shulman and Bruno Augenstein, Santa Monica, January 14, 1986; Augenstein, *A Revised Development Program*, 13. Corporal was a product of one of the seedbeds of American missile and space technology, the Jet Propulsion Laboratory of the California Institute of Technology in Pasadena. With origins in Second World War work, Corporal had by the early 1950s evolved into a short-range (up to 80 mile) missile for the U.S. Army. Its radio guidance system was similar to that developed for the V-2 (see chapter 2), but with the addition of a World War II radar to track the missile. The history and technical evolution of Corporal is described in

Proponents of radio guidance greatly outweighed those of inertial guidance in the decision-making processes central to the ICBM. Charles Stark Draper, though consulted by the von Neumann committee,[65] was not a member of it, while no fewer than three of its nine members were drawn from the California Institute of Technology, where the radio-guided short-range Corporal missile had been developed. They included Louis Dunn, Director of the Jet Propulsion Laboratory, who was actively involved in the Corporal program.[66] The experience of Ramo and Wooldridge also seemed to point in the same direction.[67] Furthermore, the Convair group, the only people who could claim to have built a prototype ICBM, had chosen radio guidance.[68]

Much more than accidental choice of personnel was involved in the resultant decision to guide ICBMs, initially at least, by radio. In the United States of the 1950s, "probably no technique is more generally associated with guided missiles in the public mind than is radar," and the prestige of radar, following its great success in World War II, was high.[69] Radio, radar, and electronic components had had a great deal of effort lavished on them, much more than inertial components. The "electronikers," as they were sometimes called by proponents of inertial guidance, were indeed well entrenched.

But they were not unopposed. Inertial guidance, too, had well-placed allies. As we saw in chapter 2, by the mid-1950s a significant number of Air Force officers had passed through Draper's MIT Instrumentation Laboratory to earn higher degrees. On their return, some of them formed what they themselves now describe as a "guidance mafia," keen to see the

Clayton R. Koppes, *JPL and the American Space Program: A History of the Jet Propulsion Laboratory* (New Haven, Conn.: Yale University Press, 1982).

65. See above, note 60.

66. Beard, *Developing the ICBM*, 157, and Koppes, *JPL*, 46–47.

67. There had been two main tasks at the Hughes electronics laboratory they headed: "one to link an airborne search radar with a Sperry gunsight fitted with a computer and the other an electronic seeker for an air-to-air missile for bomber defense." James Parton, *"Air Force Spoken Here": General Ira Eaker and the Command of the Air* (New York: Adler and Adler, 1986), 459.

68. They developed a radio guidance system called Azuza. See Consolidated Vultee Aircraft Corporation, *MX-774*, and John L. Chapman, *Atlas: The Story of a Missile* (London: Gollancz, 1960), 50–52.

69. Renne S. Julian, "Radar," in Allen E. Puckett and Simon Ramo, eds., *Guided Missile Engineering* (New York: McGraw-Hill, 1959), 379–405.

ideas they found at the Instrumentation Laboratory put into practice and willing to channel funds to ensure that technological development there continued.[70]

The "guidance mafia" had a powerful argument in the black-box nature of inertial guidance. Indeed, there seemed to be something of a consensus among Air Force decision makers that inertial guidance was the preferable technique *if* it could be proven to be "mature."[71] It was, for example, accepted as an option for use in future variants of Atlas, as the lead ICBM program had come to be called.[72]

From the beginning, too, the ballistic missile Western Development Division of the Air Force had a member of the "inertial mafia" as its guidance expert. B. Paul Blasingame had been seconded by the Air Force in 1947 to gain his doctorate at Draper's MIT Instrumentation Laboratory. Initially, he was constrained to use radio guidance. The constraint came both from the physical weight of the inertial systems of the period—none was available weighing less than 200 pounds, the maximum calculated permissible for the ICBM program—and also

70. "Mafias"—strong personal networks, crossing formal institutional boundaries, of people committed to a particular technical approach—are an interesting feature of technological politics around the U.S. Air Force, and more generally. See the description of the "fighter mafia" in Jacob Goodwin, *Brotherhood of Arms: General Dynamics and the Business of Defending America* (New York: Times Books, 1985), chapter 8.

One result of the work of the inertial guidance mafia can be seen in a 1961 report on activities at the Air Force's Ballistic Systems Division: "There is a continuing program supported at MIT's Instrumentation Laboratory to generate new ideas in inertial guidance. There is no comparable program in the radio guidance category." Anonymous, "USAF pushes Missile Systems State-of-Art," *Aviation Week and Space Technology,* (September 25, 1961), 131–133, quote on p. 133.

71. Interview with General Bernard Schriever, Washington, D.C., March 25, 1985.

72. Inertial guidance was developed for Atlas and eventually deployed on the E and F variants, not by any of the groups discussed in chapter 2, but by the Arma division of American Bosch Arma Corporation. Arma was a traditional naval gyroscope supplier. Its work for Atlas was successful, and the firm's inertial sensor work was remarkably innovative. But it was afflicted by severe organizational and managerial difficulties and never won a major contract to replace its Atlas work. Interview with Paul H. Savet, New York City, September 26, 1986. (Savet was a key technical figure in Arma's inertial work and an important contributor to the development of both accelerometer and gyroscope technology.)

from the "social weight" of the supporters of radio guidance in the von Neumann committee and Ramo-Wooldridge. But Blasingame was able to win approval for Air Force sponsorship of inertial development work at both MIT and the AC Spark Plug Division of General Motors, which had been brought in by the Air Force to help put into practice the work done on bomber navigation by Draper.[73]

A further Air Force ballistic missile program called Thor provided the opportunity to move this work from back-up to primary status. Thor was an intermediate rather than intercontinental range ballistic missile. Gardner and General Schriever initially had opposed its development by the United States, fearing it would dilute the priority of the ICBM program, and proposed instead that work on it be devolved to Britain. But as fears of a Soviet missile lead grew, and as the Army intermediate range missile program, Jupiter, gained momentum, Air Force leaders decided that their own intermediate range missile was a necessity. On November 28, 1955, the Western Development Division was instructed to proceed with it.[74]

With a range specification of 1,500 rather than 5,500 nautical miles, Thor seemed like "duck soup" compared to the Atlas ICBM. Guidance system weight was less of a constraint, and the accuracy specification—one nautical mile circular error probable—was moderate.[75] Blasingame talked "late into the night" with Draper and left convinced that inertial guidance could do the job and that AC Spark Plug, backed up by MIT, could produce working systems. Though there was still skepticism to be met, it was overcome, and Thor went ahead with inertial as the primary system and a radar system as back-up.[76]

73. Interview with B. Paul Blasingame, Santa Barbara, September 12, 1986.
74. Armacost, *Politics of Weapons*, 55–64.
75. The eventual weight of the Thor inertial guidance system was reported to be 650 to 700 pounds: Norman L. Baker, "Polaris Pioneers Future Ballistic Missile Design," *Missiles and Rockets* (February 1958), 137.
76. Interview with Blasingame. Performance data for missiles of this period have now been declassified. See the originally "top secret" presentation by General N. F. Twining, Chair of the Joint Chiefs of Staff, to the Defense Subcommittee of the House Committee on Appropriations, January 13, 1960 (Eisenhower Library, White House Office Staff Secretary Subject Series, Defense Department Subseries Box 11, Folders General Twining Posture Briefing). I am grateful to Graham Spinardi for a copy of this document. J. Scott Ferguson, *Historical Collection Projects* (Cambridge, Mass.: Charles Stark

It was an important contract for Draper's Instrumentation Laboratory. Finally, their accumulated expertise could be put to work in a major national inertial guidance program. MIT-designed gyroscopes and accelerometers were produced for Thor by AC Spark Plug, though not without considerable difficulties and an initially low yield of satisfactory instruments.[77] The onboard computer was analog, but an ingenious mathematical scheme called Q-guidance, developed at the Instrumentation Laboratory by Richard H. Battin and J. Halcombe Laning, shifted the bulk of computational requirements out of the missile.[78] The components of the "Q-matrix" could be computed well in advance of firing, and all that was needed on board was to use these components in simple repetitive calculations that could be implemented in analog hardware.

Inertial guidance worked. Thor may not actually have met its accuracy specifications: its circular error probable in test flights was only "less than 2 nautical miles."[79] But that was of relatively limited concern. The success with Thor helped the MIT-AC Spark Plug team win a place in the next generation of U.S. ICBMs. Titan II (Titan I was a more advanced analog of Atlas, also radio-guided) was designed for inertial guidance from the outset, and the MIT-AC team designed a lighter, more sophisticated and more accurate inertial system for this missile. The first of the missiles we have discussed to survive into the modern strategic world, Titan II became operational in 1963 and was phased out in the mid-1980s.[80]

The ultimate cause of the shift to inertial is quite clear: the perceived vulnerability, in the eyes of the key military decision makers involved, of radio guidance. General Schriever recalls: "Obviously the self-contained system was a hell of a lot better

Draper Laboratory, Inc., August 3, 1979, C-5249), unpaginated, reports the radio/inertial issue slightly differently: "Thor had a dual guidance system that used both inertial and radio guidance. But the Air Force's increasing confidence in all-inertial systems dictated a design change in 1957 which dropped the radio guidance prior to Thor's first test launch."
77. Interview with Robert G. Brown, Topsfield, Mass., April 8, 1985.
78. For Q-guidance, see Richard H. Battin, "Space Guidance Evolution—A Personal Narrative," *Journal of Guidance and Control*, Vol. 5, No. 2 (March-April 1982), 97–110; interviews with Battin, Cambridge, Mass., October 11, 1984 and J. H. Laning, Jr., October 3, 1984, Cambridge, Mass.
79. Twining, *Presentation*, 12.
80. There is a useful description of Titan II in Cochran et al., *Nuclear Weapons Databook*, Vol. 1, 112.

from a military standpoint . . . the radio . . . system required a very substantial ground installation which was highly vulnerable and we wanted to get rid of that as soon as we could."[81]

It is interesting, however, that the shift from radio to inertial was also seen as having a cost in loss of accuracy. Although the advocates of inertial guidance, like Blasingame, regarded it as merely a "presumption" that inertial was less accurate than radio, it was a presumption that was widely shared.[82] The radio-guided Atlas D had performed far better than its five-mile specification, well enough indeed for President Eisenhower—by then under increasing political pressure for allegedly having allowed a missile gap between the United States and the Soviet Union to have opened up—to boast of its two-mile accuracy in his 1960 State of the Union Message.[83] By 1963, Atlas D test firings were landing within a nautical mile of the target 80 percent of the time and were significantly more accurate than the inertially guided Atlas E and F, which could manage to be within only 1.5 nautical miles 80 percent of the time.[84] In a direct comparison of radio and inertial guidance, R. C. Berendsen of General Electric, a firm with involvement in both, described radio guidance as being twice as accurate as inertial. Though inertial guidance would improve, so would radio guidance, and it would keep its relative advantage.[85]

It was not to be. The inertial "guidance mafia" won a swift victory. Radio guidance was never seriously considered for an American ICBM after Atlas and Titan I. Funding of radio guidance dried up, and so the technique did not improve. Proponents of ICBM radio guidance can still be found (see chapter 7), but in advocating a technique whose development ceased thirty years ago against one whose development has

81. Interview with Schriever.
82. Interview with Blasingame.
83. "In 14 recent test launchings, at ranges of over 5,000 miles, Atlas has been striking on an average within 2 miles of the target." Dwight D. Eisenhower, in Fred C. Israel, *The State of the Union Messages of the Presidents, 1790–1966*, Vol. 3, (New York: Chelsea House Hector, 1966), 3100–3101.
84. George Alexander, "Atlas Accuracy improves as Test Program is Completed," *Aviation Week and Space Technology* (February 25, 1963), 54.
85. Berendsen's paper, to a meeting of the Institute of Radio Engineers and American Rocket Society, is summarized in Philip J. Klass, "ICBM Guidance Techniques Compared" *Aviation Week* (May 2, 1960), 159–160. General Electric was involved both in Atlas radio guidance and Polaris inertial guidance.

continued, they are at a huge disadvantage. So the initial victory of one technique in practice became irreversible.

But why the initial victory? Radio guidance supporters would have had difficulty in credibly disputing that inertial was less vulnerable: the early 1960s move of putting missiles into silos increased pressure to get rid of any remaining installations above ground. But why did the obvious counterargument—that radio was more accurate—not count for more? The answer appears to be that in the crucial period in which radio and inertial guidance were real competitors, missile accuracy was not highly prioritized. Why that was so requires us, finally, to begin to examine what the ICBM was thought to be for, its place in nuclear strategy.

Accuracy and Nuclear Strategy

If we can summarize what we have learned about the setting by the Air Force and others of ballistic missile accuracy specifications, it would be to say that there was a rough accuracy level—perhaps around two nautical miles circular error probable—that came to be considered acceptable. Early attempts to impose a much more stringent requirement are best seen as a means of preventing the missile from ever emerging as a serious competitor to the bomber rather than as a realistic hope of developing a missile with those characteristics. Once the emergence of the missile became irreversible, the pressure for great accuracy vanished, at least within the Air Force. (As we shall see later, the Army was different.)

Implicit in the setting of an accuracy specification of two miles or worse was the assumption that the primary target of the ICBM—and probably also the intermediate-range ballistic missile—would be cities. The RAND report on the ICBM made this explicit. The "major relaxation of requirements" it was advocating was in circular error probable (CEP), and thus it was "necessary to show that an acceptable target destruction capability exists with CEP's of several miles." The "target system" for which this was shown to exist was "the larger urban areas in the Soviet Union. CEP's of 2 to 3 n[autical] miles are adequate to produce tremendous damage against this target system using one-half megaton (500 KT) yield. For the largest urban areas, a CEP of even five miles still provides a significant capability."[86]

To the extent that there was serious thinking at all about its purpose—and reading the record of these years, one is struck by how little of this there was—the ICBM was originally conceived as a countercity, rather than a counterforce weapon. But even to put matters in this way is to be somewhat anachronistic, for this distinction was not made as sharply before 1960 as it has been since.

Before 1960 the Air Force had a straightforward view of nuclear war, located most strongly in the Strategic Air Command, which operated the bomber fleets and, when they were deployed, the ICBMs as well. As suggested earlier, this view was originally little more than of World War II strategic bombing conducted with atomic weapons. Of course, as the Soviets developed an operational nuclear arsenal during the early 1950s, the risk of a Soviet nuclear attack on the United States or its allies increased. Naturally, "first priority" had to be allocated to "the destruction of known targets affecting the Soviet capability to deliver atomic bombs," as the Joint Chiefs of Staff decided as early as August 1950.[87]

But this formal prioritization of counterforce, to use the modern strategic term, had in the 1950s fewer consequences than might have been expected in the shaping of the American missile arsenal. The reason did not lie in its apparent assumption that the United States would strike the first nuclear blow. Air Force leaders, particularly General LeMay, seemed confident that they would receive intelligence warning of an imminent Soviet attack and would be able to get an American strike in first. A 1954 Navy memorandum, which General LeMay later agreed contained no "major misstatements," quoted LeMay's answer to the question "How do SAC's [the Strategic Air Command's] plans fit in with the stated national policy that the United States will never strike the first blow?":

I have heard this thought stated many times and it sounds very fine. However, it is not in keeping with United States history. Just look back and note who started the Revolutionary War, the War of 1812, the Indian Wars, and the Spanish-American War. I want to make it

86. Augenstein, *Revised Development Program*, 7, 8, 10.
87. David Alan Rosenberg, "Origins of Overkill," 16–17.

clear that I am not advocating a preventive war; however, I believe that if the U.S. is pushed in the corner far enough we would not hesitate to strike first.[88]

"Preventive war," in the parlance of the time, was an attack on the Soviet Union "out of the blue." This was not without highly placed advocates, who saw a unique opportunity to strike before the Soviet Union developed a full-fledged nuclear arsenal, but it was never considered seriously as a real, immediate option by either Truman or Eisenhower. What LeMay was considering instead was "preemptive war," the crucial difference being its triggering by intelligence warning of a Soviet strike. Preemption, notes historian David Alan Rosenberg, lay in a "gray area."[89] It was never officially approved as national policy, but it was not officially ruled out either, though it certainly existed in the war plans as an option, as Chairman of the Joint Chiefs of Staff, General Lyman L. Lemnitzer, told President Kennedy in a top-secret 1961 briefing.[90]

Nor was intelligence on the location of Soviet targets necessarily a problem. Many assumed it to be so. Satellite reconnaissance did not begin until the early 1960s, and the U-2 program, with its high-altitude spy flights, did not begin until 1956 (and even many defense "insiders" would not have known of it because of the secrecy surrounding it until Soviet air defenses finally succeeded in shooting down the U-2 containing Gary Powers on May 11, 1960). But although knowledge of this was similarly restricted, the CIA and Air Force intelligence had captured crucial data from the Germans and Japanese. During the 1930s, Soviet cartographers had engaged in a major

88. Moore, "Memorandum Op-36C/jm," 23. Rosenberg, who reprints this memorandum in his "Smoking, Radiating Ruin," checked its accuracy with LeMay (p. 6).
89. Ibid., 13. For details concerning both "preventive" and "preemptive" war, see Rosenberg, "Origins of Overkill."
90. "It [the Single Integrated Operational Plan] may be executed as a total plan (1) In retaliation to a Soviet nuclear strike of the U.S., or (2) As a preemptive measure." "The JCS Single Integrated Operational Plan—1962 (SIOP-62)," briefing to the President, September 13, 1961. Reprinted in Scott D. Sagan, "SIOP-62: The Nuclear War Plan Briefing to President Kennedy," *International Security*, Vol. 12 (1987), 22–51, 41–51, quote on p. 50. The 1962 was a reference to Fiscal Year 1962; the plan came into effect on April 15, 1961.

exercise to update and improve the old czarist maps. Copies of their work were obtained by the Germans, and through them the Americans. Both German and Japanese intelligence were also in possession of large sets of photographs of Soviet military bases, and these too fell into U.S. hands.[91]

This baseline information was supplemented by a variety of means. In Project WRINGER, prisoners of war being repatriated from the Soviet Union were interrogated on a massive scale.[92] "Skyhook" high-altitude balloons were carried, along with their cameras, by upper-atmosphere winds from Western Europe to Japan (though knowing what it was they had photographed was a problem).[93] Even before the U-2 a limited number of photoreconnaissance overflights were carried out, and many electronic "ferret" missions were flown near the Soviet borders.

Whether U.S. intelligence really knew where the key Soviet counterforce targets were could still be challenged, and some Air Force leaders were skeptical.[94] But confidence that it did know was high among RAND personnel privy (most were not) to the information, and by the latter part of the 1950s even Strategic Air Command leader General LeMay was expressing optimism about the abilities of U.S. intelligence.

There also seems to have been confidence that warning of a Soviet attack would be forthcoming. To prepare a large-scale attack on the United States, the Soviet Union would have to move much of its bomber force to staging bases in the Arctic, the Americans believed.[95] They seem to have felt confident of detecting this. In these Arctic bases, Soviet bombers would be sitting ducks. "If I see that the Russians are amassing their planes for an attack," LeMay told Robert Sprague, co-chair of the high-level Gaither Committee, who was concerned about Strategic Air Command vulnerability to a Soviet preemptive

91. Gregg Herken, *Counsels of War* (New York: Knopf, 1985), 80. Rosenberg, "Origins of Overkill," paints a more pessimistic view of the target information available to the United States in this period. This difference of opinion amongst historians perhaps reflects differing assessments at the ti ·e.
92. Rosenberg, "Origins of Overkill," 21.
93. Ibid., also Prados, *The Soviet Estimate*, 29–30.
94. Reed, *Minuteman ICBM*.
95. See for example the 1959 Congressional testimony of Chair of the Joint Chiefs Nathan Twining and the 1963 testimony of Secretary of Defense Robert McNamara, quoted in Sagan, "SIOP-62," 31fn. 28.

strike, "I'm going to knock the shit out of them before they take off the ground."[96]

So a preemptive strike against Soviet nuclear forces seemed—at least to key Air Force operational planners like LeMay—to be feasible and not incompatible with national strategy. Four interrelated factors, however, prevented this from translating into a perception of the missile as a counterforce weapon (and therefore into a high accuracy requirement) and meant that the obvious argument for radio guidance—that if one was going to preempt, then its vulnerability was irrelevant while its accuracy was crucial—never seems to have been made. These factors were, first, that these planners could see the counterforce mission only as a supplement to the countercity role; second, that the bomber, not the missile, was their preferred counterforce weapon; third, that the technical specification of missiles was largely and deliberately kept separate from operational strategy; and, fourth, that counterforce targets in the 1950s and early 1960s were not "hard."

We have already touched on most of these factors. The countercity role was not merely what men like LeMay knew best. In both Europe and the United States, it had been the key justification, in the interwar period and onward, for investment in air power and for the organizational independence of air forces. "Terror bombing" had been criticized as both immoral and militarily ineffective, perhaps most stingingly by senior naval officers who wished greater funding for their aircraft carriers and carrier-based planes, rather than the Strategic Air Command's heavy bombers. The Air Force's strategy, one rear admiral asserted in public Congressional testimony in 1949, was "ruthless and barbaric . . . random mass slaughter of men, women and children . . . militarily unsound . . . morally wrong . . . contrary to our fundamental ideals."[97] But this criticism had been fought off, and the senior officer in the "admirals' revolt," Chief of Naval Operations Admiral Louis Denfield, was fired.[98] So to abandon the countercity role would

96. Quoted by Fred Kaplan, *The Wizards of Armageddon* (New York: Simon and Schuster, 1983), 134, on the basis of an interview with Sprague; Herken, *Counsels of War*, 79–81.

97. Rear Admiral Ralph Ofstie, testimony to House Armed Services Committee, October 1949, quoted in Kaplan, *Wizards*, 232.

98. David Alan Rosenberg, "American Postwar Air Doctrine and Organization: The Navy Experience," in Alfred F. Hurley and Robert C. Erhart, eds.,

have been, for Air Force leaders of that generation, to give up the essence of what they stood for and had successfully defended.

Their operational plan, therefore, was that the U.S. attack on the Soviet Union, whether preemptive or retaliatory, should be a single, massive, close to simultaneous, all-out offensive—the "Sunday punch" as it was lightly described in Strategic Air Command circles.[99] Cities, industrial plants, airfields, military bases, and nuclear weapons facilities would all be hit, as would Soviet air defenses. Historical research, notably by David Alan Rosenberg, has revealed the broad contours of the Strategic Air Command's "basic war plan" as it evolved toward the first national Single Integrated Operational Plan (SIOP), approved by the Joint Chiefs of Staff on December 2, 1960. That plan strongly reflected these Strategic Air Command preferences.[100]

Within that overall framework, the missile was not seen as a counterforce weapon. Since it could not be retargeted in flight, any uncertainty in target location would have counted against it. More surprisingly, the ballistic missile's short time of flight does not seem to have been seen as an argument for assigning counterforce targets. If anything was considered especially time-urgent, it was the air defense sites, the early obliteration of which would facilitate the bomber offensive. It seems to have been accepted by operational commanders in the 1950s that bomber forces took a long time to get airborne. What had provoked Sprague to question LeMay about Strategic Air Command vulnerability was an exercise in September 1957 in which not a single bomber, apart from a few already on a test flight, had been able to take off within six hours of receipt of the alert signal.[101] So it could be argued that there was a good chance that even slow "Sunday punch" bombers would be able

Air Power and Warfare. Proceedings of the Eighth Military History Symposium, USAF Aacademy, 1978 (Washington, D.C.: Office of Air Force History, Headquarters USAF and United States Air Force Academy, 1979), 245–278, p. 263. See also Harvey Sapolsky, *The Polaris System Development: Bureaucratic and Programmatic Success in Government* (Cambridge, Mass.: Harvard University Press, 1972), 5–6; David Alan Rosenberg, "American Atomic Strategy and the Hydrogen Bomb Decision," *Journal of American History*, Vol. 66 (1979), 62–87.

99. Herken, *Counsels of War*, 82.

100. Rosenberg, "Origins of Overkill."

101. Kaplan, *Wizards*, 132.

to catch Soviet nuclear forces on the ground, if Soviet alert levels were no better than American.

As we have seen, the Air Force's operational commanders had in any case lost control of the technical specifications of missiles. In the layered world of defense, the civilians who played a large part in setting those specifications—though certainly "insiders" possessed of security clearances—may not have been fully aware of the realities of war planning.[102] Between 1951 and 1955 even the Joint Chiefs were not given access to the details of Strategic Air Command planning and were briefed only after a formal request from Air Force Chief of Staff Nathan Twining in June 1955.[103] So at least some of those helping to set missile specifications might not have thought much beyond the predominant lay perception of nuclear war, that the United States would attack second, not first, and the attack would take the form of the destruction of Soviet cities.

Many of the Air Force officers in Schriever's Western Development Division had had enough contact with the Strategic Air Command to know that war planning was more complicated than that. But they also knew that there was little genuine interest in the ICBM at the top of the Strategic Air Command, and that to design a large-warhead, accurate, fast-reaction ICBM to match the Strategic Air Command's apparent operational requirements was, at least in the short run, an impossibility.[104] Gardner's skilled organizational engineering had given them enough autonomy to ignore those requirements. So they too accepted that specifications be set with countercity attacks in mind, as had been quite explicit in the RAND report drawn on by the von Neumann committee.[105]

Finally, even had those who set the specifications wanted to plan for counterforce attacks, this orientation might well not have affected too much what they did. The degree of "hardening" of military targets—U.S. or Soviet—during the 1950s

---

102. Robert Sprague, though not to my knowledge directly involved in the setting of missile specifications, is a good case in point. He was unquestionably an "insider"—heavily involved in the defense industry as well as acting chair of the enormously influential Gaither Committee reevaluating defense policy—but was taken aback at what he learned from LeMay. See Kaplan, *Wizards*, 134.
103. Rosenberg, "Origins of Overkill," 37.
104. Letter from B. P. Blasingame, July 30, 1989.
105. Augenstein, *A Revised Development Program*.

was negligible by the standards of later decades. Bombers were parked in the open or in relatively flimsy structures; early missile sites were all above ground, and first generation liquid-fueled missiles were enormously vulnerable. Destroying such targets, especially with the predicted new generation of hydro-gen bombs, did not require much greater accuracy than destroying cities or industrial complexes. The difficulties of locating the counterforce targets and making sure that you did indeed get your blow in first loomed larger than the accuracy of weapon delivery. The missile's ineffectiveness against hard-ened targets was raised as an argument against it and in favor of the bomber as early as 1957,[106] but this does not seem to have been a major plank of the case against it and certainly not one that required the missile's advocates to counter by giving enormous priority to accuracy.

All of these factors—the supplementary role of counterforce in dominant Air Force thinking, the allocation of the bomber rather than the missile to the counterforce role, the lack of direct influence from operational planning on the setting of missile specifications, and the relative lack of hardening of counterforce targets—were soon to change. The extent of the influence of at least the first three on the shaping of Air Force missile design, can, however, perhaps be gauged by comparing that missile work with that undertaken by the Army, especially in the most directly comparable cases—the Air Force's Thor and the Army's Jupiter.

Jupiter versus Thor, and the Army's Exit

There were many similarities between Thor and Jupiter. Both were early, inertially guided, liquid-fuel intermediate-range ballistic missiles, launched not from silos but above ground. Some of their differences were fairly obvious consequences of the different traditions of the two services and of their conflict. The Air Force was quite happy that Thor was a missile

---

106. No fewer than eighty ICBMs would be required "to achieve 50 per cent probability of destruction" of a target hardened to withstand an overpressure of 100 pounds per square inch, according to WSEG (Weapons Systems Evaluation Group) Report No. 23, *The Relative Military Advantages of Missiles and Manned Aircraft* (May 6, 1957), 22, 27. (I owe this reference to Graham Spinardi; the document is in the National Security Archive, Washington, D.C., Kaplan files.)

launched from fixed bases such as existing airfields. The Army, on the other hand, struggled to make Jupiter mobile, battling against both the missile's considerable bulk and the opposition of the Air Force, which was trying (eventually successfully) to gain operational control over it. In the words of Army General Maxwell Taylor, "a mobile missile needs Army-type troops to move, emplace, protect and fire it . . . a decision to organize mobile ballistic missile units would in logic have led to transferring the operational use of the weapon back to the Army—where it should have been all the time."[107]

The most interesting comparison, though, is in terms of attitudes to accuracy. The Army's Jupiter was planned to be the most accurate ballistic missile of its day, so accurate that a May 1957 Weapons Systems Evaluation Group report comparing missiles and manned aircraft contained an overtly skeptical footnote suggesting, in effect, that the Jupiter accuracy goal was unrealistic.[108] The German ballistic missile team was under constant pressure from the Army to increase accuracy. This pressure was never designed to block the program, as it may have been in the Air Force. "The Army would lay down a particular accuracy, and wait for our arguments whether it was possible. We had to promise a lot, but were fortunate."[109] The extent of this pressure on the designers of Jupiter's guidance system, together with the relative absence of such pressure on the designers of Thor, in all probability contributed to test results that were interpreted as suggesting that Jupiter was perhaps as much as four times as accurate as Thor.[110]

107. Quoted by Armacost, *Politics of Weapons*, 146.
108. WSEG Report No. 23, 22–23. The sanitized version of this that was released has the accuracy goals for Atlas, Titan, Thor, Jupiter and Polaris deleted, but their relative values can be deduced from the table on p. 23. The skeptical footnote is on p. 2. The Weapons Systems Evaluation Group was "the Pentagon think tank of the Joint Chiefs of Staff" (Kaplan, *Wizards*, 93).
109. Interview with Dr. Walter Haeussermann, Huntsville, Alabama, March 22, 1985.
110. Twining, *Presentation*, 12, quotes Jupiter's circular error probable in test firings as "one half-mile" as compared to Thor's "less than 2 nautical miles." Another factor may of course have been that Thor was the Instrumentation Laboratory's first ballistic missile guidance contract, while the Huntsville team was much more experienced in this application of inertial technology. In chapter 7 I discuss the process of construction of accuracy figures for missiles.

Why did the Army award higher priority to ballistic missile accuracy than the Air Force? One reason seems likely to have been simply the competition between the two services, where as the relative "underdog" the Army might have been inclined to "sell" its product hard. The Weapons Systems Evaluation Group study noted that if the accuracy predictions for Jupiter were correct, which they doubted, "they would indicate that Jupiter is the most promising weapon for development."[111]

The other likely reason for the Army's high priority on accuracy is the development within the Army in this period of an approach to nuclear strategy quite at odds with the dominant Air Force view. This approach emphasized the use of nuclear weapons in situations short of all-out war, and "If wars were to be kept limited, such weapons would have to be capable of discriminately hitting only tactical targets."[112] While the public articulation of the "limited war" approach was made by civilian strategic theorists, the most famous being Henry Kissinger, it found an eager audience in the Army, who could use it to contest the hegemony of the Strategic Air Command.[113]

But the "technical success" of Jupiter in becoming apparently the most accurate ballistic missile of its day counted for little in the end. Jupiter had a nominal range of 1,500 nautical miles but considerable "growth potential." On September 7, 1956, a Jupiter-C missile flew 3,300 miles. "News of the accomplishment—to the manifest distress of the Army—was suppressed in the interest of averting an upset in the so-called precarious balance between the services."[114] The Air Force ultimately won operational jurisdiction over Jupiter, and the next Army missile, Pershing I, was deliberately restricted to a much shorter range than Jupiter (about 400 nautical miles).[115] "The initiative [for Pershing I] was provided by [Department of] Defense officials anxious to divert Army rocket development effort away from areas of direct conflict with the Air Force."[116] With this restriction imposed on them, and with the lure of space

111. WSEG Report No. 23, 2.
112. Armacost, *Politics of Weapons*, 46.
113. Kissinger's most famous statement is his *Nuclear Weapons and Foreign Policy* (New York: Harper, 1957). See Freedman, *Nuclear Strategy*, 102, 108, for Army support of Kissinger's views.
114. Armacost, *Politics of Weapons*, 115.
115. Cochran et al., *Databook*, Vol. 1, 289.
116. Armacost, *Politics of Weapons*, 223–224.

exploration following the establishment of the National Aeronautics and Space Administration (NASA), it was not surprising that the final big project of the von Braun team should be not a missile but a space booster: NASA's Saturn, which carried the Apollo moon mission into space, guided by a Huntsville-designed system.[117]

Two decades later, the Army was to return to the intermediate-range ballistic missile business with Pershing II. Pershing II was originally intended to be a short- (again 400 nautical mile) range system, but in the late 1970s, in the context of the Intermediate Nuclear Forces deployment in Europe, its range was considerably extended.[118] This decision, "the most mysterious and controversial part of the NATO modernization decision," and certainly an extremely consequential one in terms of the nuclear politics of the 1980s, was probably the result of a change in the wider political/strategic context making a longer-range missile possible. Certain elements in the Army and in Martin Marietta, the prime contractor for Pershing I, had been seeking such a decision as early as 1960.[119]

The tale of Pershing II would, however, take us too far from our central thread. From the end of the 1950s onward, the central institutional actors in the strategic ballistic missile business in the United States were the Air Force and the Navy, not the Army. And center stage was occupied not by any of the missiles we have discussed so far but by two new ones: the Polaris submarine-launched ballistic missile and the Minuteman ICBM. To the incoming Kennedy Administration in 1961,

117. For Saturn guidance, see W. Haeussermann, "Saturn Launch Vehicle's Navigation, Guidance and Control System," *Aeronautica*, Vol. 7 (1971), 537–556 and NASA and others, *Saturn 1B News Reference* (n. p.: NASA, December 1965). The Saturn system used externally pressurized gas bearing gyroscopes and accelerometers (see chapter 6) already developed for Pershing I (interview with Haeussermann).

118. Cochran et al., *Databook*, 294; Christopher Paine, "Pershing II: The Army's Strategic Weapon," *Bulletin of the Atomic Scientists*, Vol. 36, No. 8 (October 1980), 25–31; and F. Clifton Berry, "Pershing II: First Step in NATO Theatre Nuclear Force Modernization?" *International Defense Review*, Vol. 8 (1979), 1303–1308.

119. Diana Johnstone, *The Politics of the Euromissiles: Europe's Role in America's World* (London: Verso, 1984), 14. *Aviation Week*'s "Industry Observer" reported in April 1960 that "Martin Co. is preparing a proposal for a 1,000-mi. version of the Pershing tactical missile" (*Aviation Week*, April 11, 1960, 23).

Minuteman and Polaris were weapons vastly superior to those we have discussed, and it was around them that the American strategic arsenal was constructed in the 1960s. Minuteman and Polaris, then, constituted the "second wave" of the missile revolution. As Polaris was the earlier of the two, let us discuss it first.

Enter the Navy

Just as it was not obvious that an ICBM could or should be built, so it was not obvious that a submarine-launched ballistic missile could or should exist. During the war, German rocket scientists conceived of being able to attack the United States by having a submarine tow missiles in cannisters across the Atlantic, firing them near the American coast. But the project, code-named Prüfstand XII (Test Stand 12), never left the drawing board.[120] The U.S. Navy had shown sporadic interest in missiles: for example, awarding the Sperry Gyroscope Company a contract in 1917 to develop Lawrence Sperry's idea of a pilotless "flying torpedo."[121] This interest quickened just before, during, and after World War II. But despite the "Viking" high-altitude rocket, the test firing of a V-2 from the aircraft carrier USS *Midway* in 1947, and a "Large Bombardment Rocket" program, most Navy missile programs were, like Sperry's flying torpedo, cruise missiles or pilotless planes, not ballistic missiles.[122]

In any case the Navy remained predominantly committed to traditional weapon systems, particularly aircraft carriers and other surface ships. Submarines did grow in importance, especially as a result of vigorous heterogeneous engineering by the

120. Ordway and Sharpe, *Rocket Team*, 55.
121. Thomas P. Hughes, *Elmer Sperry: Inventor and Engineer* (Baltimore: Johns Hopkins Press, 1971), chapter 9.
122. The best study of U.S. Navy missile work before Polaris is Berend Derk Bruins, *U.S. Naval Bombardment Missiles, 1940–1958: A Study of the Weapons Innovation Process* (Ph.D. thesis, Columbia University, 1981). See also Rear Admiral D. S. Fahrney (USN ret.), "Guided Missiles: U.S. Navy the Pioneer," *American Aviation Society Journal*, Vol. 27, No. 1 (Spring 1982), 15–28, and Wyndham D. Miles, "The Polaris," in Eugene E. Emme, ed., *The History of Rocket Technology: Essays on Research, Development and Utility* (Detroit: Wayne State University Press, 1964), 162–174.

"father" of nuclear propulsion, Admiral Hyman Rickover.[123] But while in retrospect it seemed a logical next step to weld together into a single technological system the nuclear-propelled submarine and the ballistic missile, at the time that seemed, even to Rickover, technologically overambitious and a potential massive drain on resources urgently needed for other tasks. "The components needed ... did not yet exist, and there was no guarantee that any amount of research and money could bring them into existence. There was no small proven nuclear warhead of sufficient yield. An accurate guidance system, an adequate fire control system, a suitable navigation system, were all missing. There wasn't even a concept as to a launching system.'"[124]

There was, however, within the Navy a "tacit and informal low-level alliance" of ballistic missile proponents.[125] On their own they had nothing like the clout needed to generate a real ballistic missile development program, rather than sporadic experimental work. But, as in the case of the ICBM, external events suddenly gave them the opportunity to be effective.

One such event was part of the continuing turmoil in the Eisenhower Administration over defense policy: the creation of a special review committee chaired by James Killian of MIT. Navy ballistic missile proponents, circumventing their superiors, channelled papers to the Killian Committee through the Navy officer assigned to work with it.[126] The Committee concluded in September 1955 that a sea-launched intermediate-range ballistic missile system should be developed in addition to the land-based systems. As the recommendation was endorsed by the National Security Council and President Eisenhower, it was powerful external backing for the ballistic missile proponents' case.[127]

It might have seemed to have been all that was needed. But that was far from the case. With two intermediate-range ballistic missile programs, Thor and Jupiter, in existence, the

123. See Norman Polmar and Thomas Allan, *Rickover* (New York: Simon and Schuster, 1982).
124. V. Davis, *The Politics of Innovation: Patterns in Navy Cases* (Denver: University of Denver Graduate School of International Studies Monograph Series in World Affairs No. 3 (1966–67), 25.
125. Ibid., 22.
126. Ibid., 23.
127. Miles, "Polaris," 478.

Department of Defense—under pressure to keep the defense budget down—was reluctant to provide new money for a third, while most of the Navy would have been very unwilling to see it funded from the existing Navy budget. The potential impasse was avoided by a change in personnel at the very top of the Navy. In August 1955 the outgoing Chief of Naval Operations, an opponent of taking on a major ballistic missile development program, was replaced by Admiral Arleigh A. Burke.

Burke's appointment leapfrogged seniority and was seen as an attempt by the civilian Secretary of the Navy to make the service less conservative.[128] Though this may have rankled, there was no doubting Burke's loyalty to his service, his desire to develop a distinctively naval contribution to national strategy for the nuclear age, and his preparedness to fight the Navy's corner in bureaucratic disputes with the Air Force. "They're smart and they're ruthless," he told a later Secretary of the Navy. "It's the same way as the Communists. It's exactly the same techniques."[129]

What Burke saw was that the Navy had to get involved in the missile revolution, on pain of permanent exclusion, and as soon as he was appointed he began to do what he felt was necessary to achieve that goal. To begin with, he needed to compromise. Given that funding could not be found for a program exclusive to the Navy, a partner had to be found. The Army, already feeling the threat of its own exclusion, was less unwilling than the Air Force. An agreement was reached for the Navy to participate in the Jupiter program, with the Army retaining responsibility for the missile itself and for its land-based launch system and the Navy taking on the task of developing a means of launching Jupiter from ships and eventually submarines.[130] This joint Army-Navy program was approved by the Secretary of Defense on November 8, 1955.[131]

128. There is a biography of Burke by David Alan Rosenberg in Robert William Love, Jr., ed., *The Chiefs of Naval Operations* (Annapolis, Maryland: Naval Institute Press, 1980).
129. Transcript, "Adm. Burke's Conversation with Secretary Franke, 12 August 1960," Arleigh Burke Papers, quoted by Kaplan, *Wizards*, 265.
130. Miles, "Polaris," 478.
131. Harvey Sapolsky, *The Polaris System Development: Bureaucratic and Programmatic Success in Government* (Cambridge, Mass.: Harvard University Press, 1972).

Sea-launching the large, liquid-fueled Jupiter was not an attractive proposition, though one Navy officer rationalized it thus: "We were prepared to take the chance that we might lose a submarine or two through accidental explosions. But, then, there are some of us [in the military] who enjoy, or at least are acclimated to, the idea of risking our lives."[132] Burke accepted the compromise because he felt speed to be of the essence. His internal Navy memorandum of December 2, 1955, giving the joint program top priority within the Navy, made this quite clear: "I think that the first service that demonstrates a[n intermediate-range ballistic missile] capability . . . is very likely to continue the project and the others may well drop out. The missile must be fired from a ship just as early as possible even though the equipment in the ships is not as desirable as can be conceived."[133]

Although many would have expected the missile work to be assigned to one of the Navy's existing technical bureaus, Burke established a new program office for it, simply named the Special Projects Office. His goal was to establish a base for the project independent of the bureaus and their rivalries. Burke chose the Office's head, naval aviator Real Admiral Raborn, shrewdly: "I did not want a technical expert because a technical expert would be too narrow-minded. I wanted an aviator because if this missile were successful it would jeopardize the aviation branch."[134]

Under Raborn's leadership, the Special Projects Office developed a distinctive style, well analyzed by political scientist Harvey Sapolsky. The Office's leaders knew that the program faced a range of possible opponents, not only outside the Navy but inside it as well—there were widespread fears that the

132. Captain R. G. Shutt, U.S. Navy Special Projects Office, in December 1961 interview with Robert E. Hunter. Quoted in Hunter, *Politics and Polaris: The Special Projects Office of the Navy as a Political Phenomenon* (Wesleyan University, Senior Honors Thesis, June 1962), 132. I am grateful to Graham Spinardi for a copy of Hunter's remarkable thesis.
133. Arleigh Burke, "Memorandum for Rear Admiral Clark (Op.51) and Rear Admiral Raborn (Office of the Secretary of the Navy), Subj: ICBM-IRBM," December 2, 1955, reprinted in Lockheed Missiles & Space Company, Inc., *The Fleet Ballistic Missile System: Polaris, Poseidon, Trident* (Sunnyvale, Calif.: Lockheed Missiles & Space Company, Inc., n.d.), 7.
134. Lockheed Missiles & Space Company, Inc., *Fleet Ballistic Missiles—25 Years* (Sunnyvale, Calif.: Lockheed Missiles & Space Company, Inc., n.d.), 1.

missile program would take funds away from other Navy activities despite a pledge extracted from Secretary of Defense Wilson that it would not. Sapolsky notes four strategies that contributed to the Special Projects Office's success in the face of these threats: differentiation of a special role, which was represented as of crucial national importance; co-optation of potential critics and disruptive elements; moderation of short-term goals to maximize long-term support; and managerial innovation to create an aura of efficiency.[135]

These strategies, it is important to note, were not limited to the realms of "politics." The "technical" decisions made at Special Projects also reflected them. Crucially, almost as soon as it was established, Special Projects began to deconstruct the unwieldy techno-political compromise that had given birth to the joint Army-Navy ballistic missile program.

By March 1956 they left behind the world of liquid fuel (all the missiles discussed so far were liquid-fueled) and made the daring decision to move to solid fuel. Using the latter meant that the missile could be manufactured with its fuel already inside it (like a child's firework); there was no need to try to pump volatile liquid fuel into a missile on a ship or, worse, in a submarine. Though the decision involved "betting" on a technology many believed was not yet ready for use in other than tactical missiles, the advantages of solid fuel seemed overwhelming to the Special Projects Office. Special Projects also set to work to shrink the missile they inherited. The Office sponsored studies that argued that a missile weighing 30,000 pounds—not Jupiter's 160,000 pounds—was feasible.[136] Special Projects was among the most eager listeners to Edward Teller's predictions of a new generation of much lighter hydrogen bombs: they would make the small missile not just feasible but militarily meaningful. And a small solid-fuel missile meant that Special Projects could move straight to submarine-launching.

So liquid fuel, Jupiter, surface ships, and the Army (by December 1956, when the year-old marriage of convenience ended in divorce) were left behind. The move underwater into the nuclear-powered submarine could have created a new dependency: on the formidable Admiral Rickover and his Naval Reactors Branch. But the small missile solved that prob-

135. Sapolsky, *Polaris System Development.*
136. Ibid., 28.

lem, too. Polaris, as the redesigned Fleet Ballistic Missile (no longer merely a sea-launched intermediate-range ballistic missile) was christened, was small enough to fit inside a modified existing type of nuclear submarine. So no big new submarine or new reactor was needed, and Rickover was kept at one remove from the Fleet Ballistic Missile.

By technical as well as political means, the Special Projects Office thus ensured for itself a remarkable degree of autonomy, one considerably greater, for example, than that enjoyed by their counterparts in the Air Force. Much of Special Projects' subsequent technical decision making had to do with preserving that autonomy.

Later, the concern for autonomy led the Special Projects Office to be classed—somewhat unjustifiably—as technically "conservative."[137] Its initial strategy was anything but conservative. Solid fuel, underwater launching, a missile that was tiny by contemporary standards, and a new generation warhead were all radical steps. From the point of view of this book, however, the most significant steps were in navigation and guidance.

Polaris Navigation and Guidance

Launching a missile from a submarine introduced a whole set of problems not faced in launching from fixed sites on land. The most obvious was simply knowing where the submarine was. But the submarine's velocity also had to be known, since a submarine, unlike a missile silo, could not be assumed to be at rest with respect to the earth. The missile guidance system would have to be oriented correctly so that the missile would fly in the right direction toward the target, but this would have to be done without the external benchmarks used for land-based missiles. Radio guidance from a submerged submarine did not seem plausible, so an inertial missile guidance system had to be used. But, given how crucial it was to make Polaris small, that system would have to be much lighter and more compact than those of Jupiter and, especially, Thor.[138]

137. See Ted Greenwood, *Making the MIRV: A Study of Defense Decision Making* (Cambridge, Mass.: Ballinger, 1975), 34.
138. According to Baker, "Polaris Pioneers Future Design," the weight of

At least in the United States, the solution to the need to know submarine position, velocity, and orientation was obvious by the mid-1950s.[139] Ballistic-missile submarines had to be equipped with black-box inertial navigators. These navigators would then provide this initial information to the guidance systems of the missiles carried by the submarines.

Charles Stark Draper's MIT Instrumentation Laboratory had worked for the Navy on applying inertial navigation to submarines well before Polaris was conceived. In 1951 the Laboratory was awarded a Navy Bureau of Ships contract to develop a prototype Ships Inertial Navigation System (SINS), which it delivered in 1954. For the Polaris program, the Laboratory's work was taken up by the Sperry Corporation, pioneers of gyroscope technology in the pre-inertial era, and a firm with especially strong links to the U.S. Navy.[140]

Although it seemed a powerful combination, matters did not go smoothly. A "little detail"of the MIT/Sperry design led to almost as much anxiety for the developers of Polaris as any other issue. Draper believed in the elegant virtues of letting gyroscopes remain in fixed orientation in inertial space (i.e., with respect to the stars), even for the navigation of an earth-bound vehicle such as a submarine. But as the earth rotated and the submarine's position changed, this meant that the gyroscopes were subjected to a varying gravity field. The slightest imbalance in their rotors would lead to significant errors. But achieving perfect or near perfect balance was an exceedingly difficult task, especially as one moved outside the laboratory to the "real world" of production. The problem dogged attempts to make an inertial navigator accurate enough to launch Polaris.

Fortunately for Polaris's tight schedule, an alternative was available. As interest in ballistic missiles grew during the 1950s, cruise missiles fell out of favor. In July 1957 the Navaho intercontinental cruise missile was abruptly canceled, and the Auto-

the Polaris system was 150 to 200 pounds, as against Jupiter's 250 to 300 pounds and Thor's 650 to 700 pounds.

139. Soviet nuclear missile submarines of this generation may not have had inertial navigators. See chapter 6.

140. W. F. Raborn and J. P. Craven, "The Significance of Draper's Work in the Development of Naval Weapons," in S. Lees, ed., *Air, Space and Instruments: Draper Anniversary Volume* (New York: McGraw-Hill, 1963), 25. On the history of Sperry, see Hughes, *Sperry*.

netics Division of North American Aviation was left with a guidance system, the XN6, with "its missile shot out from under it." [141] Autonetics moved quickly to find a new customer.

Unlike the MIT/Sperry design, the XN6 was a "local level" system, kept horizontal at all times, so the gyros were not subject to large changes in the direction of gravity. By 1958 the XN6 system was mature and reliable enough to be taken on a mission that ensured its fame and rescued it from the status of a component in a canceled system. It navigated the nuclear submarine USS *Nautilus* on its widely publicized voyage from the Pacific to the Atlantic under the ice surrounding the North Pole.

Special Projects ran sea trials on their test platform, the USS *Compass Island,* to compare the performance of the Sperry-MIT SINS, known as the Mk-1, and the adapted XN6. The Autonetics SINS performed much better, but both companies were awarded contracts to develop a SINS for Polaris. With navigation requirements so stringent by the standards of the day, Sperry supporters found it easy to argue for duplication. A modified Sperry system, the Mk-3 Gyronavigator, was to be installed in the first five submarines, the 598-class, while a new Autonetics design, the Mk-2 Autonavigator, would go in the next five 608-class. However, the Sperry SINS failed to keep to this schedule and the deployment pattern was reversed. After a few years these Sperry systems were replaced, and the Autonetics Mk-2 (in various modifications) became standard on fleet ballistic missile submarines.

A considerable mystique developed around the admittedly demanding SINS technology. Extraordinary restrictions, for example, were placed on British access to the SINS that were sold to the United Kingdom as part of the Polaris purchase. The United Kingdom, including the captains and navigators of the Polaris submarines and the Admiralty's technical specialists, were not and are not allowed navigational data from the SINS. It is transmitted directly to the American-provided fire-control system and thence to the missile, with its diversion into British (human) hands prohibited.[142]

141. J. M. Slater, *Twenty Years of Inertial Navigation at North American Rockwell* (Anaheim, Calif.: North American Rockwell, 1966), 23.
142. United Kingdom interview data.

But no one believed that even the best SINS that was likely to become available could operate wholly as a black box for an indefinite time. Periodic "resets"—updates from external sources of navigational information—were seen as needed (around every eight hours in the case of Polaris) to stop unacceptable errors from building up. Furthermore, the problem of the vertical, the need for knowledge of the gravity field through which the submarine was passing, had to be solved.

In chapter 2, we met these difficulties as objections to the very idea of black box navigation. We saw in that chapter the heterogeneous engineering required to eliminate objections. They were also defeated in the more particular form they took here, but heterogeneous engineering of a different kind was still needed. This time delicate matters of international relations were involved.

One solution to the reset problem that was free of international relations difficulties was to use the human navigator's old stand-by, the stars. A system was developed for taking star-sights through a periscope while the submarine lay submerged near the surface. In use the system "was a real dog. I mean it was a mechanical marvel, but it was a hydraulically driven, hydraulically supported periscope: it's like taking . . . sights at the top of a forty-foot pole, and you've got to remember that you've got to track and everything else while the ship is moving all over the place, and we were only too happy to get rid of it."[143] With its "usefulness . . . limited by marginal accuracy, cloud cover, daylight, alignment problems, and maintenance costs," the system, known as the Type 11 periscope, had been discarded by the time the navigation systems for the British Polaris submarines were supplied by the United States, and was finally eliminated from the U.S. fleet in 1969.[144]

Other solutions to the reset problem had to be found for the U.S. and U.K. fleets, and here the potential problems of international relations arose: French pessimism about reliably solv-

143. Interview with Thomas A. J. King, Arlington, Virginia, April 2, 1985.
144. S. A. Conigliaro, *From Polaris to Trident Navigation* (Mimeo of speech given to National Marine Meeting, Institute of Navigation, U.S. Merchant Marine Academy, Kings Point, Long Island, New York, October 23, 1973), 6; "marginal accuracy" had been added to the text by hand in the copy provided to Graham Spinardi at Unisys.

ing them has meant that French ballistic missile submarines are still equipped with star-sighting periscopes.

Three solutions were pursued.[145] The first involved surveying the sea floor with sonar and identifying distinctive features. The limited range of the first Polaris missiles meant that the submarines would not patrol in the open oceans, but in sea areas, particularly the Norwegian Sea and Mediterranean, close to the Eurasian landmass and targets in the Soviet Union. With such patrol areas having been surveyed, the submarine could then navigate from one surveyed feature to another, updating its SINS at each. So long as no extraordinary maneuvers were required, a submarine could follow surveyed features without coming near the surface, except for communications.

This form of navigation and accurate mapping of the earth's gravitational field in Polaris patrol areas required detailed surveying by surface ships. This attracted the attention of the Soviet Union, and Soviet vessels began to shadow the survey ships.[146] Even so the true purpose of the survey ships was apparently considered too sensitive to be imparted to America's NATO allies off whose coasts they were operating, except in most general terms.[147] A further difficulty was that the survey ships themselves had to navigate very exactly so as to map the sea floor accurately, but this in turn required a different source of navigational information.

This information was provided by the second solution to the reset problem, a more accurate version of Loran (Long Range Aid to Navigation) known as Loran-C.[148] Loran was developed during the Second World War at the MIT Radiation Laboratory. Time differences between the arrival of radio signals from widely spaced land-based transmitters enabled positional fixes to be made. By the late 1950s Loran-C receivers were able to provide absolute navigational accuracy of about a quarter of a

145. Interview data. A useful summary of means of submarine navigation resets can be found in Owen Wilkes and Nils Petter Gleditsch, *Loran-C and Omega: A Study of the Military Importance of Radio Navigation Aids* (Oslo: Norwegian University Press, 1987).

146. Interview data.

147. See Wilkes and Gleditsch, *Loran-C and Omega.*

148. P. J. Klass, "Computer Simplifies Loran-C Navigation," *Aviation Week and Space Technology,* June 15, 1964, 95–97.

mile at a thousand mile range and were sensitive to differences of thirty to forty feet.[149]

But even the thousand-mile range meant that Loran-C transmitters had to be relatively close to where Polaris submarines were going to patrol. Stations had to be built in Greenland, Norway, the United Kingdom, Spain, Italy, Turkey, Libya, and elsewhere. Potential controversy was avoided by the simple tactic of not disclosing their role in Polaris navigation, with, at most, a few trusted top leaders in these countries made aware of it. In future years, however, stations designed—or believed to be designed—for submarine-launched ballistic missile navigation were to lead to open political dispute in New Zealand, Australia, and Norway.[150]

Free from such risks, and also at least in the immediate future safe from possible Soviet attack, was the third reset system. This was the world's first satellite navigation system, Transit. Although the Johns Hopkins University Applied Physics Laboratory, where the idea emerged, had a division devoted to the Polaris program, Transit did not arise directly from it. Two researchers in the laboratory had developed a technique for tracking the orbit of earth satellites such as Sputnik using the changes in frequency in satellite signals as they passed over a ground station of known location. Their department head, who was aware of the needs of the Polaris program, then realized that by reversing the process a navigational fix could be obtained when a satellite of known orbit passed overhead. In December 1958 funding was received, and, remarkably, the first experimental Transit satellite was in orbit by September 1959.[151]

As the Transit system became operational in the mid-1960s it provided an alternative solution to the reset problem (other methods were not discarded, however, if only because Transit fixes are not continuously available, since they depend on a satellite being in an appropriate position). Transit also helped

149. Interview of Special Projects officer by Graham Spinardi.
150. Wilkes and Gleditsch, *Loran-C and Omega*, 63, and chapter 11. These disputes largely concerned Omega, a global radio navigation aid that can be received underwater, but which is less accurate than Loran-C and appears therefore not actually to have been used for missile submarine navigation.
151. Thomas A. Stansell, Jr., "Transit, the Navy Navigation Satellite System," *Navigation: Journal of the Institute of Navigation*, Vol. 18, No. 1 (Spring 1971), 93–109.

enormously with the problem of the vertical. Tracking Transit satellite orbits enabled a new, highly accurate model of the earth's gravitational field to be constructed. The task was laborious, requiring the collection, editing, and processing of vast amounts of data: "Magnetic tapes are literally worn out and must be replaced during the process [of computation]."[152] But by 1964 the Applied Physics Laboratory's gravity model was ready and "was sufficiently accurate to make possible [their] goal of better than 0.1 mile navigation at sea."[153]

Along with the roughly contemporaneous move from airborne to spaceborne reconnaissance, Transit marked the beginning of the use of space to avoid what might otherwise be troublesome ("political" rather than "technical") problems of earth-based systems. Earthbound links remained, of course, but the operationally crucial ones for Transit could all be on U.S. territory. Others links were less direct and could be presented as either purely scientific or of civilian benefit.[154] For the construction of the gravity model, for example, thirteen Transit tracking stations had to be built worldwide.[155] In at least one case, the site in Misawa, Japan, the link to a nuclear system would have been controversial, had it been identified. But, to my knowledge, it never was.

The Ships Inertial Navigation System, reset by sonar fixes, Loran-C, or Transit and supported by the new earth gravity model, was seen as solving the submarine navigation problem. But that was, of course, only the start of the problem of achieving the delivery of the missile's warhead to its target. The missile itself had to be guided in flight.

Once the Navy was free to develop its own missile, it was no longer tied to the German guidance specialists working for the Army at Huntsville. Draper's MIT Instrumentation Laboratory was well placed to offer itself as an alternative. It was involved in the SINS work and had also received a contract to study stabilization problems in ship-launch of Jupiter. Ralph Ragan of the Laboratory remembers how its Polaris guidance work

152. Ibid., 101.
153. R. B. Kershner, "Technical Innovations in the APL Space Department," *APL Technical Digest* (Silver Spring, Maryland: Johns Hopkins University Applied Physics Laboratory), Vol 2, No 1. (Jan.–March 1981), 269.
154. Civilian navigational use of Transit did indeed develop.
155. Stansell, "Transit," 100.

began even before the formal start of the program. Again, one of the officers who had passed through the Instrumentation Laboratory was a key intermediary:

It looked twenty years . . . before they [would] ever put a ballistic missile in a submarine. Submarines did not want [liquid] ballistic rockets, and . . . the [Jupiter] missile was just plain . . . too big and too unsafe for the submarines to put up with . . . I said to Sam Forter [Commander Forter, a former Instrumentation Laboratory student now in SPO Fire Control Branch] . . . "What we really ought to be doing for you is . . . doing studies on a smaller . . . ballistic missile for a submarine . . . we know how to make you a small one." . . . So he took me in to talk with then Captain [Levering] Smith [of SPO].

A contract was arranged with Navy Bureau of Ordnance money (according to Ragan, "the Navy didn't care too much as long as we didn't embezzle it"). So the Instrumentation Laboratory "had six months' head start on a design for a ballistic missile for submarines before the Polaris program was signed and given a name."[156]

In addition to their general inertial expertise and their familiarity to many in the Navy, Instrumentation Laboratory staff could offer another crucial "technical detail": a mathematical guidance formulation, already being used in the Thor program, which seemed particularly well suited to Polaris, Q-guidance. Q-guidance shifted much of the computation not only outside of the missile, but also out of the submarine. Crucial data such as the components of the Q-matrix could be calculated onshore by the powerful computers at the Naval Ordnance Station in Dahlgren, Virginia. So even though launch conditions could not be known in advance as they could be for land-based missiles, only fairly simple tasks were left for the submarine's fire control system and the missile's on-board computer. SPO also saw Q-guidance as being particularly suitable for guiding solid-fuel missiles.[157] Admiral Levering Smith, then Technical Director of SPO, recalls that MIT was chosen over

156. Interview with Ralph Ragan, Cambridge, Mass., October 4, 1984.
157. The magnitude of thrust of a solid-fuel rocket motor could not be controlled, only its direction could be. While this was also in practice the case with the large liquid-fuel rocket motors of this generation, low thrust "vernier rockets" were added to Atlas and Titan to permit finer control after main engine shut-down. Letters to author from Levering Smith (October 13, 1986) and B. P. Blasingame (December 31, 1988).

the German team at Huntsville "primarily because of the Q-guidance. It did appear that we could work more closely with Draper than with Huntsville, partly because I thought the [Draper] fluid-floated gyro would adapt easier to the solid motor accelerations, but to my way of thinking it was driven more by Q-guidance than anything else."[158]

The culmination of all this was that on October 10, 1956, Admiral Raborn and some of his staff at SPO "visited Draper to elicit his interest in developing the Polaris inertial guidance. The result was a direct contract for its development . . . The General Electric Company was selected to provide industrial support and to build the resulting guidance system."[159] Thus was set the organizational pattern that has persisted to this day for the development of Fleet Ballistic Missile guidance systems: a direct contract awarded to the Draper Laboratory, with the systems designed by Draper being produced by major industrial firms.

Working on a very tight schedule, which became even tighter in December 1957 following the launch of Sputnik, the Instrumentation Laboratory designed and developed the Mk-1 guidance system for Polaris. Three floated gyroscopes and three PIGA accelerometers were carried on a stable platform, the whole system tiny in comparison to the previous generation of inertial systems. A simplified digital computer—not a general purpose machine, but one specially designed for the few, repetitive calculations it had to perform—completed the guidance system.[160]

With the pressure on for fast production, but tight constraints on weight and volume, the Mk-1 guidance system was a nightmare to make, with difficulties in "every aspect of production."[161] With the computer there was "a lot of trouble with the memory cores . . . trouble in the wiring and testing . . . and there was trouble with the transistors." Participants still remember the "purple plague": "At the junction of the lead to the transistor they would start to rot, mold, or whatever you

158. Interview with Vice Admiral Levering Smith, San Diego, Ca., February 23, 1985.
159. Raborn and Craven, "Significance of Draper's Work."
160. B. O. Olson, "History of FBM Guidance at C[harles] S[tark] D[raper] L[aboratory]," (March 10, 1975, typescript), 2. I am grateful to Mr. Olson for giving me a copy of this useful document.
161. Interviews with Draper personnel.

call it, and under the right light, or maybe by bare eye, it turned purple, and you lost connection . . . it was like a disease that went through all the early transistors." Gyroscopes also were very difficult to produce to the standard required: "You'd make a batch of bearings that would work phenomenally . . . and had life times of 100,000 hours. And then a year later all the production people . . . were getting poor gyros. All the bearings would go bad in all the production lines."[162]

Nevertheless, the Mk-1 guidance system was ready in time to meet Polaris's schedule, and it could be made to work, although it suffered more failures and required more attention than might have been wished. Similarly delicate was the preparation of the Mk-1 for launch: "The guidance interface was quite complex in that all the loops . . . were closed through fire control, so that alignment and erection was done through fire control. . . . All of that interface . . . connected through resolvers, was extraordinarily complex, very touchy. Everything had to be done just right for both alignment and erection. Quite honestly I look back on it and it's a miracle it ever worked, but it did, and does."[163]

Polaris, Accuracy, and Strategy

If everything worked, the combined output of navigation, fire control, and guidance was a missile trajectory that would bring the Polaris warhead to the vicinity of its target. Even as Polaris A1 was being deployed, successor systems were being developed in which the weak points that might stop things from working were strengthened. Rather than discuss the detail of these here (successor systems to Polaris will be examined in chapter 5), it is more appropriate to focus on two questions I have not yet raised: how close to its target was Polaris supposed to come?, and what was that target? The answers again demonstrate the intertwining of technology, organization, and strategy.

As with the ICBM, Polaris accuracy was not a priority. An accuracy goal was set, now revealed in declassified Navy documents to have been 4,000 yards (almost two nautical miles or

162. Ragan interview.
163. Interview with Special Projects Office personnel.

rather more than three-and-a-half kilometers). This was not pinpoint accuracy, even by the standards of the day: the roughly contemporary Navy Triton cruise missile program had an accuracy goal of 600 yards.[164] Further, 4,000 yards was a goal, not a requirement, implying that it would not be considered disastrous if it was not quite met. Participants recall that Polaris did meet the goal "as far as one could tell": the extensive facilities later developed for determining missile accuracies were not then in place.[165]

This relaxed attitude to accuracy was a product of two factors. First, there was, as in the ICBM case, the technological politics of developing a system that had no close previous analogs and that was widely seen as straining the "state of the art" in several respects. What mattered was demonstrating feasibility—that it could be built, that the components would function individually and collectively—and having it available quickly, for bureaucratic as well as national reasons. As the Naval Warfare Analysis Group's first study of the Fleet Ballistic Missile, distributed in January 1957, put it: "Requirements for yield and accuracy should be subordinated to early availability of the weapon."[166]

The second factor leading to deemphasizing accuracy was the assumption that the Polaris's prime targets would be cities and industrial complexes. As we have seen, this was also true for the early Air Force missiles. But the processes leading to the assumption, and its longevity, were quite different.

Polaris was both cause and means for a determined Navy attack, led by Chief of Naval Operations, Admiral Arleigh Burke, on dominant Air Force thinking about nuclear war.[167] As we have seen, the Strategic Air Command was programmed for a single massive nuclear attack on both cities and counterforce targets and planned to conduct that attack preemptively. Polaris made possible and would benefit from a quite different strategy. That strategy, which Burke began to elaborate after

164. Bruins, *Bombardment Missiles*, 285.
165. Interviews. For how missile accuracies are determined, see chapter 7.
166. *NAVWAG Study No.1—Introduction of the Fleet Ballistic Missile into Service* (Washington, D.C.: Naval Warfare Analysis Group, January 1957), 7. The author of this now declassified document, John Coyle, kindly provided a copy to Graham Spinardi.
167. See Rosenberg, "Burke."

becoming Chief of Naval Operations, was what we would now call finite deterrence. If adopted, it would transform Polaris from a marginal weapon, threatened with absorption into the Air Force's command structure, into the centerpiece of national defense.

Finite deterrence involved abandoning the hope of preemption and even the subordinate elements of counterforce in the Air Force's plan. All the United States needed, according to Burke, was a nuclear force that was invulnerable (unlike the Strategic Air Command's bomber bases) and sufficient to destroy Soviet cities. It could be quite small, its size determined by, in Burke's words, "an objective of generous *adequacy for deterrence alone* (i.e., for an ability to destroy major urban areas), not by the false goal of adequacy for 'winning.'"[168]

The nation, in other words, needed Polaris, and perhaps, in the nuclear sphere, not much else. The Navy projected that a fleet of 45 Polaris submarines, with 29 deployed at all times, would be "sufficient to destroy all of Russia." To Eisenhower's Budget Director, Maurice Stans, this raised the obvious question as to why the United States needed "other IRBMs [intermediate-range ballistic missiles] or ICBMs, SAC [Strategic Air Command] aircraft, and overseas bases." While Navy leaders liked the conclusion that these forces could be cut back and money better spent on conventional weapons such as ships and submarines, they were not about to call explicitly for the virtual elimination of the Strategic Air Command; that, they said, "was somebody else's problem."[169]

Understandably, Air Force leaders saw Polaris as a threat. They fiercely resisted the idea of finite deterrence, and their weight within the processes of war planning forced Burke and his Army allies—for the Army, like the Navy, saw the advantages in increased spending on conventional forces that would flow from the new strategy—to present it not as a full-fledged replacement of the dominant plan but as an "alternative undertaking" providing for the possibility of "general war initiated under disadvantageous conditions" (i.e., with the Soviet Union having struck first).[170]

168. Quoted in Rosenberg, "Origins of Overkill," 56–57.
169. Ibid., 56.
170. Ibid., 53.

The Fleet Ballistic Missile was, it is important to note, not originally justified by this role of invulnerable countercity deterrent. To begin with, a Navy missile was argued for on the quite different grounds of the need to hit what were called "targets of naval opportunity," such as submarine pens and port facilities.[171] Only as the Polaris program was under way was it agreed that "population or industrial targets should be specified by the CNO [Chief of Naval Operations] as the target for the initial FBM [Fleet Ballistic Missile] capability."[172] Nor did this new role accord well with traditional Navy preferences, for it had been precisely the Air Force's emphasis on "population and industrial targets" that the Navy had criticized as immoral and ineffective in the "admirals' revolt" of the previous decade. Burke, though not implicated in the failed "revolt," had been the driving force behind a secret 1948 report in which these points had been made.[173]

It was, in short, Polaris that created the perception within the Navy of the need for an invulnerable, countercity "finite deterrent," rather than vice versa. It was never, even within the Navy, a unanimous vision: as early as the winter of 1956-1957 traces can be found of an interest in increasing Polaris's accuracy so as to make it more effective against the military "targets of naval interest."[174] But the countercity deterrent vision triumphed, with its consequence of low priority to accuracy.

As time went by, the status of that vision moved from effect to cause. It was adopted not merely as the best justification for a program already embarked on, but became held by many involved in the Fleet Ballistic Missile program, including its top leadership, as the proper guiding principle for shaping the program. To the extent that the wider public in the United States also believed that the purpose of nuclear weapons was countercity deterrence, the visions of "outsiders" and "insiders" were thus at their closest with Polaris and its immediate successors. The transformation of this vision of the Fleet Bal-

171. Sapolsky, *Polaris System Development,* 44.
172. Coyle, *NAVWAG Study,* 2.
173. Rosenberg, "Postwar Air Doctrine," 257.
174. Bruins, *Bombardment Missiles,* 307.

listic Missile, documented in chapter 5, was thus a cause of unease to some "insiders" as well as many "outsiders."[175]

But that transformation lay two decades into the future. To complete my account of the missile revolution, I must now turn to the other "modern" missile it generated, Minuteman.

Minuteman

Polaris had no direct analogs or rivals. As an ICBM, Minuteman did. What differentiated it from the early Atlas and Titan ICBMs was, in the first instance, simply its fuel: it was conceived as a solid-fuel, rather than liquid-fuel, ICBM. Its key advocate was a relatively low-ranking Air Force technical officer, Colonel Edward N. Hall, who from 1956 was in charge of the solid-propellant research program in General Schriever's ballistic missile Western Development Division of the Air Force.[176]

There was nothing new about solid propellants. They had always been seen as appropriate for short-range missiles, Polaris had just become committed to them, and the research program Hall headed existed because Hall's superiors were interested in using them for a new medium-range land-based missile, probably to be deployed in Europe. But could solid-fuel motors large enough, and powerful enough, for ICBM use be built? Unlike many of his colleagues, Hall believed they could be built, and he produced a proposal for a solid-fuel missile with a full ICBM range of 5,500 nautical miles.

The use of solid fuel alone would not have been sufficient to justify yet another ICBM program. The risks of liquid fuel were not as pressing an issue for the Air Force as for the Navy, and storable liquid fuels were being developed that eliminated the delay caused by the need to fuel up before launch. But Hall integrated his preferred fuel into a distinctive picture of a new ICBM. Many elements of Hall's picture did not survive. Following Hall's experiencing "friction with others" in the Western Development Division, Schriever transferred him to Great Britain in 1958 to command the deployment of Thor

175. Resignations apparently took place from the program over the issue.
176. Hall also had responsibility for Thor development, but solid fuel was where his true interests seemed to lie. For Hall, see Reed, *Minuteman ICBM,* and Roy Neal, *Ace in the Hole: The Story of the Minuteman Missile* (Garden City, N.Y.: Doubleday, 1962).

missiles there, thus removing him from a continuing role in ICBM development.[177] But enough of the original scheme survived, leaving a permanent stamp on the American ICBM arsenal, to make it worth describing.

More radical in its way than the proposal to use solid fuel was the central criterion Hall applied to his missile: it should be cheap. Had he succeeded, Hall would have become the Henry Ford of the strategic missile world, turning the missile from a hand-tooled military luxury into a robust mass-market product.

Hall knew well that the costs of a missile lay as much in its deployment as in its production: each of the early ICBMs required literally dozens of people continuously to maintain and oversee it. He was also aware of the hope that continuous-flow production, controlled by computers, would greatly reduce costs in industries such as oil and chemicals.[178]

Hall envisaged the analog for missiles: a single site where as many as 1,000 to 1,500 missiles would be assembled, deployed in dispersed silos in constant readiness for firing, and their parts recycled if computers continuously monitoring their status indicated failure. Solid fuel was of course an integral part of the scheme. In his vision there was no time for pre-launch fueling or other complicated preparations: the missile's chosen name of Minuteman conveyed instant launch as well as Revolutionary pedigree. The need for human intervention was kept as low as possible: people would still initiate launch procedures, but for salvos of many missiles at a time, not missile by missile.[179]

The goal of cheapness and the need to fit into this ambitious scheme for a huge automated missile "farm" structured Hall's detailed design for his missile. Together with limits on the physical size of solid motors that it was possible to cast, they dictated that Minuteman would be small. Its eventual diameter was 71 inches, rather than the 120 inches of Atlas, Titan, and Thor.[180] All technical desiderata should be subordinated to low cost: "the basis of the weapon's merit was its low cost per

177. Ibid., 87.
178. Reed, *Minuteman ICBM*, 85.
179. See ibid., also Neal, *Ace*, especially 152.
180. Figures from Schwiebert, *Ballistic Missiles*, 243–250.

completed mission; all other factors—accuracy, vulnerability, and reliability—were secondary."[181]

Hall was not a sophisticated strategic thinker. Paradoxically, though, his apparent lack of insight into the contradictions glossed over by the Eisenhower Administration's stated doctrine of "massive retaliation" may have helped him defend his proposal in the simplest possible terms. Thus when having to oppose recommendations from Ramo-Wooldridge that Minuteman be "developed with a view toward accuracy," he was able simply to note, in the words of a later analyst, "that under current strategic doctrine the missile's mission would be to attack enemy cities." Hall argued: "A force which provides numerical superiority over the enemy will provide a much stronger deterrent than a numerically inferior force of greater accuracy."[182]

In later years, this simple judgment would seem questionable. Unwittingly, however, in emphasizing economy Hall was laying the basis for Minuteman's success once the 1961 Kennedy Administration was to bring the tools of cost-benefit analysis to bear on the world of strategic weapons. Without Hall— he departed just as full-scale development was beginning—his more radical cost-cutting ideas, such as integrated assembly, deployment, and recycling in a single site, were abandoned. But relative to any competitor, Minuteman remained a cheap missile. So it was not surprising that economy-minded analysts should conclude that it should become the foundation of the land-based arsenal. Earlier ICBMs, the Kennedy Administration decided, should be scrapped, and the deployment of Minuteman's contemporary, Titan II, sharply curtailed.

Only after it was made did the ramifications of this rapid, apparently "obvious" decision, which led to an ICBM force consisting of 1,000 Minuteman missiles and 54 Titan IIs, become clear.[183] It left the United States with an ICBM force,

181. Reed, *Minuteman ICBM*, 79; Reed is paraphrasing a 1958 report by Hall on "Future Trends and Potentials of Solid Rockets for Ballistic Missiles."
182. Reed, *Minuteman ICBM*, 58, 77; Hall, Memorandum to Col. C. H. Terhane, December 26, 1957, quoted ibid., 59.
183. The Kennedy Administration took up office in January 1961, and the curtailment of Titan II was announced by Kennedy on March 28. See Ball, *Politics and Force Levels*, for the full story of how the size of the arsenal was decided.

the vast bulk of which was designed, to the extent that any clear strategic purpose can be imputed to the process of the design of Minuteman I, for a single, massive, and relatively indiscriminate attack.

This was soon to be regretted, both by influential civilian strategists and Air Force officers.[184] But for many years those who wished other capabilities from ICBMs were unable to start wholly afresh with an entirely new missile. Instead, they had to seek to build what they wanted into successive generations of Minuteman. Since what they wanted was, increasingly, counterforce capability and therefore high accuracy, it is appropriate to end this chapter by examining the Minuteman guidance system.

Minuteman Guidance

Minuteman's advertized readiness for instant use and highly automated deployment brought to the fore an issue in guidance system design not faced in the early ICBMs, intermediate-range ballistic missiles, or Polaris: keeping a guidance system in constant operation. There seems to have been no question that Minuteman was to be inertially guided (if nothing else, its proposed launch in salvos from underground silos pointed in that direction). But inertial guidance systems of the time needed substantial preparation before launch. If the gyros were floated gyros, and had been allowed to cool, they had to be heated until the fluid reached the correct temperature, a process that could introduce sources of error as well as cost time. The system would then have to "erect," that is achieve the correct orientation with respect to the direction of gravity. It would also have to be "aligned," positioned correctly in the horizontal plane. This last process, in the early land-based missiles, required human intervention. Altogether, tens of minutes, and quite possibly an hour or more, might be needed before the guidance system was ready. All this could, however, be avoided if the guidance system were kept continously run-

184. See James R. Schlesinger, "The Changing Environment for Systems Analysis," in Stephen Enke, ed., *Defense Management* (Englewood Cliffs., N.J.: Prentice Hall, 1967), 89–112.

ning, and that became a key aspect of the Minuteman guidance system.

The task of designing and producing the guidance system for Minuteman was put out for competitive contract. It was a big, crucial contract for the nascent inertial industry. Just how big was not immediately clear: one manager of the firm making the successful bid reckons his organization eventually earned $6,000 million of business from Minuteman guidance and related work.[185] The key figure in an unsuccessful contender vividly recalls the bidding process.

I can remember writing a proposal night and day. By then it had gotten to where you had to have a huge stack of documents . . . some 40 or 50 volumes. In a deadline it's got to be at 10 a.m. on a certain date. That means you're going to be flying the night before with this proposal, and it's got to get the airplane, the last airplane, and everyone has to do his part of the thing, that means everyone is up all night for the last two or three nights in a row.

I remember the Air Force made each of the proposers go in and give a big . . . management presentation. I had to go and Tex [Thornton, co-founder of Litton Industries] had to be there. AC Spark Plug Division of General Motors had their proposal . . . and they had the President of General Motors come from Detroit to attend . . . but North American got that job.[186]

What might on the face of it appear a mere technical detail won the contract for the Autonetics Division of North American Aviation, at least in the view of Autonetics staff.[187] The technical detail concerned the bearing on which gyro wheels spun. Conventionally, that was a ball bearing, and a great deal of effort had been devoted at Draper's MIT Instrumentation Laboratory to its improvement. Autonetics proposed to do away with the ball bearing altogether, and replace it with a self-activating gas bearing, a technology that had been under development for some years.

In a self-activating (or "hydrodynamic") gas bearing there is no mechanical contact. The stationary and rotating parts of the bearing are separated by a thin film of gas and are designed in such a way that rotation causes flow patterns in the gas that produce the necessary supporting forces. Although the self-

185. Interview with J. S. Gasper, Anaheim, Calif., September 9, 1986.
186. Interview with Henry Singleton, Los Angeles, March 4, 1985.
187. Gasper interview.

activating gas bearing had a variety of uses, Autonetics pioneered its application to gyros.[188]

What was crucial about the self-activating gas bearings was not, in terms of Minuteman, greater gyro accuracy than was possible with ball bearings, although the proponents of the former did claim high accuracy.[189] More important was that the lack of mechanical contact in a gas bearing seemed to imply that a gyro using the technique might run for years without wear and tear. So the goal of a guidance system permanently in operation seemed feasible. Proponents of ball bearings could argue that long lifetimes were possible from these too, and the guidance systems of Titan II missiles, with their Draper-designed ball-bearing gyros, were kept continuously running after their deployment in 1963.[190] But, if nothing else, Autonetics had a dramatic demonstration of their case to point to: a prototype gas bearing that in 1957 had been in continuous operation since 1952.[191]

Gas bearings offered other advantage as well. Because they could be made in house, they solved what for many gyro firms was a troublesome problem of external dependence on ball bearing manufacturers: "As long as he [a gyro firm] was buying ball bearings from the bearing manufacturers there was a big unknown there. He didn't know from batch to batch just what he was going to get."[192] Ball bearings, too, were a typical area

188. Gas had been used before in gyros—it was the trademark of the German group at Huntsville—but in quite a different way. In the Huntsville gyros pressure was created by an external gas bottle or pump rather than by the bearing's own rotation, and the technique was applied not to the spin-axis bearings (which were still conventional ball bearings) but to the bearings supporting the output axis around which the gyro precessed relatively slowly. Figure 6.1 shows an externally pressurized gas-bearing gyro, and the difference between output and spin axes can be seen in figure 2.15.

189. Tests at Autonetics in the winter of 1955–56 were interpreted as implying a "probable drift rate" of 0.0022 degrees/hour in an experimental gas-bearing gyro, comfortably better than the 0.01 degrees/hour regarded as "inertial grade" performance. See Anonymous, *Autonavigator System Design and Gyroscope Development Program for Marine Use: Quarterly Progress Report No. 6, November 1955–February 29, 1956* (Downey, Calif.: Autonetics Division of North American Aviation, February 29, 1956, AD-107 998/7ST), v, 5. I am grateful to Nils Petter Gleditsch for access to this document.

190. Letter to author from B. P. Blasingame, December 31, 1988.

191. J. S. Ausman and M. Wildman, "How to Design Hydrodynamic Gas Bearings," *Product Engineering* (November 25, 1957), 103–106.

192. Interview with Michele Sapuppo, Cambridge, Mass., October 10, 1984.

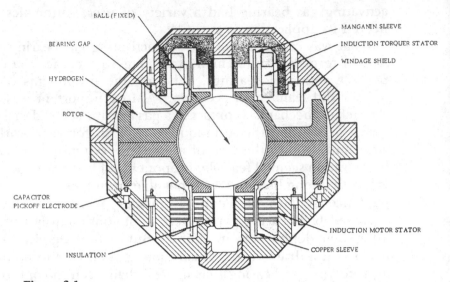

Figure 3.1
Free-Rotor, Self-Activating Gas-Bearing Gyroscope of Type Used in
Minuteman Guidance
Source: J. M. Slater, *Twenty Years of Inertial Navigation at North American
Aviation* (Anaheim, Calif.: North American Aviation, Autonetics Division,
1966), 37.

of gyro production where "tacit knowledge" seemed hard to
eliminate. "Ball bearings seem to be more of an art than a
science,"[193] though the hope of escaping this with gas bearings
proved in part illusory: "The bearing in principle is simplicity
itself—a cylinder spinning on a rod, with grooved end-plates
for thrust support—though economical and reliable fabrication
methods took considerable effort to achieve."[194]

For Minuteman, Autonetics built the self-activating gas bearing into a geometrically novel gyro design. The rotor spins (the
present tense is appropriate, for detail modifications aside, this
is the gyroscope still to be found in the Minuteman force) on
a central ball, the size of a "large marble," kept apart from the
ball by the gas film, "a smidgin of gas."[195] The basic design is
shown in figure 3.1. Because the rotor is (within limits) free to
reorient itself around the ball, it is sensitive to rotations not

193. Interview with Peter Palmer, Cambridge, Mass., October 10, 1984.
194. J. M. Slater, *Twenty Years of Inertial Navigation at North American Rockwell*
(Anaheim, Calif.: North American Rockwell, 1966), 37.
195. Ibid., 36.

just around one axis but around two. The Minuteman gyro is thus a "two-degree-of-freedom gyro," not a one-degree-of-freedom device as developed by Draper at MIT. So only two gyros are needed per guidance system (though a third has in fact been added for reasons we will return to in chapter 4).

This gyro design owed much to the imagination of John Slater, then a key figure in inertial instrument design at Autonetics.[196] Slater's preferences can also be seen in another feature of the design of the Minuteman I guidance system.[197] Like the German-designed systems of the time (see figures 2.8 and 2.9), but unlike those designed at MIT and elsewhere, it was what was called an "inside-out" platform. Instead of being held within supporting gimbals, the gyros and accelerometers were clustered around a central ball joint.

That was not a feature that was to survive into the next generation of Minuteman. Also not to survive was the use of a distinctive Autonetics accelerometer, the "velocity meter," designed by Jack Wuerth.[198] In chapter 4, I discuss the significance of the removal of these features in Minuteman II.

A fourth feature of the Minuteman I guidance system was an unequivocal harbinger of the future, not just for Minuteman but for all other U.S. strategic missiles. Minuteman was equipped with an on-board digital computer: a general-purpose computer, not just the limited special-purpose "digital calculator" used in Polaris. There had been plans for use of an on-board digital computer from remarkably early on in the history of missile guidance: one of the first U.S. digital computers, the BINAC, was commissioned by Northrop in 1947, apparently for on-board guidance of the Snark cruise missile.[199] But BINAC never flew, and building a small, reliable

196. See J. M. Slater, *Inertial Guidance Sensors* (New York: Reinhold, 1964).
197. See J. M. Slater, "The North American Gyro-Stabilized Platform," in Scientific Advisory Board, *Seminar,* Vol. 1, 139–145.
198. Interviews with J. M. Wuerth and John Slater, Fullerton, Calif., February 23, 1985, and letter to author from Jack Wuerth, January 6, 1989. Wuerth had also invented a double integrating accelerometer called, therefore, a "distance meter," that was used on earlier Autonetics systems, such as the N-6 that navigated the USS *Nautilus* under the North Pole. See John M. Wuerth, "Accelerometer and Integrator" (U.S. Patent 2,882,034, April 14, 1959, filed November 1, 1948).
199. Nancy Stern, *From ENIAC to UNIVAC: An Appraisal of the Eckert-Mauchly Computers* (Bedford, Mass.: Digital Press, 1981), chapter 6.

on-board digital computer, especially for the harsh environment and weight constraints of a ballistic missile, remained a
daunting task, even as solid-state transistors became available
to replace vacuum tubes. The on-board guidance calculations
necessary for "first generation" ballistic missiles were performed on analog, not digital, hardware.

The innovation in Minuteman was not simply in using a
general-purpose digital computer for guidance calculations.
Earlier Air Force missiles had employed an autopilot to stabilize
the missile in flight as well as a guidance system; in Minuteman
the two functions were combined in an "integrated digital guidance and flight control system." And, in an echo of Hall's
original ideas, the computer was also assigned the task of continuously monitoring the status of key components of the missile as it sat, ready to fire, in its silo.[200]

Building not one but hundreds of computers that could
operate unattended and unmaintained for years on end was a
major challenge to the electronic technology of 1960. Jack
Wuerth, a key figure in Minuteman guidance, explains the
problem faced: "By the late 1950's, the average failure-free
life, or Mean-Time-Between-Failures . . . of individual resistors,
capacitors, transistors, and diodes was already measured in
years. However, many thousand such components were needed
in a single system. Probability theory therefore revealed that
at least one such component could be expected to fail at frequent intervals."[201]

With generous funding from the Air Force, Autonetics established a program to improve the reliability of electronic components "by a factor of 100," awarding contracts of around $1
million to $2 million to at least thirteen electronics compa-
nies.[202] The program was a resounding success. "Minuteman
high rel parts," as they were known, achieved a good reputation

200. See J. M. Wuerth, "The Evolution of Minuteman Guidance and Control," *Navigation: Journal of the Institute of Navigation,* Vol. 23, No. 1 (Spring
1976), 64–73.

201. J. M. Wuerth, 'The Impact of Guidance Technology on Automated
Navigation,' *Navigation: Journal of the Institute of Navigation,* Vol. 14, No. 3
(Fall 1967), 328–339.

202. Anonymous, "Autonetics picks Suppliers," *Aviation Week* (June 13,
1960), 95; anonymous, "Minuteman Avionics Reliability Increases," *Aviation
Week* (December 12, 1960), 99–105.

not only in the missile itself but more widely, as the electronics suppliers involved made this characterization a selling point.

Conclusion

By 1967, the consequences of the missile revolution were in place. Though the first-generation land-based missiles, never deployed in large numbers, had been withdrawn, 1,000 Minuteman ICBMs and 54 Titan II ICBMs now sat in underground silos dispersed, miles apart, over nine Air Force reservations in Montana, the Dakotas, Missouri, Wyoming, Arizona, Kansas, and Arkansas. The U.S. Navy deployed 656 Polaris submarine-launched missiles in 41 submarines, a substantial number of which were, at any given point in time, patrolling silently under the oceans. The U.S. Army's longest-range missile was the 400 nautical mile Pershing I, which was deployed jointly with the armed forces of West Germany until its removal in the late 1980s under the Intermediate Nuclear Forces agreement.

The processes leading to that result were complex, Here I have examined only some of the processes affecting the nature of the missile arsenal, not those shaping its size, and have not even attempted to discuss the evolution of the strategic bomber force, the host of "tactical" nuclear weapons, or the command-and-control systems that attempt to manage this plethora. And while the overall size and shape of the U.S. ballistic missile arsenal has remained remarkably stable since the mid-1960s, vitally important qualitative changes were already under way by 1967.

No single person or organization decided what the American nuclear arsenal should look like. The 1961–1963 Kennedy Administration, with its famous Secretary of Defense, Robert McNamara, might seem to have done so, since the forces that were in place by 1967 were largely the result of decisions taken in the Kennedy years. But the position of Kennedy and McNamara at the head of government by no means translated into the capacity simply to make the choices that they thought best reflected the national interest, and, even had that been the case, those choices were between options that had largely been shaped before they came to power.[203] By January 1961, Atlas, Titan, Thor, Jupiter, Polaris, and Minuteman already existed in a form that was hard to change in any fundamental way.

Nor did the actions of the Soviet Union determine in any detail the course of the missile revolution. True, the systemic conflict between the United States and the Soviet Union was the precondition for the whole business of building a nuclear arsenal. And, as we have seen, a potential "missile gap" was an important resource for U.S. proponents of the ballistic missile. But a program such as Minuteman or Polaris was in no simple sense a reaction to Soviet actions, and certainly in no sense at all a copy of anything the Soviets or anyone else had done. Even decisions about the size of the American missile arsenal are hard to explain in terms of Soviet actions. The arsenal grew fastest from 1961 to 1967, when it was much larger than the Soviet arsenal, and then stopped growing completely just at the point of fastest expansion of the Soviet force.[204]

More generally, it can be concluded that the United States built its missile arsenal without any agreed understanding— even within élite circles, much less among the general population—of why it was doing so. Polaris is the best example, because there was later seen to be a clear and urgent need for an invulnerable, ultimate countercity deterrent. But the program was not begun for this reason or even with this as justification. It is only slightly unfair to say that the Navy began the program for a reason no more sophisticated than that ballistic missiles were the coming thing, and the Navy had to have one.

With political leaders by no means simply in command, the Soviet Union a shadowy mirror reflecting American fears rather than a well-understood foe, and nuclear strategy often rationalization after the fact rather than genuine guiding principle, the initiative for new missile systems came largely from below. Particularly important were technologists with career investment in and enthusiasm for particular lines of techno-

203. See Ball, *Politics and Force Levels* for the classic study of the constraints under which Kennedy and McNamara operated.
204. Ball, ibid., documents the extent to which the Kennedy Administration's decisions to expand greatly the American missile force were taken by people who no longer believed that the "missile gap" was a reality.

logical development. But they did not operate in a vacuum. Without the support of key civilian officials such as Trevor Gardner or military officers such as Arleigh Burke, who had concerns and interests of their own, their activities would have come to nothing. And they could succeed only when conditions were ripe, as they were in the winter of 1953–1954.

Guidance technology was both central and marginal during the missile revolution. It was central in that it was seen as one of the most daunting tasks in building a long-range ballistic missile. Disagreement over accuracy specifications was one form taken by conflict over whether massive resources should be invested in such an enterprise. Those who did not wish this to happen placed tight specifications that could not credibly be met, while the proponents of the missile struggled to relax accuracy specifications.

But in a different sense, guidance was also marginal. The particular level of accuracy actually achieved was not a major priority in most of the missile programs we discussed, at least as long as it was within loose bounds of what was adequate— around two nautical miles average error. That this was not simply a consequence of an immature state of the art can be seen from the one clear exception among first generation U.S. ballistic missiles. The German guidance specialists working on Jupiter were under constant pressure to obtain the greatest accuracy possible and seem to have achieved half a mile average error. Organizational circumstances, such as the relative "underdog" status of the Army, and strategic questions, such as the assumption that the ICBM's targets would be cities, shaped issues of guidance even in the early phase of U.S. ballistic missile development.

This interaction of guidance technology, organizational priorities, and nuclear strategy soon moved to center stage. It did so not only within the overall legacy of the missile revolution for the size, shape, and organizational framework of the U.S. arsenal, but also within its particular legacy for guidance. Inertial guidance was left the irreversible victor of the struggle with the radio guidance of strategic ballistic missiles. In the years to come, those who wanted greater accuracy from ballistic missiles sought to refine inertial guidance, not to revert to radio guidance, barring a small and powerless minority of continuing proponents of the latter.

The victory of inertial guidance raises an issue about the nature of technological change I have not had to address in the discussion so far. Once a basic form of technology (a "paradigm," some would call it) has emerged as dominant, is its future path of development then determined? Do we need to refer to social factors to explain the growth after 1960 of the accuracy of ballistic missile inertial guidance? Or is it just a natural trajectory of technology, with no need for further explanation? That is the topic of the next chapter.

The Beryllium Baby and the Technological Trajectory

The metaphor that seems to come to mind most naturally is the embryo in the womb. "The system, Duffy said, treated its recording instruments with the tenderness of a mother carrying her unborn child: AIRS was an embryo, a beryllium baby, nestled in a womb."[1]

AIRS is the Advanced Inertial Reference Sphere, the inertial measurement unit of the MX/Peacekeeper missile. It is held by many to represent the ultimate in accuracy achievable in an inertially guided ICBM. The astonishing accuracy specification of MX referred to at the start of this book is due in large part to AIRS.

I shall return to the striking metaphor of the embryo in the womb. But let me now focus the concern of this chapter more quantitatively. Consider figure 4.1. This authoritative graph shows the increase in guidance and control accuracy[2] of the three generations of Minuteman. For security reasons it lacks a unit on the vertical axis, but the pattern is clear—an improvement of better than tenfold in less than two decades.[3] AIRS was only at the stage of testing in 1976, but represents at least a further twofold improvement. Though sources of error other than guidance and control have to be taken into account as

1. Michael Parfit, *The Boys behind the Bombs* (Boston: Little, Brown, 1983), p. 85. When interviewed by Parfit, Robert Duffy, a former Air Force Brigadier General, was President of the Charles Stark Draper Laboratory, Inc.
2. In the terms we shall learn in chapter 7, figure 4.1 shows the guidance and control portion of the error budget, not the overall circular error probable. I am grateful to Jack Wuerth for confirming this latter point in a letter to me of March 22, 1988.
3. The broken line represented a prediction in 1976, but as far as I can tell, that prediction was borne out.

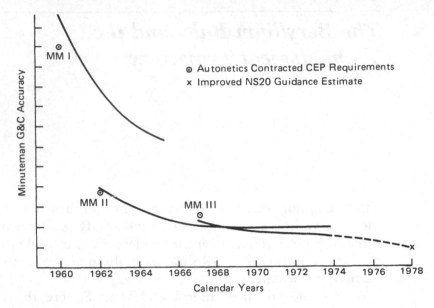

Figure 4.1
Development of Minuteman Guidance and Control Accuracy
Source: J. M. Wuerth, "The Evolution of Minuteman Guidance and Control," *Navigation: Journal of the Institute of Navigation,* Vol. 23, No. 1 (Spring 1976), 69.

well, we have here probably the largest single component of the increase in overall missile accuracy charted in table 4.1.

What sort of explanation does the growth in missile accuracy over the last three decades require? Conventional wisdom suggests that explanation resides in an internal logic of technological change: "technology creep" or a "technological imperative."[4] Missile accuracy increased because it was natural for it do so, natural for those working on guidance and the other contributors to accuracy to seek and find improvements.[5]

Although they have not considered the particular case of missile accuracy, more general recent studies of technology

4. For explicit statements of what has been a widely held assumption, see Deborah Shapley, "Technology Creep and the Arms Race: ICBM Problem a Sleeper," *Science,* Vol. 201 (September 22, 1978), 1102–1105, and Dietrich Schroeer, "Quantifying Technological Imperatives in the Arms Race," in David Carlton and Carlo Schaerf, eds., *Reassessing Arms Control* (London: Macmillan, 1985), 60–71.
5. See the authors quoted in chapter 1.

Table 4.1
The Main Missiles Discussed in this Chapter

Year of first deployment	Name	No. of warheads	Guidance system designer	Estimated accuracy
1962	Minuteman 1	1	Autonetics	1.1
1966	Minuteman 2	1	Autonetics	0.26
1970	Minuteman 3	3	Autonetics	0.12
1986	MX/Peacekeeper	10	Draper/Northrop	0.06
?	Small ICBM	1	Draper/Northrop	0.06

The above are inertially guided U.S. Air Force intercontinental ballistic missiles. Minuteman 2, Minuteman 3, and MX/Peacekeeper are all currently in service.
"Estimate accuracy" is my estimate of circular error probable at time of first deployment, expressed in nautical miles. See appendix A.

offer us a similar notion of a "technological trajectory."[6] Ambiguities and inconsistencies in the use of the term abound. Crudely put, however, a technological trajectory is a direction of technical development that is simply natural, not created by social interests but corresponding to the inherent possibilities of the technology. Some authors explicitly use the phase "natural trajectory."[7]

One can see why the notion is attractive. It has been adopted, particularly by economists, as a counterweight to the notion that technological change is somehow purely plastic, a putty to be formed into any shape by social forces, market forces in particular.

In this chapter, however, we find that there is something quite misleading in the mechanical metaphor of "trajectory" and in the conventional explanation of the growth of missile accuracy, an explanation that roughly corresponds to the notion of a technological trajectory. What is wrong is the fundamental idea that technological change can be self-sustaining, that its direction and form can be explained in isolation from the social circumstances in which it takes place.

6. See, for example, Giovanni Dosi, "Technological Paradigms and Technological Trajectories: A Suggested Interpretation of the Determinants of Technical Change," *Research Policy*, Vol. 11 (1982), 147–162.
7. See Richard R. Nelson and Sidney G. Winter, *An Evolutionary Theory of Economic Change* (Cambridge, Mass.: Harvard University Press, 1982), 258–262.

I do not deny that trajectories exist in the sense of persistent patterns of technological change. They certainly do, and the growth of missile accuracy is an example. Their very persistence points to the key missing element in how they are usually understood. What turns the potential anarchy of technological change into a persistent trajectory is the way social interests are created in its continuing to take a particular form. It is not simply that people want it to take that form—though sometimes they do—but that they build into their calculations that it will take that form. They invest money, careers, and credibility in being part of "progress," and in doing so help create progress of the predicted form. Finding their predictions correct, they continue to make them with confidence.

A technological trajectory, we might say, is in this sense an institutionalized form of technological change. Look at any putative natural trajectory—the "increasing mechanization of operations that have been done by hand," to take one example from the literature—and one can see that social interests cluster around it and that those who oppose it can be silenced by the argument that if it does not happen in one factory it will happen in others, and the factories where it does not happen will go out of business.[8]

Why, then, does a trajectory ever appear natural? The answer, I think, is that there is a sense in which the trajectory is a self-fulfilling prophecy. Those lines of technical development that do not get pursued do not improve; those that get pursued often do.[9] So, in retrospect, it is easy for the latter to appear to correspond to the inherent possibilities of a technology.

None of this justifies a view of technological change as wholly plastic. Configurations of social interests, however powerful,

8. Ibid., 259.

9. This argument has been interestingly put by W. Brian Arthur, "Competing Technologies and Economic Prediction," *Options*, April 1984, 10–13. Arthur writes: "Very often, technologies show increasing returns to adoption—the more they are adopted the more they are improved. . . . When two or more increasing-returns technologies compete for adopters, insignificant chance events may give one of the technologies an initial adoption advantage. Then more experience is gained with the technology and so it improves; it is then further adopted, and in turn it further improves. Thus the technology that by 'chance' gets off to a good start may eventually corner the market of potential adopters, with the other technologies gradually being shut out."

will not always be able to give technology a desired shape. Some extremely widely held technological prophecies turn out to be false. The material world is obdurate.

But although the apparent virtue of the trajectory-type notions is the way they take this into account, the reality can be very different. If the growth of missile accuracy is taken to be a technological trajectory, then it is easy to extrapolate that growth into the future. Even those who would like to see missile accuracies stop growing do so.[10] As we shall see at the end of this chapter, however, those actually involved in seeking to increase the accuracy of U.S. ballistic missiles are far less confident. They believe they have a met a barrier that may be insurmountable, though they dispute whether its nature is physical or social.

If that is the case, if this apparently permanent pattern of technological change has come to an end, it is a matter of some importance, especially for arms control. We can see here why the fatalism of the metaphor of trajectory constitutes a crucial flaw. For while the barrier to increased accuracy may not be surmounted, it may be circumvented by the adoption of new forms of guidance. Those who wish to stop missile accuracies from increasing could focus their efforts on preventing these becoming a reality. But they will not do so if they believe that missile accuracies will naturally continue to increase.

The Twofold World of Inertial Sensors: The Inertial Market

Fortunately, in the case of missile accuracy we are not restricted to abstract argument against the notion of a "natural trajectory" of technological change. Here we find not merely a potential but an actual rival form of technological change. Its existence shows there is nothing "natural" about increasing missile accuracy. Comparison of the two worlds of inertial technology reveals the different interests and institutional structures that sustain the two different forms of technical change. The bifurcation is not absolute: the two forms interact and cross-fertilize and occasionally still inhabit the same organizations. But the differences are clear.

The form of technical change alternative to ever-increasing accuracy has by no means been the less "high-tech." It has seen

10. Schroeer, "Technological Imperatives."

considerable changes in gyroscope technology: first the replacement of floated mechanical gyroscopes by sophisticated fluidless "dry" designs; and then the replacement of mechanical gyroscopes altogether by laser gyroscopes in which rotation is detected using beams of coherent light. Even the basic stable-platform configuration of an inertial system has been challenged successfully.

What ultimately has sustained this form of technical change has been the creation since the 1960s of a market for black-box navigation well beyond strategic nuclear systems. "Creation" is indeed the appropriate word: the market certainly did not smoothly and automatically emerge. This is clearest in the most interesting development: civil air inertial navigation. Now commonplace in long-range air transport, it was brought into being only with great difficulty.

The case for it was in good part economic. Navigation systems that were not black boxes could not individually be relied upon sufficiently to be made automatic. Weather, for example, could interfere with both radio signals and star sightings. So long flights carried human navigators who continuously collated information from all sources available to them. Their skill and experience meant that even in the early days of the 1930s, "no transoceanic airliner was lost . . . because of faulty navigation."[11]

But these skilled people were expensive. Could they be replaced by black boxes—inertial navigators? During the 1960s, this question was asked more and more seriously. It was made more urgent by the widespread assumption that a new generation of supersonic airliners would appear over the next decade, and they would fly too fast to give the human navigators the time to ply their trade effectively.[12]

The key issues were whether inertial navigators could be made safe enough and cheap enough to replace humans. Both questions ultimately came down to reliability: the cost of manufacturing an inertial system, while not inconsiderable, was less salient than the cost of a wordwide network of replacement

11. Monte Duane Wright, *Most Probable Position: A History of Aerial Navigation to 1941* (Lawrence, Kans.: University Press of Kansas, 1972), 167.
12. For some of the reasons why this assumption turned out to be wrong, see Mel Horwitz, *Clipped Wings: The American SST Conflict* (Cambridge, Mass.: MIT Press, 1982).

inertial systems and skilled maintenance and repair personnel. The military might be prepared to pay for this, but airlines would make sure its burden fell on the manufacturers, rather than on them.

So could the black box genuinely be made autonomous, not in the protected environment of a missile silo but in the world of the airline? The first, bold move to create a civil air inertial navigation market came from Sperry Gyroscope, the firm that had pioneered gyro culture in the United States.[13] By then a division of Sperry Rand, it contracted in 1964 to supply inertial navigators for Pan American's long-range Boeing 707s. The airline's decision was "prompted by its look-ahead to the supersonic transport," and it also expected "to reduce its flight crew size from four to three without labor problems because all of its crew members are pilot-navigators and they will be needed to meet expanding fleet needs."[14] But serious difficulties with inertial navigator reliability ensued, and Sperry was forced to withdraw from the contract bearing major financial losses.[15] It was a setback from which Sperry Gyroscope's attempt to translate its formidable gyro expertise into the new world of inertial navigation never fully recovered.

Reliability was also a desperate problem for the first successful supplier of civil air inertial systems, the AC Spark Plug Division (later Delco Electronics Division) of General Motors, which had worked with Charles Stark Draper's MIT Instrumentation Laboratory on projects such as Thor and Titan II. Their crucial contract was to supply inertial systems for the new Boeing 747 jumbo jets. Each 747 was to be equipped with three, so that failure in one could be detected by comparing its readings with those of the others. Given the popularity of the 747, it was a major contract.

Having committed themselves to provide systems with a mean time between failure of 2,000 hours, AC Delco found the first systems failing ten times as often as that. B. Paul Blasingame, who had resigned from the Air Force to become Technical Director at AC Spark Plug, recalls that it was a "larger effort to redesign them to get reliability and learn how

13. See chapter 2.
14. See Philip J. Klass, "707 Navaid Stresses Maintenance Ease," *Aviation Week and Space Technology* (September 28, 1964), 45, 53.
15. Interview with Philip J. Klass, Washington, D.C., October 19–21, 1984.

to manufacture them than to design the system in the first place."[16] But with the financial backing of General Motors behind them, AC survived. After a decade of losses, it finally broke even in its commercial inertial navigator business—Blasingame recalls the "mortgage burning" ceremony that took place—and then, with a mature product, began to make profits.

Although these traumatic experiences reinforced the centrality of reliability, the accuracy of an inertial system—its performance when working normally—was much less of a problem. The key single event in the creation of the civil air inertial market was the agreement in 1966, through the remarkable system that exists to harmonize the needs and conflicting interests of airlines and the suppliers of equipment,[17] of a standard definition of an inertial navigator, Characteristic 561.[18]

In effect, Characteristic 561 laid down that the probability of position errors greater than 25 nautical miles should be no more than 5 percent, so as to maintain safe separation of aircraft on routes such as those over the North Atlantic.[19]

16. Interview with B. Paul Blasingame, Santa Barbara, Calif., September 12, 1986.

17. The central role was played by an unusual company called ARINC (Aeronautical Radio, Incorporated), and its Airlines Electronic Engineering Committee. ARINC was created in December 1929 by the U.S. airlines to provide radio communications with aircraft, but by the 1960s it had moved well beyond this original remit. Its shareholders are those airlines, joined by major non-U.S. carriers. The role of the Airlines Electronic Engineering Committee is to "define form, fit and function standards such that equipment built by different manufacturers can be accommodated in a common aircraft installation and perform as intended." See David H. Featherstone, "AEEC— The Committee that Works!" Airlines Electronic Engineering Committee Letter 82–000/ADM-218, October 20, 1982. Insight into how the harmonization of interests works in practice can be found in a book by this committee's former chair William T. Carnes: *Effective Meetings for Busy People: Let's Decide it and Go Home* (New York: McGraw-Hill, 1980).

18. ARINC Characteristic 561,"Air Transport Inertial Navigation System," was approved by the Airlines Electronic Engineering Committee on December 2, 1966. The version I have in my possession is the eleventh, ARINC Characteristic 561–11, *Air Transport Inertial Navigation System* (Annapolis, Maryland: Aeronautical Radio, Inc., January 17, 1975).

19. "An accuracy which has been deemed appropriate for use over the North Atlantic under present (and probably also under presently contemplated) separation standards is obtained by limiting Cross-Track error to a maximum of ± 20 nautical miles and Along-Track error to a maximum of ± 25 nautical

Compared to the goal of one-mile error that the Air Force had presented the inertial pioneers of the 1940s, this was not too demanding. A system with an average error of a nautical mile per hour of flying time would easily meet it. Though civil air navigation specifications have evolved over the near quarter century since the first version of Characteristic 561, a mile-per-hour accuracy would still satisfy them.

Perhaps suprisingly, creating a market for military air navigators and then competing in it was not wholly different in its demands, though it was a process that began earlier. A black-box navigator could be argued to be useful for all military aircraft, including tactical bombers and even fighters, not just strategic bombers. But the inertial systems of the 1950s had to be changed radically before that could become a reality. They were too big and heavy, and even though by the early 1960s that was being vigorously addressed, there were still problems of "system complexity, low reliability, and high costs."[20]

In 1963, the U.S. Air Force launched a concerted campaign for a "low-cost" inertial navigator, in collaboration with one of the most colorful characters of the early years of inertial navigation, Wladimir Reichel. Born in St. Petersburg in 1891, Reichel was another long-standing member of gyro culture who "felt, in 1946, that we were in a position to think seriously of actually achieving the dream of every navigator, whether in the air or otherwise, from time immemorial, i.e., a self-contained system of navigation independent of outside points of reference—stars, ground observation, radio beam, etc." In 1945 he joined the Kearfott Company, originally a supplier of marine windows, and played an important part in its rise to become a major inertial navigation firm.[21]

miles, determined on a 95% probability basis for flights of typical durations on selected routes and at appropriate latitudes" (Ibid., 43B).

20. W. A. Reichel and Captain Russell E. Weaver, Jr., "The Air Force Design and Engineering Approach to Low-Cost Navigation Systems," read to Institute of Navigation, *Low-Cost Navigation Symposium*, Los Angeles, November 6–7, 1963. I am grateful to Leonard Sugerman, then an Air Force Major, who worked with Reichel on the low-cost inertial project, for sight of the proceedings of this symposium.

21. Wladimir A. Reichel, "Additional Background," Addendum to "Career Executive Roster" (June 4, 1963), 18. Again, I am grateful to Leonard Sugerman for providing me with a copy of this fascinating autobiographical document. Before the revolution Reichel worked in his family's naval gunsight firm. He arrived in the United States in 1923, where he worked on

The direct approach taken by the Air Force to solving the problem of cost was a special Air Force organization under the technical direction of Reichel, charged with designing an inertial navigation system costing no more than $25,000.[22] It failed. But gradually a cost norm of around $100,000 per system was established, one that has risen only gradually with inflation. Several of the firms that tried to establish themselves in the military air inertial navigation market failed to do so. But three principal U.S. firms remained and competition between them, sometimes fierce, has generally acted to keep a downward pressure on prices.[23] The firms are Kearfott, which became a division of the diversified multinational Singer; Litton Industries, an early 1950s "start-up" firm, which gained some of Autonetics's inertial expertise, played a major part in the development of the military air inertial market, and became AC Delco's original main competitor in the civil air market; and Honeywell, the Minneapolis thermostat company, which during World War II employed its knowledge of feedback mechanisms in autopilot manufacture.

What is most important for our argument is that the accuracy norm for military aircraft inertial navigators became established in the 1950s and 1960s at roughly one nautical mile per hour, and has not, for most applications, grown significantly.[24] It is not that more accurate systems are impossible: at least two substantially more accurate inertial navigators were marketed. These were the early 1960s Hipernas (High Performance Navigation System) of Bell Aerospace and the 1970s SPN/GEANS

aircraft instruments. Fluent in German and French as well as Russian and English, Reichel developed international gyro culture contacts, such as Captain Altvater, and knew of the work of Boykow.
22. Interview with Colonel Leonard R. Sugerman (U.S. Air Force, rtd.), Las Cruces, New Mexico, September 18, 1986. Sugerman was another important member of the guidance mafia, seconded by the Air Force to MIT in 1953–1955, who became Director of the Central Inertial Guidance Test Facility, Holloman Air Force Base, New Mexico. There is a history of the low-cost aircraft inertial navigation system effort in Lloyd H. Cornett, Jr., *History of Air Force Missile Development Center, Fiscal Year 1969* (Holloman Air Force Base, New Mexico: Air Force Missile Development Center, Historical Division, 1969).
23. Q.v. Jacques S. Gansler, *The Defense Industry* (Cambridge, Mass.: MIT Press, 1982).
24. Now 0.8 nautical miles per hour seems to be a typical specification.

(Standard Precision Navigator/Gimbaled Electrostatic gyro Aircraft Navigation System) of Honeywell.

Only a handful of the former were ever produced, but the story of the latter is of some interest because it reveals the existence of major barriers to increasing the accuracy specifications in military aircraft inertial navigators. SPN/GEANS grew out of work in the 1950s by Professor Arnold Nordsieck of the University of Illinois on a novel gyroscope design: the electrostatically suspended gyroscope, or ESG (figure 4.2 shows Nordsieck holding an early experimental model). In the ESG, the rotating mass is a spherical ball supported in a vacuum by an electrostatic field.

The corporation that took up Nordsieck's work most actively was Honeywell. The Minneapolis firm owed most of its inertial systems experience to "black programs," notably the SR-71 strategic reconnaissance aircraft, for which it designed what was for its day (the early to mid 1960s) an unusually accurate (quarter to half a nautical mile per hour error) inertial system.[25] Honeywell failed to gain a stake in the emerging market for $100,000 "one mile per hour" systems, however, apparently in large part because the very high secrecy of "black programs" meant that even Honeywell's own marketing management could not be informed of what the corporation had to sell.[26] The ESG, as "the world's most perfect gyro" (see figure 4.3), potentially offered Honeywell a way back in to the military aircraft navigation market.

But it could do so only if accuracy requirements were tightened. There was little hope of selling the ESG on the grounds of superior reliability or lower cost; its key claim was that an aircraft inertial navigator based on it was accurate, not to a nautical mile per hour, but to a tenth of a nautical mile per hour.[27] But persuading the Air Force of the need for this proved difficult: "we did an awful lot of marketing work on the Strategic Air Command to try to convince them of the advantages of a tenth of a mile per hour system" for strategic bombers; Honeywell tried "awfully hard for four years to get

25. As we saw in chapter 2, there is evidence that the SR-71 was also equipped with a Northrop stellar-inertial system.

26. Interview.

27. Interviews and Anonymous, *U.S. Air Force: Standard Precision Navigator* (St. Petersburg, Florida: Honeywell Aerospace Division, n.d.).

Figure 4.2
Professor Arnold Nordsieck Holding Early Electrostatically Suspended
Gyroscope (ESG)
Note the spherical rotor in Nordsieck's right hand. Photograph courtesy
University of Illinois at Urbana-Champaign.

Because of the virtually frictionless environment of its rotating element, the Honeywell ESG (Electrically Suspended Gyro) offers very low drift rate, long life and high reliability. Eliminating conventional mechanical suspension, it operates in a coasting condition for long periods of time with optical pickoffs sensing its orientation. Like the stars, this gyro will serve as a precise reference for long range navigation.

The Honeywell ESG was originally developed with funding by the Navy Special Projects Office and Wright Air Development Division. Further developments, including the MEG (Miniature Electrostatic Gyro), are being made with the technical support and funding of the USAF Aeronautical Systems Division, as well as company sponsorship. The MEG demonstrates Honeywell's advanced technology in the field of inertial systems and sensors.

The ESG and MEG are based upon Honeywell's experience in developing and producing more than 35,000 inertial gyros and accelerometers. These inertial devices have been used on 62 of America's successful orbital shots and on such missile programs as Sergeant and Polaris.

For further information on Honeywell's inertial sensor capabilities, from research through manufacturing, write Dept. AW-11-88, Minneapolis-Honeywell, Minneapolis 40, Minnesota.

Honeywell

H *Military Products Group*

Figure 4.3
Honeywell Advertisement for Electrostatically Suspended Gyroscope, 1962
Courtesy Honeywell, Inc.

the Services to start to tighten the requirements."[28] They did eventually succeed in having SPN/GEANS adopted for upgrades of the B-52 bomber force during the 1970s, but only with a high-risk maneuver: bypassing the normal acquisitions procedures and taking an unsolicited proposal directly to higher levels in the Pentagon. This success turned out to be only temporary, perhaps in part because of resentment on the part of those who had been circumvented, and the B-1 was equipped with an inertial navigator significantly less accurate than SPN/GEANS.[29]

So, while accuracy demands have in some cases been higher in military than in civil air navigation, there is no sign here either of a technological trajectory leading inexorably to greater inertial accuracy.[30] The root cause seems to be the assumption that, at least at certain selected crucial moments, inertial accuracy can be supplemented by non-black-box inputs such as visual identification, radar, and "homing" missiles. In the case of the B-52 and B-1 bombers, for example, it seems to have been assumed that navigation would not wholly be black box. If such a bomber were entering Soviet air space, as it passed over the Soviet coast it would take a rapid, downward radar "landfall fix," which would be used to update its inertial system to enable the latter to navigate it accurately down its narrow "penetration corridor."[31] The more accurate the inertial system, the less sophisticated and expensive the radar

28. Interview with John Bailey, Minneapolis, March 7, 1985.
29. This is the Kearfott system known as HAINS (High Accuracy Inertial Navigation System), which has an accuracy of around a quarter to a fifth of a nautical mile per hour. See Thomas Shanahan and James E. McCarthy, "High-Accuracy Inertial Navigation," *Proceedings of the IEEE,* Vol. 71 (October 1983), 1166–70. At the time of writing, no details have been revealed of the navigation system of the B-2 "Stealth" bomber.
30. It is worth noting that positional accuracy is not the only characteristic of a military air inertial navigator that is of interest to customers. The accuracy of velocity information, largely irrelevant in itself to civil customers, matters in military applications when the inertial system is used to assist bomb-aiming or the firing of tactical missiles. Furthermore, military aircraft, especially fighters, do not always fly the gentle flight paths that civil transports do, imposing a further demand on military inertial navigators.
31. See the slide "Strategic Bomber Performance Analysis with SPN/GEANS," *Proceedings, Ninth Data Exchange for Inertial Systems, Life Cycle Cost Workshop* (Macdill Air Force Base and Honeywell Aerospace Division, St. Petersburg, Florida, November 18–19, 1975), 343.

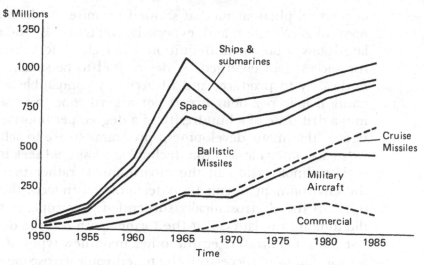

Figure 4.4
Western Inertial Market
Annual volume in then-year U.S. dollars. Rough estimates only. Data supplied by Litton Guidance and Control Systems.

needed for the "landfall fix," argued Honeywell. On the other hand, if the highly accurate inertial system was also expensive and unreliable (as SPN/GEANS's opponents argued that it was), then it could be seen as making sense to stick with an only modestly accurate inertial navigator and spend the savings on the radar.

The Inertial Market and Innovation in Inertial Sensors

The overall evolution of the inertial market is summarized in figure 4.4. As shown there, the ballistic missile inertial market peaked in its volume around the mid-1960s, at the height of the massive expansion agreed by the Kennedy Administration. From 1965 onward, growth in a nonexpanding market lay in military and commercial air navigation and the revived cruise missile field. For reasons to be discussed below, this last came to resemble in its effects on the inertial market air navigation more than ballistic missile guidance.

The creation, stabilization, and growth of the air navigation inertial market played a major part in shaping technological innovation in inertial sensors. The late 1950s and early 1960s saw an explosion of interest in exploiting an enormous range of physical phenomena that showed promise for the measurement of acceleration and, especially, rotation.[32] By the mid-to-late 1960s, a dominant definition of the characteristics needed for success had emerged. A device had to be small, reliable, fairly easy to produce, and of accuracy compatible with one nautical mile per hour error. For a gyroscope this translated into a drift rate of a hundredth of a degree per hour, a "magic figure" that many development programs strove to achieve.

No absolute rule said that technologists should seek to shape their technology to suit the environment, rather than shape the environment to suit their technology. Indeed, the developers of the electrostatically suspended gyroscope, as we saw, did attempt the latter. But the former path was the dominant one, and it characterized the other main new types of inertial sensor that were successful: the tuned-rotor gyroscope and the laser gyroscope.

The tuned-rotor gyroscope is a "dry" gyroscope, employing neither fluid nor gas. There was nothing in itself new about being dry in this sense—most pre-1945 gyroscopes were dry. The crucial difference is a sophisticated mechanical design that allows the gyro rotor to become in effect decoupled from its surroundings. In a dry, tuned-rotor gyroscope the gyro wheel is mounted on a shaft with a specially designed flexure or hinge, in a "mushroom" configuration (see figure 4.5): the wheel and flexure are designed so that the wheel's rotation cancels out the "spring effect" of the support.[33]

The laser gyroscope is unlike all the other gyroscopes described in this book in that it does not contain a spinning mass. It employs two coherent laser light beams travelling in different directions in a cavity in a solid block (see figure 4.6). If the gyro turns,[34] the frequencies of the clockwise and counterclockwise beams divergence by an amount determined by the speed of rotation. That divergence in frequency produces

32. A useful survey is Robert C. Langford, "Unconventional Inertial Sensors," *Astronautica Acta*, Vol. 12 (1966), 294–314.

33. This paragraph glosses over complex questions of technical history and design, which cannot be explored here. I am grateful to interviewees, especially John Stiles, Harold Erdley, Paul Savet, Walter Krupick, Stanley Wyse, Jerome S. Lipman, and D. A. Ulrich, for discussing these with me.

34. That is, if the device rotates around an axis perpendicular to the plane of the cavity.

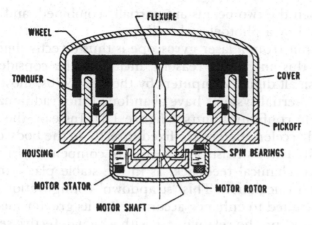

Figure 4.5
Dry Tuned-Rotor Gyroscope (schematized)
Courtesy Singer Corporation, Kearfott Division.

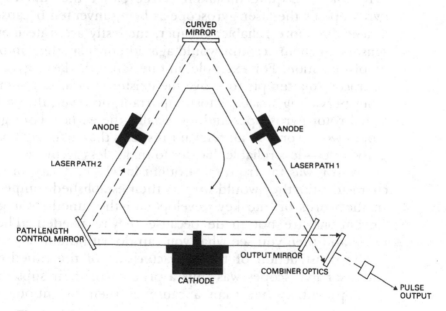

Figure 4.6
Laser Gyroscope (schematized)

"beats" when the two beams are partially combined, and these are detected by a photoelectric cell.[35]

The output from a laser gyroscope is thus directly digital in form. For this and other reasons, including the considerable power of small digital computers by the late 1970s, most laser gyroscope inertial systems have abandoned the traditional stable-platform configuration (see figure 1.1). Instead, the gyroscopes and accelerometers are fixed directly to the body of the vehicle. The latter's twists and turns are compensated for not by electromechanical feedback, as in the stable platform, but by computer calculation. This "strapdown" configuration is not generally argued to enhance accuracy. But its greater mechanical simplicity and the relative ease with which defective sensors can be unbolted and replaced are seen as increasing reliability and reducing the costs of manufacture and maintenance.

It would be quite mistaken to see either the tuned-rotor gyroscope or the laser gyroscope as being invented because of a need for more reliable, cheaper, modestly accurate inertial sensors. Such an account would again underplay the complexity of invention. For example, in one sense the laser gyro can be traced to attempts to prove the existence of a luminiferous ether pervading space.[36] Most important, however, the earliest tuned-rotor gyroscopes and, especially, the earliest laser gyroscopes were worse in all relevant respects than existing floated gyroscopes—less reliable, harder to make, less accurate, and so on—and it was far from self-evident in which, if any, of these characteristics they would surpass their established competitor. In the words of one key developer of the tuned-rotor gyro, "inventions are not made because they're needed. They're made and then you see what you can do with them."[37]

The construction of the characteristics of the tuned-rotor and laser gyroscopes was thus a product of their subsequent development, rather than a cause of their invention. That

---

35. A useful technical history of the laser gyro does exist: C.V. Heer, "History of the Laser Gyro," *SPIE* [Society of Photo-Optical Instrumentation Engineers], Vol. 487 (1984) (*Physics of Optical Ring Gyros*), 2–12. For helpful discussions of the development of the laser gyroscope, I am grateful to interviewees Frederick Aronowitz, Joseph D. Coccoli, Clifford V. Heer, Stephan Helfant, Tom Hutchings, Joseph Killpatrick, Warren Macek, Bernard de Salaberry, and John Stiles.
36. See Heer, "History."
37. Interview with John C. Stiles, Wayne, N.J., September 25, 1986.

development certainly did not simply follow the will of the developers. The intractabilities of the material world played a major part: no one could discount these intractabilities after listening to interviewees describing the long and painful process of turning the prototype laser gyroscopes of the early 1960s into the hundredth-of-a-degree per hour air navigation equipment of the late 1970s and 1980s.

Unfortunately for any simple account of technological change, however, it is not self-evident, at least in advance of attempts to tackle them, what precisely the intractabilities of the material world are and how they bear upon technological development. It is one thing to use quantum theory, as one proponent of the laser gyro did in its early years, to argue that its fundamental limiting errors were as small as 10^{-7} to 10^{-9} degrees per hour, far superior to the 10^{-2} degrees per hour of air navigation, and better than even the best Draper floated gyro.[38] It was quite another to get others to believe you, to persuade them that the fundamental limits could in practice be approached, and that it would satisfy goals of theirs if they provided you with resources to undertake the effort.[39]

Technological trajectories leading to successful tuned-rotor or laser gyro development could thus be created only where credible promises could be made to those who had both the resources to make it possible to fulfill them and also an interest in having them fulfilled. These promises seem to have concerned producibility and reliability rather than accuracy. Thus it was known that keeping the flotation fluid clean and free of bubbles was a major problem in conventional gyroscope production, and the tuned-rotor gyro combined absence of fluid with what could be seen as a less complex structure. The laser gyroscope's great attraction was that it was apparently free of moving parts and all the reliability problems they bring.[40]

38. Joseph Killpatrick, "The Laser Gyro," *IEEE Spectrum*, Vol. 4, No. 7 (October 1967), 51.
39. This is no longer believed of the laser gyro, even by the author of the claim: interview with Joseph Killpatrick, Minneapolis, March 7, 1985. A figure of 10^{-4} degrees per square root of hours of operation, for laser gyros of reasonable size, is quoted in Terry A. Dorschner et al., "Laser Gyro at Quantum Limit," *IEEE Journal of Quantum Electronics*, Vol. QE-16, No. 12 (December 1980), 1378.
40. This promise has not to date been fulfilled. At low rotation rates, the

There was nothing absolute about the selection of producibility and reliability rather than accuracy: proponents of the two devices held, and hold, that accuracy superior to the floated gyro could be achieved from them. But the selection did correspond to the environment in which tuned-rotor and floated gyro development was conducted—large corporations with an eye primarily to the air navigation market—while Draper's Instrumentation Laboratory, where accuracy, as we shall see, was a paramount priority, did not systematically develop either device. This "fit" of technology and environment was not automatic however. Persuasive champions of the technology were needed to mediate between the technological enthusiasm of development teams and the caution and skepticism of corporate management.

The tuned-rotor gyroscope became an important tool of Kearfott's and Litton's competition in the "mile per hour" market, though the former in particular has also used it in successful attempts to gain entry to market niches demanding greater accuracy (see chapter 5). The corporation that more than any other supported the lengthy and painful process of laser gyroscope development was Honeywell, probably because it saw in the new device a route into the "mile per hour" market from which it was excluded. Its faith eventually paid off, at least in market share if not in profit.[41] And as other firms saw the beginnings of laser gyroscope success at Honeywell in the late 1970s, they hastened to invest major resources of their own in the device, fearing eventual exclusion from the main inertial market if they did not.

Participants still dispute whether the laser gyroscope's success is through inherent worth or "technology charisma," but by the mid-1980s the success seemed irreversible. Only the original pioneer of civil air inertial navigation, Delco, did not have a major laser gyro program, and was developing instead

two laser beams "lock in"—refuse to demonstrate a frequency difference—and the device is, inelegantly, dithered rapidly to avoid this happening.

41. Honeywell interviewees would not be drawn on the financial aspects of the laser gyroscope work, but the consensus of others interviewed in the mid-1980s was that the firm had yet to recoup its investment in laser gyroscope development and production facilities. The first operating laser gyro was not at Honeywell but at Sperry Gyroscope. Honeywell devoted greater resources to it and developed it more successfully.

a different device, the hemispherical resonator gyro (see figure 4.7).[42]

The Twofold World of Inertial Sensors: The Generations of Charles Stark Draper

Use of the power of the digital computer, sophisticated mechanical designs, and electro-optical technology to build more reliable and cheaper inertial systems has thus proven a sustainable technological trajectory. These developments were, however, viewed with ambivalence in Draper's MIT Instrumentation Laboratory, or Charles Stark Draper Laboratory, Inc. as it became when divested from MIT in the early 1970s. There certainly were staff there who were excited by them and wished to become part of them. But they did not fit within Charles Stark Draper's vision of technological change in inertial guidance. That vision, and the social and technological framework sustaining it, retained sufficient power to keep the Laboratory on a course quite different from that being mapped in corporations such as Kearfott, Litton, and Honeywell.

Draper certainly believed increased inertial accuracy to be a natural trajectory of technology. In the 1960s, when others were moving in the directions described above, he created a way of periodizing the history of inertial technology that expresses well this point of view.[43] Two features of this stand out: that the focus is on the basic inertial sensors, gyroscopes, and accelerometers; and that the periodization is in terms of

42. Edward J. Loper and David D. Lynch. "The HRG: A New Low-Noise Inertial Rotation Sensor," paper presented at the Sixteenth Joint Services Data Exchange for Inertial Systems, Los Angeles, November 16, 1982; interview with David D. Lynch, Goleta, Calif., September 11, 1986.

Another interesting inertial sensor development has been the emergence of fiber-optic gyroscopes. These are currently targeted not on the air navigation market, but on the growing market for simple, low accuracy inertial systems for the midcourse guidance of tactical missiles. One interesting development is that there is now serious work on fiber-optic gyroscopes at the Draper Laboratory; its seriousness may possibly be related to the situation of the Laboratory's traditional gyro work.

43. The fullest expression of this is Charles Stark Draper, *Importance of Research directed toward the Development of Ultimate Performance for Inertial System Components* (Cambridge, Mass.: Charles Stark Draper Laboratory, Inc., January 1975; P-030), but the Draper was using the terminology of "generations" well before then.

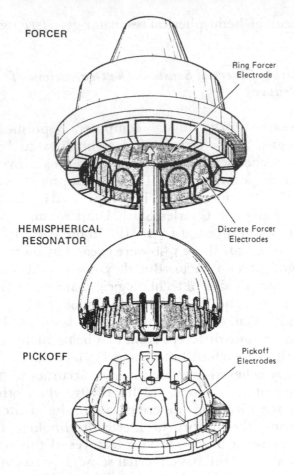

FORCER

Ring Forcer
Electrode

HEMISPHERICAL
RESONATOR

Discrete Forcer
Electrodes

PICKOFF

Pickoff
Electrodes

Figure 4.7
Hemispherical Resonator Gyroscope
Courtesy David Lynch, Delco Systems Operations, General Motors
Corporation.

their accuracy, not any other characteristics such as cost or reliability.

According to Draper, first generation inertial sensors were the aircraft instruments such as the artificial horizon and directional gyro dating from the interwar period. The "second generation" was the generation that made inertial guidance and navigation possible, sensors of the type used in the missiles described in chapter 3 and in air navigation systems. A typical second generation performance would be a gyroscope drift rate of a hundredth of a degree per hour.

"Third generation", performance would be drift rates not of 10^{-2} but 1.5×10^{-5} or 1.5×10^{-6} degrees per hour. Although second generation performance translated into position errors of 1,000 to 10,000 feet after an hour's elapsed time, third generation instruments would contribute only an average error of one to ten feet after an hour. Beyond this, Draper could envisage a fourth generation with performances of the order of 1.5×10^{-7} to 1.5×10^{-9} degrees per hour. Though this would correspond to an error in position of a about an inch or less after an hour's elapsed time, Draper believed that it did not "appear to be forbidden by any basic properties of materials or fundamental laws of physics."[44]

Considering the laser gyroscope reveals how sharply this view of the trajectory of inertial technology differed from that of others. This 1980s device, for all its "high-tech" basis in quantum electronics and the hundreds of millions of dollars spent on its development, would remain in Draper's terms a second generation technology, in the same category as the mechanical gyroscopes of the 1950s.

Draper's uncompromising attitude to accuracy came to be adopted by at least some in his laboratory as an defining an appropriate niche for the laboratory in an inertial industry populated by large, resource-rich corporations. Robert Duffy, whose metaphor of the beryllium baby I quoted at the start of this chapter, spelled out this argument in October 1972, shortly before he took over the Laboratory's presidency:

I sense a very strong shift in emphasis from performance oriented requirements to one of high reliability, or in commercial terms, low

---

44. Ibid., 26. The paper made similar predictions for accelerometer development.

cost of ownership. Although we do not interpret these requirements to be in conflict in all cases, this shift has affected our Laboratory in an unusual way. Our performance orientation has been sharpened rather than decreased, because the market, particularly the commercial market, has drawn the industry itself towards cost/user utility goals, leaving the door to performance somewhat unguarded.[45]

Draper's viewpoint was certainly not mere rhetoric or empty programmatics. It was an expression of the technological practice of his Laboratory. He set his face against novel types of inertial sensor, believing them to be a diversion from the true path of technological progress. During the 1960s, some staff at the Instrumentation Laboratory developed an interest in laser gyroscopes, built one, and secured external development funds.[46] With no support—almost opposition—from Draper, the work did not develop. In my interviews with him in 1984, I asked him why he had not supported the laser gyroscope work. He replied that "the laser path . . . by and large doesn't lead to better accuracy . . . the laser business sort of by-passed people away from accuracy."[47]

The path that led to accuracy, according to Draper, was the intensive, continuous refinement of his single-degree-of-freedom floated gyroscope and the analogous accelerometers. The method was in essence simple, and was a continuation into the 1960s and beyond of how Draper had created his original second generation sensors.[48] Painstakingly determine the dominant error sources in current versions of the floated gyro and accelerometers (the reverse salients, in Hughes's sense), and focus inventive effort on identifying and solving the critical problems that will eliminate them. Then repeat the process for the new devices.

Though the method was simple, its concrete manifestation was not. Space prohibits full examination of it, but the following quotation from an eyewitness to the process conveys some of the favor: "The art of gyros and accelerometers has pushed

45. R. A. Duffy, "Summary of New Developments at the Draper Laboratory," in *Inertial Navigation Components and Systems, AGARD Conference Proceedings 116* (Neuilly-sur-Seine: NATO Advisory Group for Aerospace Research and Development, 1973), 3.1.

46. Draper Laboratory interviews.

47. Interview with Charles Stark Draper, Cambridge, Mass., October 12, 1984.

48. See chapter 2.

developments in precision machining and measurement; it has forced new knowledge in the dimensional stability of materials; it has demanded unusual new materials with special properties; it has developed new bearings of very low and zero friction and high-speed bearings of enormous stiffness, stability and life; it caused the development and refinement of many new electromechanical transducers of phenomenal accuracy; it has required highly specialized assembly, calibration, and test techniques in special clean-room facilities."[49]

By the 1970s, Draper believed third generation performance to have been achieved. Figures disappear behind the cloak of classification, but the gyroscope designed by the Laboratory for MX's beryllium baby was called the third generation gyroscope, and such fragments of information as are available on its performance are consistent with this designation.[50] The accelerometers designed by the Laboratory for MX and Trident II also appear to be third generation. So the form of technological change practiced at Draper's Laboratory was, in terms of its own goals, a powerful one. Over a period of 40 years, the accuracy of the paradigmatic Draper floated gyroscope had been increased by around five orders of magnitude—a hundred-thousand-fold.[51]

How was this path of technical development sustained? Part of the answer lies in Draper's sheer skill as a heterogeneous engineer, his extensive contacts, and his solid reputation as someone who delivered on his promises. It is interesting, in particular, that a significant proportion of the funding for "third generation instruments" came from an agency that has never used such an instrument, the National Aeronautics and Space Administration (NASA).[52]

49. D. G. Hoag, "Ballistic-missile Guidance," in B. T. Feld et al., eds., *Impact of New Technologies on the Arms Race* (Cambridge, Mass.: MIT Press, 1971), 83–84.

50. Thus one set of interviewees, not at Draper, told me that for high-accuracy gyrocompassing as performed by the MX Advanced Inertial Reference Sphere, a "resolution" of 10^{-5} degrees per hour was required from the gyroscopes.

51. I have here taken Draper's 1947 estimate of "the lowest drift rate in good existing equipment" of 30 milliradians, or 1.7 degrees, per hour and the upper boundary of his 1975 definition of "third generation" performance, 1.5×10^{-5} degrees per hour.

52. The contract, for "development of an advanced gyro and accelerometer," ran from June 1967 to June 1970, and amounted to close to $8 million. See

To understand this, one must understand Draper's relations to NASA, the site of his most prestigious public success—the Instrumentation Laboratory's design of the navigation system for the Apollo moon mission. Draper had a key contact in NASA's top official or "administrator," James Webb: "Jim Webb was the Treasurer of the Sperry Gyroscope Corporation in World War Two and he remembers this little professor from MIT who designed the gunsight that made them millions of dollars, and he knew this guy could do anything."[53] Webb's confidence helped Draper win, without formal competition, the contract for the guidance system for the Apollo command module and lunar excursion module. An extensive proposal had been written, and the Instrumentation Laboratory was riding high on the success of Polaris, but what may have been the crucial moment was highly personal:

In September [1961], a crisp, cold day, we were called out to have dinner in Jim Webb's mansion in Washington. They had Jerry Wiesner, who was Jack Kennedy's Science Advisor, and a lot of top brass.

After dinner was over, Jim Webb said, "well, let's go into the living room and have some brandy." So we walked in and there was a big fire going in the fireplace. . . . After everyone was served brandy, Webb says to Doc [Draper], "Can you design a guidance system that will take us to the Moon and back?" and Doc says "Yes." Silence. And Jim says, "Well when can you have it?" Doc says, "I'll have it when you need it." Silence. And Jim said, "Well, how will I know it will work?" And Doc says, "I'll go along with it and make sure it does."[54]

Having "delivered" on his promises to Webb, Draper was in a good position to secure funding for the work that appears to have been closest to his heart: the development of inertial

Instrumentation Laboratory, (Untitled) Annual Report (Cambridge, Mass.: MIT Instrumentation Laboratory, April 1968), and memorandum to Professor Hill from J. F. O'Connor, January 13, 1970 (Hill Papers, 83–40, Box 3, MIT Archives and Special Collections).

53. Interview with Ralph Ragan, October 4, 1984. James Webb confirms that he had developed confidence in Draper from his Sperry days (Webb interview, July 22, 1983, Space Astronomy Oral History Project, Nationa Air and Space Museum, Washington, D.C.).

54. Ragan interview. Draper of course never flew on Apollo, but a letter from him officially volunteering to do so ("I realize that my age of 60 years is a negative factor in considering my request") is on file at NASA: Draper to Robert C. Seamans, November 21, 1961 (NASA Headquarters History Office Archives, Washington, D.C., Biography file—Draper).

instruments. "Doc used to tell us that he got the funds for the NASA program [on] third generation instruments by handshake with Jim Webb . . . in a bar, I think in Germany."[55]

Some of the funding for third generation instruments was thus not tied to any particular application. Draper, too, seemed committed to finding a nonmilitary rationale for them. His laboratory had become by 1970 a primary target for antiwar protest in the Boston area—weekly pickets of it, for example, had begun, which were still continuing when I visited it in the mid-1980s—and this pressure was a factor in the decision by MIT to divest itself of the Instrumentation Laboratory.[56] In any case, as we shall see, it was not at all clear around 1970 that "third generation" instruments would find a home in military systems.

55. Interview with William G. Denhard, Cambridge, Mass., February 18, 1985.

56. See Dorothy Nelkin, *The University and Military Research* (Ithaca, N.Y.: Cornell University Press, 1972). It is clear from the material concerning the Instrumentation Laboratory in the MIT Archives that opposition to the Instrumentation Laboratory, especially to its sheer size with its two thousand or so staff, predated the protests. A 1965 "Memorandum to: The Files. Subject: the Poseidon Contract," by MIT Vice-President James McCormack, for example refers to "MIT's reluctance to expand the size or dollar volume of Instrumentation Laboratory" (MIT Archives and Special Collections, Hill Papers, 83–40, Box 4). During the early 1960s the Instrumentation Laboratory had begun to be seen as a direct competitor by corporations involved in navigation and guidance, and the award of the Apollo guidance contract to it, without a formal bidding or competition process, sparked public criticism from guidance and navigation firms such as Arma. In apparent response MIT President James Stratton warned the Instrumentation Laboratory (and also the MIT Lincoln Laboratory) to "exercise strict self discipline against any tendency to expand our part in [government] programs beyond the area where our services are clearly necessary" (memoradum of December 15, 1961, made public in "Laboratories warned on Over-Expansion," *Aviation Week and Space Technology*, January 22, 1962, 38). For industry protests about the Apollo contract, see Philip J. Klass, "Apollo Guidance Bidders Protest NASA Choice of Non-Profit Firm," *Aviation Week and Space Technology* (January 8, 1962), 23–24. MIT officials were certainly sensitive to the charge of competition with industry (see the letter from General B.A. Schriever to James R. Killian of the MIT Corporation, June 25, 1963, Hill Papers, 83–40, Box 4) and as early as 1962 at least limited divestment of some Instrumentation Laboratory functions was being considered: V. A. Fulmer, "Memorandum to J. A. Stratton, Subject: the M.I.T. Instrumentation Laboratory," October 15, 1962, ibid.

The minutes of the Board of Directors of the newly founded not-for-profit corporation, the Charles Stark Draper Laboratory, Inc., show that in the early-to-mid 1970s Draper and others in the Laboratory pushed the case that civil aircraft navigation required third generation instruments.[57] The paper that most clearly outlines Draper's view of the trajectory of inertial technology, for example, argued that "[I]mprovements of three to four orders of magnitude in performance of angular deviation sensors and specific force sensors are desirable if the one foot location inaccuracy performance essential for safe, reliable and free-from-weather-interference aircraft landing operations is to be realized in practice."[58]

The reference to landing is significant. Draper's belief was that inertial navigation could be used not only to navigate and maintain safe separation of aircraft over the oceans, but—with his predicted one foot error—even during an aircraft's final, possibly blind, approach to the runway.

Draper's hopes never came near realization. In civil aviation he lacked the extensive contacts with those in command of resources that he had in the military sphere, and his accuracy prediction, had it been listened to at all, would no doubt have sounded like an empty boast. In any case, the extent of existing investment, financial and conceptual, in ground-based navigation and air traffic control systems was enormous. The task of persuading the civil aviation business that they needed inertial instruments with third or fourth generation accuracy was wholly beyond him.

Accuracy at the target measured at least in tens of feet was, however, a goal of the new segment of the inertial market that opened with the revival in the 1970s and 1980s of the cruise missile. No longer the large, high-flying devices of the 1940s and 1950s, the new systems were designed to be small, highly accurate, and able to evade detection and interception by flying

57. See the Minutes of the Board of Directors of the Charles Stark Draper Laboratory, in the Albert G. Hill papers, box 2, files 1/5, 2/5, 3/5, 4/5, 1/3, 2/3, 3/3 (MIT Archives and Special Collections), also W. G. Denhard, "Technology of Tomorrow's Commercial Air Traffic Control," in *Proceedings of the Institute of Navigation National Air Meeting* (Washington, D.C.: Institute of Navigation, 1971), 1–21.
58. Draper, *Ultimate Performance*, 33.

low.[59] Draper attempted, through his position on the Guidance and Control Panel of the Air Force Scientific Advisory Board, to tie the development of the cruise missile to his view of the technological trajectory of guidance technology.[60] How better to achieve high accuracy than to equip the new cruise missiles with third generation inertial sensors?

Again, though, Draper's attempt was a failure. The form of cruise missile guidance chosen was an attempt to combine cheapness and high accuracy by "opening the black box." Cruise is equipped with what is essentially a standard "one nautical mile per hour" inertial navigator. This steers this missile into preplanned "boxes," where a radar altimeter generates an altitude profile that is compared with a computer-stored map and used to correct accumulated errors in the inertial system.[61]

There were seen to be risks in this way of doing things, particularly that the radar altimeter's operation would be detectable and that snowfalls might distort the altitude profile of the "boxes." Satellite mapping of the "boxes" has also been an extremely expensive operation. But Draper was unable to turn these admitted disadvantages into acceptance of high accuracy cruise missile inertial navigation.

What seems to have swung the matter against Draper, at least in the Scientific Advisory Board, was an old enemy, the problem of the vertical. Because cruise missiles fly very low, their inertial systems would be strongly affected by the gravitational anomalies caused by such features as mountain ranges. Global gravity models of the sort developed using Transit satellites (see chapter 3) would not be good enough for very high accuracy. Gravity maps based upon ground-level surveys existed for some regions, but the Soviet Union and China, realizing the new military significance of gravity, were not prepared to divulge detailed gravity data.[62] So it was argued

59. See Kenneth P. Werrell, *The Evolution of the Cruise Missile* (Maxwell Air Force Base, Alabama: Air University Press, 1985).
60. Interview.
61. See, for example, John C. Toomay, "Technical Characteristics," in Richard K. Betts, ed., *Cruise Missiles: Technology, Strategy, Politics* (Washington, D.C.: Brookings Institution, 1981), 37–38.
62. My knowledge of this is drawn from conversations with Dr. Roger Hipkin, Department of Geophysics, University of Edinburgh.

against Draper in the Scientific Advisory Board that inertial instrument performance was no longer the key "reverse salient" to the achievement of accuracy. Knowledge of the gravity field was, and without this even inertial instruments with zero errors could not provide the requisite accuracy.[63]

The Draper Laboratory had a new solution to "the problem of the vertical" under development, an inertial instrument to permit onboard calculation of the gravity field, known as a gravity gradiometer.[64] But to equip each cruise missile with an third-generation inertial guidance system and gravity gradiometer would add weight and make what was put forward as a cheap weapon system much more expensive.[65]

With Draper thus having failed to root his desired technological trajectory in any of the expanding areas of the inertial business, its future thus became almost wholly dependent on the traditional area of ballistic missile guidance.[66] And that area did indeed sustain it, perhaps not as generously and predictably as Draper would have liked, but adequately. In the 1960s substantial development funds for "continuing development of components and inertial guidance systems for ballistic missiles" were provided by the Air Force.[67] A Navy missile "Improved

63. Interview. The phrase "reverse salient" is, of course, not the interviewee's, but conveys well his argument.

64. It is discussed further in chapter 5.

65. The current Advanced Cruise Missile program is "black" (protected by exceptional security classification), so nothing can be said with any certainty about its guidance aspects. It may be, however, that a stellar-inertial navigation system (see chapter 2) is being considered as a means of overcoming the perceived disadvantages of the radar altimeter.

66. There was a considerable felt need for accuracy in ballistic-missile submarine inertial navigators, but this was (a) in any case in a sense derivative of the need for missile accuracy, (b) an area where the electrostatically suspended gyroscope would be a powerful competitor (it tended to be ruled out for ballistic missile use because of fears that it could not easily survive high accelerations), and (c) a niche occupied by a major corporate actor, Autonetics. The remaining major market segment, space, was also "canvassed" by the Laboratory to no effect. See Scott Matthew Britten, *Space Tug Applications of Fourth Generation Instruments* (M.Sc. dissertation, MIT, 1977). Currently, the precision pointing and tracking requirements of the Strategic Defense Initiative are another possible avenue, but one whose outcome is profoundly unclear.

67. In 1962, for example, $6.1 million of a total Laboratory budget of $22.6 million was provided by an Air Force contract with this purpose, the bulk of the remainder being accounted for by the Polaris and Apollo programs. J.

Accuracy Program" was also important in the 1970s. The major missile guidance design contracts won by the Laboratory supported the process of turning the inertial sensor development work into producible, working devices and paid for much of the Laboratory's infrastructure. It also indirectly supported further development work. The type of military contract under which the Laboratory typically works provides a certain proportion of funds for independent research and development (IR&D). This, for example, has been used to support the Laboratory's fourth generation work.

But Draper's technological trajectory was tied by all this not just to the ballistic missile, but to the counterforce ballistic missile. Second generation instruments could, it was widely agreed, guide a missile perfectly adequately for the purpose of destroying cities. Only if missiles were to be aimed at hard, point targets such as missile silos was there a need for third generation instruments.

Remarkable predictions of possible missile accuracies were made. Even in the mid-1950s, with ICBM inertial guidance not yet a reality, Draper was "predicting extreme accuracies down the road."[68] One interviewee recalls Draper predicting in a meeting of the Air Force Scientific Advisory Board, probably in the 1960s, that he could "guide a missile to within a hundred feet at 6,000 miles."[69] I have found no documentary record of such a prediction by Draper himself, but David Hoag of the Draper Laboratory argued in a 1970 paper that no inherent barrier prevented ICBM accuracies of 30 meters, and perhaps even an order of magnitude better: "Such performance is possible. It will become reality if the government and military incentive cause resources to be committed to the challenge."[70]

So the histories of the Draper Laboratory, third generation instruments and counterforce nuclear strategies became wholly interwoven from the mid-1960s onwards. In chapter 5 I examine this interweaving as it manifested itself in the development

B. Feldman to General J. McCormack, "Instrumentation Laboratory Contract Summary," May 10, 1962 (Hill Papers, 83–40, Box 4, MIT Archives and Special Collections).

68. Herbert York, *Race to Oblivion: A Participant's View of the Arms Race* (New York: Simon and Schuster, 1970), 54.

69. Interview with John Brett, Wayne, N.J., January 15, 1987.

70. Hoag, "Ballistic Missile Guidance," 81.

of U.S. Navy missiles. Here, we must now turn to look in greater detail at the evolving strategic thinking of the U.S. Air Force, and at the path that led ICBM guidance to MX's beryllium baby.

The Air Force, McNamara, and Counterforce

In the 1950s, dominant tendencies in the U.S. Air Force were ambivalent about counterforce. Though not opposed to it, and still committed to counterforce as part of a preemptive nuclear attack, they tended to see it as an adjunct to their primary countercity role. By the 1960s and thereafter, however, this pattern of priorities had reversed: counterforce was primary, and countercity attacks an adjunct.

The central cause of the change was Polaris, and the challenge mounted by the Navy in the late 1950s to existing Air Force strategies. If the Navy won the argument that national nuclear strategy should rest primarily on the "ultimate deterrent" threat of the destruction of cities, then it was going to be difficult to dispute that Polaris, because of its invulnerability, was the best nuclear weapon system the United States possessed. The Air Force did briefly seek to challenge the claim that Polaris was invulnerable: Air Force Chief of Staff Thomas D. White told a subcommittee of the House of Representatives early in 1960 that Polaris was "certainly not invulnerable," because Soviet submarines could simply shadow Polaris submarines, waiting to destroy them.[71]

That, however, was an argument that was increasingly difficult to make credible. So the more long-term Air Force response to Polaris was to emphasize the aspects of nuclear strategy not encompassed by the ultimate deterrent countercity threat. By the end of 1960, General White had come close to embracing counterforce, and, after some initial resistance, his colleagues were soon to follow.[72]

A central document in the switch in the Air Force position was a February 1960 memo entitled, significantly, "The Puzzle

71. Quoted in Cecil Brownlow, "Defense Leaders Dispute Weapon Roles," *Aviation Week*, February 15, 1960, 29.
72. George A. Reed, *U.S. Defense Policy, U.S. Air Force Doctrine and Strategic Nuclear Weapon Systems, 1958–1964: The Case of the Minuteman ICBM* (Ph.D. dissertation: Duke University, 1986), 297.

of Polaris." The author was a RAND Corporation analyst, William W. Kaufmann. During the mid and late 1950s, Kaufmann and others at RAND had pondered the central problem of the Eisenhower Administration's professed "massive retaliation" strategy. Given the growing nuclear strength of the Soviet Union, would it make sense to launch the Strategic Air Command's "Sunday punch" in response, say, to a Soviet conventional attack on Western Europe? Increasingly, they concluded it would not, given the devastating retaliation against the United States that would follow. Some more limited response—conventional or tactical nuclear—was necessary instead. Or perhaps, as Kaufmann began to argue in the late 1950s, it would be possible to conduct a purely counterforce attack on the Soviet Union. With Soviet cities still hostage, the Soviet Union might be forced to desist from responding by using their surviving nuclear weapons to destroy the cities of the United States.[73]

Kaufmann's earlier ideas on limited war had appealed to the Army rather than the Air Force—"partially informed" was how an aide to Air Force General White had described Kaufmann to his boss.[74] But in "The Puzzle of Polaris," Kaufmann pointed out how the Fleet Ballistic Missile, because it lacked hard-target kill capability, could only "constitute a valuable supplement to our land-based strategic force." It would not permit the United States "to pursue meaningful counterforce and damage-limiting strategies."[75]

This time the response was quite different. The head of RAND's Washington office circulated "The Puzzle of Polaris" to Air Force leaders, including Chief of Staff White. White liked what he read, had Kaufmann write up his ideas for private circulation in Washington, and in September 1960 touched on the matter in a speech to the Air Force Association.[76] There was stern resistance to some aspects of what

73. Fred Kaplan, *The Wizards of Armageddon* (New York: Simon and Schuster, 1983), chapters 12 and 13.
74. Memo, Colonel L. F. Paul to General White, "Princeton Report on Massive Retaliation," n.d., quoted in ibid., 195.
75. W. W. Kaufmann, "The Puzzle of Polaris," February 1, 1960, quoted in ibid., 237–238.
76. Alfred Goldberg, *A Brief Survey of the Evolution of Ideas about Counterforce* (Santa Monica, Calif.: RAND Corporation, October 1967, revised March 1981; RM-5431–PR), 18–19.

Kaufmann was saying—the General commanding the Strategic Air Command, Thomas Power, reportedly reacted to the idea of sparing Soviet cities by bellowing "Restraint! Why are you so concerned with saving *their* lives? The whole idea is to kill the bastards!"—and in the face of it White retreated.[77] But the message of the episode, that the counterforce mission could be used to counter the challenge of Polaris and the Navy, was not lost.

The attractiveness of this line of argument grew when in 1961–62 the incoming Kennedy Administration began to move toward adopting counterforce as national strategy. The new Secretary of Defense, Robert McNamara, and the band of young civilian analysts he gathered around him, were appalled by what they considered to be the indiscriminate and irrational nature of the nuclear war plans they inherited. The "rational" and "limited" approach embodied in counterforce appealed. The official adoption of a counterforce strategy was soon unveiled by McNamara, first in a speech (written by William Kaufmann), to a secret session of NATO in Athens on May 5, 1962, and then publicly on June 6 in the incongruous form of a graduation address at the University of Michigan at Ann Arbor. The Michigan graduates were told:

The U.S. has come to the conclusion that to the extent feasible, basic military strategy in a possible general nuclear war should be approached in much the same way that more conventional military operations have been regarded in the past. That is to say, principal military objectives, in the event of a nuclear war stemming from a major attack on the Alliance, should be the destruction of the enemy's military forces, not of his civilian population.[78]

For a brief moment, then, the interests of the Air Force and of the administration appeared to be aligned. But almost as soon as the strategy of counterforce was announced, it began to fall apart. To the administration's analysts, it was a way of

77. Quoted by Kaplan, *Wizards*, 246. Emphasis in original.
78. Ann Arbor address, quoted in William W. Kaufmann, *The McNamara Strategy* (New York: Harper and Row, 1964), 116. The Ann Arbor address was written by Daniel Ellsberg, on the basis of Kaufmann's text for McNamara's Athens speech. See Desmond Ball, *Politics and Force Levels: The Strategic Missile Program of the Kennedy Administration* (Berkeley, Calif.: University of California Press, 1980), 197fn.

attempting to limit and fight "rationally" a nuclear war. To at least some in the Air Force, it was carte blanche to develop as large an arsenal as possible with a view to being able, in a preemptive strike, to disarm the Soviet Union. On November 21, 1962, McNamara wrote to Kennedy: "It has become clear to me that the Air Force proposals . . . are based on the objective of achieving a first-strike capability."[79]

The date of the memorandum, right after the Cuban missile crisis, may be significant. Faced with a situation that could easily have led to serious consideration of implementing a preemptive counterforce strike if it had deteriorated further, McNamara seems to have lost confidence in counterforce.[80] He backed away from public advocacy of counterforce to a position that sounded not dissimilar to the Navy's idea of a "finite deterrent." Assured destruction, he called it, because of the fundamental tenet that U.S. nuclear forces must be able to inflict unacceptable damage on the people and industry of the Soviet Union, even if the Soviet Union had struck first.

Aside from the possible effects on McNamara of his having had to contemplate what initiating a counterforce strike would, humanly, mean, assured destruction was also useful to the administration as a basis for combating the Air Force's interpretation of counterforce and its demands for more and more weapons with which to implement counterforce targeting.[81] It permitted an apparently objective answer to the fraught question of "how much is enough?"

An attack of 400 "equivalent megatons," targeting all the major cities of the Soviet Union, would kill about 30 percent of the people of the Soviet Union and destroy half of Soviet industry, while increases in the size of attack beyond 400 megatons would increase deaths and damage only slowly.[82] If the

79. Memorandum for the President, November 21, 1962, quoted in Robert Scheer, "Interview with McNamara," *Los Angeles Times,* (April 8, 1982), 1, 13 14, quote on p. 13.
80. Gregg Herken, *Counsels of War* (New York: Knopf, 1985), 169–70.
81. See Desmond Ball, "Déjà Vu: the Return to Counterforce in the Nixon Administration," California Seminar on Arms Control and Foreign Policy, December 1974, especially 14–17.
82. Kaplan, *Wizards,* 317. "Equivalent megatonnage" is a measure of the area destroyed by a nuclear warhead. It is based on the assumption that this area does not increase linearly with the yield of the warhead, but only as the two-

Soviets knew that whatever they did first, the United States would be able to inflict as much damage as this, they would be deterred, went the reasoning, while they would not be substantially more deterred by larger forces. By the late 1960s, the size of U.S. strategic forces was such that, even allowing for substantial losses of them in a Soviet strike, the 400 equivalent megaton criterion was easy to meet.

It was, it is important to note, a change in "stated posture," rather than in actual targeting. The nuclear war plans, the SIOP, or Single Integrated Operational Plan, which had been revised to allow a separate attack on Soviet counterforce targets, were not changed to reflect the new position. Number one of the five main options in the 1962 SIOP (known as SIOP-63 because it came into effect during Fiscal Year 1963) was an attack on "Soviet strategic nuclear forces, including missile sites, bomber bases and submarine tenders."[83] That option remained in place throughout the years when assured destruction was official policy. An Assistant Secretary of Defense in the Johnson Administration wrote to author Desmond Ball in 1971:

The SIOP remains essentially unchanged since then [McNamara's Ann Arbor speech of 16 June 1962]. There have been two developments, however: (1) it has become more difficult to execute the pure-counterforce option, and its value is considered to be diminishing and, (2) all public officials have learned to talk in public only about deterrence and city attacks. Too many critics can make too much trouble (no-cities talk weakens deterrence, the argument goes), so public officials have run for cover. That included me when I was one of them. But the targeting philosophy, the options and the order of choice remain unchanged from the McNamara speech.[84]

thirds power of the yield. So a warhead with a yield of y megatons has an "equivalent megatonnage" of $y^{2/3}$.

83. Desmond Ball, *Targeting for Strategic Deterrence*, Adelphi Paper No. 185 (London: International Institute for Strategic Studies, 1983), 11. The other main options were: "II. Other elements of Soviet military forces and military resources, located away from cities—for example, air defenses covering U.S. bomber routes. III. Soviet military forces and military resources near cities. IV. Soviet command and control centers and systems. V. If necessary, all-out urban-industrial attack" (ibid.).

84. Anon. to Desmond Ball, February 16, 1971, quoted in Ball, "Déjà Vu," 16–17.

The failure to alter the SIOP to reflect changing "stated posture" was not mere bureaucratic inertia, though the war in Vietnam, which came to be the central preoccupation of McNamara's last years as Secretary of Defense, did to some extent divert attention away from nuclear issues.[85] As we shall see, McNamara's Department of Defense continued to press for counterforce capabilities in weapons despite the adoption of assured destruction. In effect, what happened was that a divergence opened up between, on the one hand, the nature of at least some of the weapons that were acquired and what it was planned to do with them in the event of war, and, on the other hand, the way nuclear weapons were talked about and publicly legitimated.

This should not, however, be seen as commitment by the Department of Defense to a secret first-strike plan. The crucial question was the attitude one should take to the fact that "it has become more difficult to execute the pure-counterforce option." This was a consequence of the Soviet Union expanding and protecting its nuclear force. In retrospect, 1961–62 looked like the high tide of Soviet vulnerability to a U.S. counterforce strike. Soviet land-based missiles and bombers would be slow to launch, and were in vulnerable above-ground bases, its radar early-warning network was patchy, and its few ballistic missile submarines had to come to the surface to fire. In August and early September 1961, with the Berlin crisis in the background, a Kennedy Administration team headed by RAND analyst Henry Rowen used the latest intelligence information to plan in detail a first-strike counterforce attack on the Soviet Union, and calculate the number of American deaths that would result from the Soviets retaliating with their surviving forces. The "best case" answer was two to three million; the "worst case," ten to fifteen million.[86]

The perceived "diminishing value" of the counterforce option in the SIOP meant, in effect, that those figures kept rising as the 1960s went on. Reactions to this bifurcated sharply. On the one hand, there were those who saw it as no bad thing that assured destruction was becoming mutual assured destruction (though the extra word, forming the

85. Herken, *Counsels*, 225.
86. Kaplan, *Wizards*, 294–301; interview with Carl Kaysen (one of the participants in the study), Cambridge, Mass., November 14, 1984.

famous acronym, MAD, was actually added in 1969 by a "hawk-ish" critic of the strategy[87]). Mutual assured destruction would promote "crisis stability." With no need to fear being disarmed by a U.S. first strike, the Soviet Union would have no incentive to launch preemptively. Neither the United States nor the Soviet Union would be placed in a "use 'em or lose 'em" situation, and so both could afford to act cautiously in a crisis. Secretary McNamara and many civilian analysts, liberal elements in Congress, and, as we shall see in chapter 5, a significant section of the U.S. Navy took this view. The analysis prepared the way ideologically for the 1972 Strategic Arms Limitation Talks (SALT) and Anti-Ballistic Missile treaties, which formalized the "balance of terror" between the two superpowers.

The other reaction was to see it as foolish for the United States not to seek to develop its counterforce capability to match the growing number and hardness of Soviet targets. As a civilian analysts' response, this was relatively quiescent during the 1960s. It was certainly present during the SALT debates of 1969–1972, when mutual assured destruction was criticized from the right as immoral in its assumption that "killing people is good, killing weapons is bad." But "the prompt ratification of the SALT treaties in the summer of 1972 indicated that the moral argument against mutually assured destruction was more confusing than compelling to a Congress now plainly enthusiastic about arms control."[88] Criticism of mutual assured destruction was to emerge in full public force only later in the 1970s and 1980s.

It was the Air Force's dominant response throughout the period. Initially, it was not merely a general reaction to the "puzzle of Polaris," but intimately tied up with the single dominant concern of the Air Force: to forge a convincing strategic rationale for the manned bomber, even in the wake of the missile revolution. In May 1957 the Pentagon Weapons Systems Evaluation Group spelled out what could be seen as the military advantages of the manned bomber in the age of the missile: "a. Operational flexibility, b. Accuracy of delivery, c. High pay-

87. The critic was the strategist Donald Brennan. See Herken, *Counsels*, 248.
88. Herken, *Counsels*, 249; the first quotation is from one of the critics, Fred Iklé.

load capability, d. Established reliability, e. Reconnaissance capability."[89]

By 1960 these claimed characteristics of the bomber formed the keystone of the Air Force's case for a new plane to follow-on from the B-52, first called the B-70 then renamed RS[Reconnaissance-Strike]-70. "Advertized as more flexible, reliable and precise than the inaccurate missile, the RS-70 was the centerpiece of the Air Force's counterforce doctrine."[90] This case did not convince McNamara, who, despite great pressure from the Air Force and its supporters in Congress, succeeded in stopping the program. But it was repeated in the Air Force's 1963 look to the future, "Project Forecast." In the words of a later writer:

While it [the Project Forecast Report] acknowledged McNamara's argument that mutual deterrence was emerging, the report argued that this was only true at the level of mutual attacks on cities, called countervalue targeting. This did not mean that the U.S. could not achieve military superiority in the ability to attack enemy military facilities, or counterforce targeting . . . Missiles, while well suited for attacks against cities and other area targets, caused extensive collateral damage and were not tailored for counterforce operations. . . . Aircraft . . . had superior accuracy [to missiles].[91]

Counterforce and the ICBM: From Minuteman I to Minuteman II

This emphasis on accuracy and counterforce was written into the 1964 version of the *United States Air Force Basic Doctrine.*[92] But while initially connected to the (unsuccessful) struggle for a new manned bomber, it was an emphasis that from the early 1960s onwards had its effects on Air Force missiles. Air Force missile advocates were not prepared to accept that the ICBM was inherently less accurate than the bomber. The Guidance Panel of Project Forecast was headed by the same Robert Duffy I quoted at the head of this chapter, then an Air Force colonel.

89. WSEG Report No. 23, *The Relative Military Advantages of Missiles and Manned Aircraft*, May 6, 1957 (National Security Archive, Washington, D.C., Kaplan files).
90. Reed, *Minuteman ICBM*, 298.
91. Ibid., 280–282.
92. Ibid., 289–291.

Duffy had been one of the Air Force students at the MIT Instrumentation Laboratory in 1952,[93] and was "the sort of leader" of the guidance mafia.[94]

Duffy shared Draper's view of the technological trajectory of inertial guidance. Others in Project Forecast were skeptical: "the accuracies that they [the Guidance Panel] were projecting . . . had everyone saying they must be smoking opium."[95] But Duffy and the guidance mafia were committed to proving that extreme accuracy in an ICBM was not a pipedream.

Sheer technological enthusiasm certainly played a role in this, but so did the need to demonstrate an area where ICBMs were clearly "better" than Polaris. Increasingly, too, it was harder to be able confidently to allocate the "Option I" counterforce targets in the SIOP to manned bombers. As early as 1958, the belief that the Soviets were constructing a large ICBM force led to worries that "even though the manned bombers get there, they may not get there in time" to destroy the ICBMs before they were fired.[96] During the 1961 Berlin crisis, it had still been believed that low-flying American bombers could catch Soviet forces on the ground.[97] But as the decade wore on, the argument grew stronger that counterforce targets were so "time urgent" that ballistic missile warheads had to be allocated to them.

The paradox was that the United States had an ICBM— Titan II—that in many ways was suitable for the task, but there were insufficient numbers. Unlike the early ICBMs, whose characteristics were only weakly connected to the realities of Strategic Air Command war planning (see chapter 3), "Titan II was designed to meet SAC's known operational 'requirements'."[98] With a warhead rated at an enormous nine mega-

93. Interview with Robert Duffy, Cambridge, Mass., October 1, 1984.

94. Interview with Major General John W. Hepfer (U.S. Air Force, rtd.), Washington, D.C., August 29, 1985.

95. Interview with General Bernard Schriever (U.S. Air Force, rtd.), Washington, D.C., March 25, 1985.

96. Weapons Systems Evaluation Group, *First Annual Review of WSEG Report No. 23: The Relative Military Advantages of Missiles and Manned Aircraft* (August 8, 1958), 7. (Copy provided by Graham Spinardi.)

97. See above, p. 201.

98. Letter to author from B. P. Blasingame, July 30, 1989. Blasingame's role as Air Force guidance specialist in the ballistic missile program is discussed in chapter 3.

tons, and an accuracy of two-thirds of a nautical mile, Titan II posed a significant threat to the hardest Soviet target. Even bomber advocate General LeMay, who expressed his disdain for the early ballistic missiles by pointedly turning his hearing aid off when the status of the Atlas ICBM was being discussed, "turned it back on when Titan II status was being reported."[99] But one of the first acts of the Kennedy Adminstration had been to curtail the deployment of Titan II to 54 missiles, apparently on the grounds of the lesser cost of Minuteman, and that decision seems to have been treated as irreversible.[100]

So those who wished large-scale counterforce capability from the U.S. ICBM force had no other option, at least in the near term, than to seek to turn Minuteman, with its much smaller warhead but much larger numbers, into a more effective counterforce weapon. That was not the role for which it had been designed. As we saw, Colonel Hall, who played the major role in the early emergence of Minuteman, fought against having a stringent accuracy requirement placed on it. In doing so, he had even embraced the doctrine of "finite deterrence" that was growing in popularity in the Navy: "the Air Force should avoid claims of high accuracy because many people would not believe them. The weapon should instead be used against cities. Hall suggested that the ability to destroy 135 Soviet cities, even if the Russians struck first, would be enough to deter a general war."[101]

Minuteman I had an accuracy specification of 1.1 to 1.5 nautical miles circular error probable (CEP).[102] In the development process, as more was learned about the causes of inaccuracy, the guidance and control portion of the overall error was gradually reduced (see figure 4.1). Assuming the other

99. The anecdote was told by Titan Program Director, Colonel "Red" Wetzel, and reported ibid.
100. It is possible that the enormous collateral damage that would be caused in a counterforce attack by a nine-megaton warhead may also have been a factor.
101. Reed, *Minuteman ICBM*, 77.
102. Slide accompanying General N. F. Twining, "Presentation before the Department of Defense Subcommittee of the House Committee on Appropriations," January 13, 1960, originally classified "Top Secret" (my thanks to Graham Spinardi for a copy of this document, which is in Box 11 of the Defense Department Subseries, White House Office Staff Secretary papers, Dwight D. Eisenhower Library).

sources of error such as inaccuracy in reentry and errors in the model of the earth's gravitational field were reduced proportionately, this would have left Minuteman I with a circular error probable of 0.6 to 0.8 nautical miles (1.1 to 1.5 kilometers).

With a warhead in the range 0.4 to one megaton, Minuteman I was a formidable countercity weapon.[103] But once the Soviets started to follow the Americans and place their missiles in silos, Minuteman I would not be an effective counterforce weapon. Assume the most optimistic circular error probable and yield (0.6 nautical miles and a megaton) for Minuteman I, and assume its target to be a silo hardened to withstand 300 lbs per square inch overpressure from a nuclear explosion, a typical 1960s hardness figure. The standard mathematics of such attacks (appendix B) suggests that the chance of the missile destroying its target was only about one in four.

Minuteman I was too close to deployment for the changes in strategic environment of the early 1960s to permit major changes in the direction of greater counterforce capability, though as a result of the deliberations of a review committee set up in mid-1961 by the Secretary of the Air Force, Eugene Zuckert, some alterations were made to the program, notably to provide the capability to launch missiles individually, not just in salvos.[104]

But already a successor to the original Minuteman was on its way. Minuteman II provided a much bigger opportunity for the proponents of counterforce, because its specifications were set only in 1962. Though Autonetics, designers of the Minuteman I guidance system, kept the role of guidance contractor for Minuteman II, the changes in guidance that took place between the two generations of the missile were dramatic.

Most dramatic was the decision to move from discrete transistors to integrated circuits in the on-board computer. The resulting greater power of the computer meant "greater accuracy through the use of more refined trajectory computa-

103. Twining, "Presentation."
104. Ball, *Politics and Force Levels*, 193–94. The committee was known as the Fletcher Committee after its chair, Dr. James C. Fletcher. It also recommended changes to increase the range of the missile and to provide greater targeting flexibility.

tions."[105] But it was a bold decision, since integrated circuit technology was a mere three years old, and only just entering the market. Indeed, in large part because of the Minuteman program, the early years of microcircuit production were dominated by military purchases. The military bought all the integrated circuits produced in the United States in 1962, and by 1965 were still buying over 70 percent of them.[106] It was also a decision that was to cost dear. As Minuteman II was deployed, the integrated circuits started to fail at an alarming rate—"the silicon was physically cracking"—and a large percentage of Minuteman II missiles were out of service while a "recovery team" replaced the electronics of the entire force.[107]

Another change made with Minuteman II concerned the method of aligning the guidance system correctly in the horizontal plane, a process that, though it took place before launch, was as crucial to accuracy as anything that happened in flight. The system for doing this on Minuteman I could make it impossible to change the target the missile was to be fired at without sending a human crew into the unattended silo to make adjustments which took several hours and could involve physically rotating the missile. There was also a source of potential vulnerability in the automatic optical system, periodically adjusted by human crews, that kept the guidance system in alignment: that system could easily be disturbed by a nearby nuclear explosion. To increase flexibility and reduce vulnerability, the alignment system was made much more automatic for Minuteman II. Dependence on alignment data from "outside the black box" was reduced by adding an ultra-accurate single-degree-of-freedom floated gyro, designed by Lester R. Grohe, a former leading engineer at Draper's Instrumentation Laboratory.[108] This extra gyro permitted

105. Anonymous, "133B Minuteman," *Missiles and Rockets,* October 7, 1963.
106. Ernest Braun and Stuart Macdonald, *Revolution in Miniature: The History and Impact of Semiconductor Electronics,* second edition (Cambridge: Cambridge University Press, 1982), 98.
107. Interview with John S. Gasper, Anaheim, Calif., September 9, 1961.
108. This gyro was Nortronics G1-T1-B, designed for the Nortronics Division of Northrop by Grohe (interview with David Ferguson, Edinburgh, June 6, 1986). Reckoned by his colleagues to be a brilliant designer, Grohe was, like Draper, committed to floated gyros and PIGAs, but took an independent position on the detail design of them, which led to friction between him and Draper.

alignment to be "remembered" even if the optical system failed.[109]

The design of the stable platform was altered radically. Minuteman I had the "inside-out" or "knuckle-joint" platform also found in the systems designed by the German team at Huntsville (see figure 2.8). There were advantages in not having the gyroscopes and the accelerometers buried within the gimbal system of a conventional platform, such as ease of access for maintenance and repair. But the knuckle-joint platform had only limited freedom of rotation in the horizontal plane, an acceptable restriction when the target was known long in advance, but a potential constraint if one wanted to change targets rapidly.[110] Perhaps most importantly, there was a suspicion that its lack of rigidity (it has been called a "rubber platform") was a barrier to enhanced accuracy.[111] To begin with, the plan had been to supplement it with a separate gyrocompass to help with alignment, but eventually it was decided to replace the knuckle-joint platform with a conventional design in which the platform was supported by external gimbals.[112]

The material out of which the platform was made was changed, from aluminium to beryllium. This metallic element,

109. See J. M. Wuerth, "The Evolution of Minuteman Guidance and Control," *Navigation: Journal of the Institute of Navigation*, Vol. 23, No. 1 (Spring 1976), 71–74, for a most helpful discussion of the evolution of Minuteman alignment techniques. Minuteman II was only partially black box in this respect, since optical alignment crews were still needed, but with Minuteman III a full "self-alignment technique" was developed. The reason for the partial solution in Minuteman II may have been that "SAC [the Strategic Air Command] could not believe that a gyro would be able to align as well as those people could outside" (Hepfer interview), but limitations in the Minuteman II computer capacity also prevented the adoption of the Minuteman III self-alignment technique for Minuteman II (Wuerth, "Minuteman Guidance and Control," 74).

110. "Taking advantage of the fact that a ballistic missile trajectory lies in a plane, and the orientation of the plane is known in advance," the Minuteman I platform had only "very limited yaw freedom" (ibid., 69).

111. Gasper interview.

112. A key actor in this was John Gasper of Autonetics, who convinced first his own management and then Robert Duffy, at that point Director of Guidance and Control for the Air Force's Ballistic Missile Division, that the change was desirable, even though it could be seen as giving up an Autonetics "trademark." Letter from Gasper, January 13, 1989.

originally of little practical interest, had begun to be used for a number of specialized purposes, for example as a source and reflector of neutrons in nuclear weapons. It had considerable disadvantages: it was extremely expensive and when machined produces a dust that is highly toxic. But guidance engineers, especially Lester Grohe, became interested in it during the 1950s, in Grohe's case as a solution to a particular reverse salient in gyroscope development.[113] Beryllium's lightness and rigidity came to be seen widely within the inertial business as outweighing its disadvantages. Hence its selection for the stable platform of Minuteman II, and of course for the later MX beryllium baby.

All these changes in the overall design of the Minuteman II-guidance system were important, and they reveal a general concern to increase accuracy. But they tell us little about the institutional dynamics involved. To find out more about these, we have to go first one, then another, step deeper into the black box.

The component of the black box seen as most crucial to accuracy was the accelerometer: "You had an error budget with . . . maybe a hundred terms. The largest contributor was the accelerometer, that was the major error contributor."[114] The Air Force managers of the program were unhappy with the distinctive Autonetics accelerometer (the VM4A "velocity meter") used in Minuteman I. Even for that program they sought to have it replaced with an early version of an acceler-

113. Grohe headed a group at the Instrumentation Laboratory seeking to design small, light but highly accurate inertial instruments. They started to make these out of aluminium, rather than the originally more typical Invar (an alloy of iron, nickel, and carbon widely used in precision instrument making), because of the lesser density of the former, but were finding problems because aluminium expanded faster than other components of the gyro, such as the steel of the ball bearings, as the temperature rose. Searching for a different material, Grohe found that beryllium had a much more suitable thermal expansion coefficient, and other advantages of the material soon became evident. Interview with Albert Freeman, Cambridge, Mass., October 3, 1984.

114. Hepfer interview: see chapter 7 below for the meaning of "error budget." The Autonetics G6B4 gyro from Minuteman I was retained for Minuteman II, though the gas used was switched from helium to hydrogen to enhance accuracy by reducing drag (Wuerth, "Minuteman Guidance and Control," 70).

ometer designed at Draper's Instrumentation Laboratory: the 16 PIGA. This was a pendulous integrating gyro accelerometer, designed on the same basic principle as that developed for the V-2 program (see figure 2.6), but made much more accurate by the results of years of development of floated gyros (the 16 refers to the 1.6 inch diameter of the floated gyro it contained).

The early 16 PIGA turned out not to be properly ready in time for use in Minuteman I, and a second non-Autonetics alternative had to be rejected because an incompatibility with the design of the stable platform (an incompatibility that some Air Force officers suspected, incorrectly, had been deliberately created by Autonetics) meant it would not work well and might even break.[115] So Minuteman I was left with the Autonetics velocity meter.

For Minuteman II the Air Force's Ballistic Missile Office "directed," this time successfully, that the velocity meter be replaced by a Draper accelerometer: the 16 PIGA-G.[116] The decision reveals something of the balance of power between the different organizations involved. The Ballistic Missile Office could formulate preferences in the light of its goals (in this case enhanced accuracy). In the absence of a compelling argument to the contrary, such as the risk of accelerometers breaking, it could impose those preferences.

The perceived additional accuracy of the 16 PIGA-G cost the Air Force dear, both in money and in worry. "The MIT design is a very difficult thing to build," recalls Major General John Hepfer, former Commander of the Air Force Ballistic Missile Office. Though the initial contract went to AC Spark Plug, that firm was unable to produce them to the reliability the Air Force

115. The second alternative was a novel accelerometer design, the Vibrating String Accelerometer, originating from Atlas inertial guidance contractors Arma. A natural resonance frequency of the Minuteman I stable platform was close to one of the accelerometer, leading to severe problems, including broken strings. The Air Force officers' suspicion was that the platform had deliberately been designed with this in mind, so as to rule out the competing accelerometer. However, the platform design actually predated the issue of the Vibrating String Accelerometer. Letter to author from John M. Wuerth, January 6, 1989.
116. Hepfer interview. As noted above, Schriever's Western Development Division underwent several changes of name before it became the Ballistic Missile Office. For simplicity I shall use the latter name throughout.

wanted in order to sustain continuous alert of the Minuteman force, and two other firms, Bendix and Honeywell, were involved: "Honeywell saved us on the 16 PIGA." But the cost was high: together with the redesign of the guidance system electronics it cost "probably $300–$400 million."[117]

It might be argued, however, that the selection of the Draper accelerometer represented, not the Air Force Ballistic Missile Office's ability to impose its will on its contractors in its pursuit of accuracy, but rather its "capture" by a different guidance contactor, Draper's Instrumentation Laboratory. Given the influence of Draper on the Air Force's guidance mafia, that is on the face of it not an unreasonable assumption.

A second step into the black box reveals, however, that the assumption is, in this case at least, false. That step involves us examining the bearings on which the gyro wheel of the Draper accelerometer spun. Traditionally, all gyro wheels spun on ball bearings, but there was a widespread feeling that imperfections in these were a barrier to accuracy. As we saw in chapter 3, Autonetics pioneered a self-activating gas bearing to replace ball bearings altogether. "Ball bearings simply do not cut the mustard in precision gyros," they concluded.[118]

Draper felt that with time and effort the imperfections of ball bearings could be removed, and his Laboratory had invested a great deal of both in seeking to do so. Machining tolerances were reduced from 50 to 200 micro-inches to 10 micro-inches by 1952, and great attention was given to overall bearing design, retainer materials, dirt, excess oil, and the assembly process. This work was supported by contracts from the Air Force, Navy, and NASA, and involved other parts of MIT and firms such as the New Departures Division of General Motors and the Barden Corporation. Considerable progress was made, and the Laboratory's report on this work was confident that, with some further effort, ball bearing design and production could be improved such that "at least 50 percent of those produced would give the requisite high performance for up to 25,000 hours of operating life."[119]

117. Ibid.
118. John Slater, *Twenty Years of Inertial Navigation at North American Aviation* (Anaheim, Calif.: Autonetics Division, North American Aviation, 1966), 36.
119. William G. Denhard, *Evolution of Precision Gyroscope Ball Bearings* (Cambridge, Mass.: MIT Instrumentation Laboratory, March 11, 1966, E-1928);

Again, Draper felt that here was the correct path of technical advance, and gas bearings a diversion: "Doc fought them [gas bearings] like mad."[120] But within the Air Force Ballistic Missile Office, even among the guidance mafia, the feeling had grown that the gas bearing was the better road to both reliability and accuracy.

A lieutenant colonel from the Ballistic Missile Office went directly to Michele Sapuppo, the engineer in the Instrumentation Laboratory responsible for accelerometer design: "Bob Savage . . . came in one night and asked me how I felt about designing gas bearings." Sapuppo had also come to the conclusion that what Draper felt to be the technological trajectory of bearing design might be a blind alley: "my people had struggled for several years to control the retainer/lubricant system in our ball bearings in an effort to prevent performance anomalies. After studying some 75 different configurations, we simply gave up." So with Sapuppo's immediate superior agreeing to provide an "umbrella" against Draper's potential wrath, the design change was made, and the 16 PIGA-G provided with a gas rather than ball bearing.[121]

The decision spelt the end for large-scale funding of gyroscope ball bearing development. It rankled with those at the Instrumentation Laboratory who had "spent all this time and effort really becoming experts in the ball bearing" and felt "about to make something that's very, very worthwhile." But the prophecy that they were wrong was self-fulfilling. "If you ask those fellows [proponents of the ball bearing] 'could you do that job with ball bearings that now has gas bearings?' they feel quite confident that they could but the problem is that they don't really, really know because the funding has been so

Albert P. Freeman, "Gyro Ball Bearings—Technology Today," in *Inertial Navigation—Systems and Components,* AGARD Conference Proceedings No. 43 (Neuilly-sur-Seine: NATO Advisory Group for Aerospace Research and Development, 1968); Anonymous, *Final Report on the Gyro Spin-Axis Program, Performed under Air Force Sponsorship under Contract AF 33(657)-7463* (Cambridge, Mass.: MIT Instrumentation Laboratory, September 1963, R-418). The quotation is from the last of these, p. 1.
120. Duffy interview.
121. Interview with Michele Sapuppo, Cambridge, Mass., October 10, 1984; letter to author from Sapuppo, January 23, 1989.

limited for so many years." The triumph of the gas bearing had become irreversible.[122]

As well as thus revealing something of the dynamics of technological trajectories, the episode shows how the Air Force could impose its wishes: "what happened was the guy in a blue suit [i.e., an Air Force officer] said he wanted gas bearings . . . so he insisted we do them."[123] Despite the existence of the guidance mafia, Air Force officers were by no means simply the captives of Draper. Their desire to make Minuteman II as accurate as possible, and their judgments as to how best to achieve this, were an independent factor in the situation.

The extent to which they were successful, despite the $300 million difficulties with the electronics and the PIGA, can be seen in figure 4.1. The guidance and control accuracy specification set for Minuteman II was three times tighter than for Minuteman I and was achieved. Given that only four years separate the systems, it was probably the sharpest single "quantum leap" that U.S. ICBM accuracy has seen.

It made Minuteman II into a counterforce weapon. A conservative estimate of its circular error probable would be 0.34 nautical miles.[124] With the missile's 1.2 megaton warhead, this would imply a probability of destroying a silo hardened to 300 pounds per square inch of nearly two-thirds, significantly better than Minuteman I's one in four or worse. Soviet silos are believed to have been made much harder since the 1960s, and so Minuteman II's counterforce capacity eroded over the two decades and more since its deployment. An attempt to maintain it by replacing its guidance system with the improved version used on Minuteman III was cancelled early in the tenure of the Carter Administration, perhaps because of the extent of opposition to counterforce in the political system.[125] But at the time of its deployment Minuteman II was a major threat to even the hardest targets in the Soviet Union. And that, we have

122. Interview with Peter Palmer, Cambridge, Mass., October 10, 1984.
123. Ibid.
124. Thomas B. Cochran et al., *Nuclear Weapons Databook*, Vol. 1, *U.S. Nuclear Forces and Capabilities* (Cambridge, Mass.: Ballinger, 1984), 114. As indicated in appendix A, I believe an estimate of around 0.26 nautical miles to be more likely than 0.34. The former would of course mean an even higher "kill probability."
125. Anonymous, "Improved Minuteman 2 Guidance Halted," *Aviation Week and Space Technology* (July 11, 1977), 15.

seen, was no accident, but the result of the most detailed shaping of its guidance system design.

MIRV and Minuteman III

Its successor, Minuteman III, broke in one crucial respect from all previous ICBMs. Each missile was equipped with three different warheads that could be dispatched to different targets, in the system known as MIRV (multiple independently targetable reentry vehicles). The idea of MIRVing ballistic missiles emerged in the early 1960s, with at least "five quasi-independent inventors."[126] It was a technology that displayed remarkable "interpretative flexibility," not simply meaning different things to the different "inventors," but also being seen by different groups as a solution to quite different problems.[127] There was opposition to MIRV, notably within the Air Force from those who felt that, whatever the "mathematics" said, several small warheads just had to be a poor substitute for one big one in attacking hard targets, who feared it would undercut the case for more missiles and who felt that any new development in the missile field would take money away from manned bombers.[128] But MIRV's "something for everyone" characteristic made it unstoppable:

To SP [the Navy's Special Projects Office], BSD [the Air Force Ballistic Systems Division], and the technical community MIRV provided a challenging new development effort and a means of extending their programs beyond the missiles then authorized. It also offered a solution to the technical problems involved in providing a high-confidence and cost-effective means of penetrating ABM [Anti-ballistic Missile] systems. To the rest of the Air Force MIRV was a way of increasing the number of deliverable warheads, thereby permitting coverage of the expanding target list generated by the counterforce targets. . . . It provided a war-fighting capability for precise surgical strikes against military targets within or near urban areas and for symbolic countervalue attacks against dams, nuclear facilities or other nonurban but important targets. . . . In OSD [the Office of the

126. Ted Greenwood, *Making the MIRV: A Study of Defense Decision Making* (Cambridge, Mass.: Ballinger, 1975), chapter 2 and appendices A and B.
127. See Trevor J. Pinch and Wiebe E. Bijker, "The Social Construction of Facts and Artefacts: or How the Sociology of Science and the Sociology of Technology Might Benefit Each Other," *Social Studies of Science*, Vol. 14 (1984), 399–441.
128. Greenwood, *MIRV*, 37–40.

Secretary of Defense] the target flexibility, counterforce and war-fighting capability were desirable attributes of MIRV, but its useful-ness as a hedge against the possible deployment of a larger Soviet ABM system became more important. McNamara and his cost-conscious Office of Systems Analysis found MIRV particularly useful in their struggles to restrain the growth of the size and budget of the strategic forces. . . . These perspectives [were] quite different and in some cases opposed, but it mattered little whether the different power centers could agree on underlying policy or priorities so long as they were unanimous in support of initiating and continuing development. That they were."[129]

A genie was being released from its bottle. It would haunt the next decades and largely undermine the value of negotiating equal numerical limits on U.S. and Soviet strategic forces. With single warhead missiles, equal forces would render any attempt at "first strike" largely pointless. Even with 100 percent effective counterforce weapons, the best that could be done was to destroy missiles "one for one." But with MIRV, a single missile might be able to destroy several missiles. So even with the roughly equal arsenals negotiated in the Strategic Arms Limitation Talks (SALT), there was always the possibility to be reckoned with of one side using only a portion of its strategic force to destroy at least the land-based portion of the other side's force.

Several analysts and arms control specialists realized in advance this consequence of MIRV. Some of them, notably Jack Ruina and George Rathjens of MIT, pressed the National Security Adviser in the new Nixon Administration, Henry Kissinger, whom they knew personally, to cease MIRV tests so that a ban on the technology could be negotiated, and there also was pressure in Congress.[130] But by then, 1969, it was far too late. Too much momentum had built up behind MIRV.

The other changes made in moving from Minuteman II to Minuteman III were much less striking. Greater accuracy was sought, but the shift was not as sharp as from Minuteman I to Minuteman II (see figure 4.1). The overall stable platform design and the inertial instruments remained essentially unchanged, with minor modifications to increase accuracy.

129. Ibid., 49–50.
130. Ibid., chapter 5, and Seymour M. Hersh, *The Price of Power: Kissinger in the Nixon White House* (New York: Summit, 1983), chapter 12.

Computer memory was almost doubled,[131] which facilitated the more complex trajectory calculations required for MIRV.

Against this stable background, however, a constant pressure to increase accuracy existed. The "users," the Strategic Air Command, maintained a permanent delegation at the Ballistic Missile Office at Norton Air Force Base outside of San Bernadino, California. The Strategic Air Command "wanted as much as they could get in terms of accuracy."[132] With the lower yield warheads necessitated by MIRV (Minuteman III's original Mk-12 warhead has a yield of 170 kilotons, as against the 1.2 megatons of Minuteman II), the "targeteers" needed all the accuracy possible out of Minuteman III to implement the counterforce options in the SIOP, especially as intelligence estimates of the hardness of Soviet targets grew.[133] Furthermore, from the late 1970s on the comfortable lead in accuracy that Air Force missiles had enjoyed relative to Navy missiles started to erode and "to stay better than submarines" in terms of accuracy became a major preoccupation.[134]

The method of increasing accuracy was very much one of reverse salients. "The vast majority of the accuracy growth . . . has resulted from painstaking attention to the analysis, measurement and eventual elimination of a myriad of small error sources. Each time a predominant contributor to error is significantly reduced some other contributor becomes predominant and receives the spotlight."[135] Given relatively stable hardware, most of this improvement was implemented in the form of software changes. This made for undramatic and relatively cheap changes, which nevertheless were highly consequential.

An instance of this that did attract some attention was a program of software alterations to the Minuteman III NS-20 guidance system in the later 1970s. This cost only $30 million, but may have come close to doubling the accuracy of Minuteman III.[136] Though many of the changes might well have been

131. Wuerth, "Minuteman Guidance and Control," 74–75.
132. Hepfer interview.
133. Cochran et al., *Databook*, 117.
134. Hepfer interview.
135. Wuerth, "Minuteman Guidance and Control," 69.
136. See the "Improved NS20 Guidance Estimate" in figure 4.1, also Deborah Shapley, "Missile Accuracies: Overlooked Program could undermine SALT," *Science*, Vol. 196 (June 10, 1977), 1185–1186.

made anyway, they were packaged into a program by Secretary of Defense James Schlesinger in 1974. Schlesinger in that year began the process by which stated U.S. nuclear posture moved decisively away from assured destruction.[137] He also had a particular concern about the factual status of existing missile accuracy figures (see chapter 7).

The major "hardware" change to Minuteman III that did take place in the late 1970s was also initiated by Schlesinger in 1974: the introduction of a new, more powerful warhead, the W78 hydrogen bomb, in the Mk12A reentry vehicle. This doubled the yield of Minuteman III's original warheads.[138] The gradual increase in nuclear warhead yield-to-weight ratios that took place throughout this period made this possible; it seems, nevertheless, that Schlesinger's particular concerns over counterforce capability and the possibility that operational missile accuracies might be much worse than specifications were important in initiating the shift.

All these "small" changes kept Minuteman III a powerful counterforce weapon. Since the mid-1960s, however, many in the Air Force had looked forward to a wholly new ICBM, free of even the residual constraints of ancestry to be found in Minuteman III. During the early 1970s that hope began to take concrete shape in a missile that, despite the later attempts to christen it "Peacekeeper," was to remain better known by the name it went under during its long period of gestation: MX. It was to be equipped with a guidance system that many, though, in apparent paradox, not its designers, believed to represent the ultimate in accuracy achievable from a black-box guidance system. And it was to cause controversy far beyond that caused by any of its predecessors.

Suckling the Beryllium Baby

At least in its guidance aspects, which give it its formidable and controversial counterforce capabilities, MX can be seen as the final fruit of the "guidance mafia" and the intimate connection

137. For the Schlesinger doctrine, see Ball, "Déjà Vu."
138. Cochran et al., *Databook*, Vol. 1, 116. The Mk 12A is somewhat heavier than the Mk 12, and so not all the Minuteman III force has been fitted with it, apparently because if it were then Minuteman III missiles would not have the range to strike the ICBM fields in the southern Soviet Union.

between the Air Force and Draper's Instrumentation Laboratory.[139] Its link to the guidance mafia was personified by the key role of Major General John Hepfer in MX. Hepfer, Commander of the Ballistic Missile Office, was central in "calling the numbers" on MX, that is, in negotiating MX's specifications.[140] Though Hepfer had never been a student at the Instrumentation Laboratory, he was in other ways part of the core of the guidance mafia.[141]

The most striking aspect of the design of the MX Advanced Inertial Reference System (AIRS) is that captured by the metaphor of the beryllium baby. There are no gimbals. The MX gyroscopes and accelerometers are enclosed in a sphere, which is then floated in fluorocarbon fluid within an outer sphere (see figures 4.8 and 4.9). The inner sphere is kept in stable orientation by feedback from the gyroscopes, but instead of this being achieved by servo motors, it is done by three pairs of hydraulic thrust valves powered by a turbopump at the center of the inner sphere. Very precise temperature control is maintained by transfer of heat from the fluid through "power shells" to freon-cooled heat exchangers.[142]

The idea goes back to the Instrumentation Laboratory at the end of the 1950s and a man who was never satisfied with the philosophy of increasing accuracy by evolutionary improvement of existing designs: Philip Bowditch. Bowditch worked on the mechanical design of the Titan II guidance system, and was well acquainted with the problems of existing systems: particularly "gimbal lock," where two of the gimbals in a three-gimbal inertial system become parallel, with loss of an axis of sensitivity and probable disastrous loss of stabilization. Gimbal lock could be avoided by restricting a missile's trajectory so that it never could happen, or by addition of a fourth gimbal. The name first given to the idea of the fluid floated sphere, flimbal, reflects its origins in these reflections on the deficiencies of

139. I shall not attempt to deal here with the full story of MX's origins, which are the subject of current research by Lynn Eden.
140. Interviews with Major Gregory Parnell (U.S. Air Force), Stanford, Calif., February 20, 1985, and Major General Aloysius Casey (U.S. Air Force), San Bernadino, Calif., February 25, 1985; letter to author from Major General John W. Hepfer (U.S. Air Force, rtd.), December 27, 1988.
141. Hepfer interview.
142. Anonymous, *Peacekeeper Advanced Inertial Reference Sphere (AIRS)* (Hawthorne, Calif.: Northrop Corporation Electronics Division, n.d.).

ADVANCED INERTIAL REFERENCE SPHERE — EXPLODED VIEW

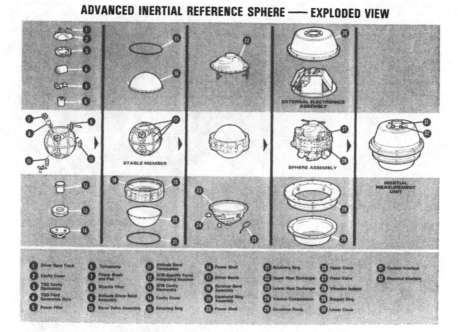

Figure 4.8
Advanced Inertial Reference Sphere—Exploded View
Courtesy Northrop Corporation, Electronics Division.

existing gimbal systems (figure 4.10). Furthermore, as anyone in the Instrumentation Laboratory would very well know, because of its work on floated gyros, flotation could be seen as bringing additional advantages, for example in terms of protection from the affects of acceleration and a generally "benign" environment.[143]

But there were immediately obvious difficulties, despite the attractions of the idea. How could the inner sphere be kept central? How could it be rotated? How could power be got in and signals got out? Solving these problems took many years of work. Bowditch moved on: "I'm an impatient person, I like to be on the front end of the development and I don't like to follow the thing through with all the minute details." Draper did not bend his formidable skills in heterogeneous engineer-

143. Interview with Philip Bowditch, Cambridge, Mass., October 11, 1984. The idea of a floated sphere inertial navigation system was also developed, apparently independently, at Northrop, but it was out of the Instrumentation Laboratory work that AIRS developed.

Figure 4.9
The "Beryllium Baby": Inner Sphere of MX Advanced Inertial Reference Unit
Removal of the two subassemblies allows one of the three third generation gyroscopes to be seen (left). One of the three accelerometers (specific force integrating receivers) can also be seen (right). Photograph courtesy Northrop Corporation, Electronics Division.

Figure 4.10
The Flimbal System in 1961
Courtesy Charles Stark Draper Laboratory, Inc.

ing to aid the flimbal: "he basically built a wall around that point solution [the floated single-degree-of-freedom gyro]. He didn't have much use for the flimbal as a thing, it was just an oddity—he didn't object to it, but he didn't enthuse about it."[144] So the responsibility for what neither Bowditch nor Draper would do—turning flimbal from an idea to a reality—passed to another member of the Instrumentation Laboratory, Kenneth Fertig. Fertig knew the importance of solving all the detail problems, but he also realized the role of what he calls "showmanship"—"we would lift it and drop it to show the benefits of the womb"—in what for many years seemed like a "stay-alive exercise" in terms of funding.[145]

Luckily for Fertig, the guidance mafia in the Air Force, especially in the Ballistic Missile Office, were impressed. They were smarting at the emphasis in Project Forecast on the manned bomber, and wanted to show just what the missile could do in terms of accuracy. They took the idea of flimbal, and built around it a program known as SABRE, to design a guidance system that would *Self Align* (doing away forever with vulnerable optical systems and messy human intervention) and guide the missile not just in *Boost* phase but also through *RE*-entry, eliminating as far as possible all the sources of inaccuracy.

It took off, and we got contracts, and then [someone] said, "look, we've already spent $12 million and we don't have authorization. . . ." So of course we all got together . . . and we briefed [General] Schriever, who always took these things as a little nuisance, and he said, "go over and tell them [the Department of Defense] you need money." We got this Fubini [Eugene G. Fubini, Director of Defense Research and Engineering], who was the top guy in the DOD at that time, and we briefed him on this need—in fact I gave the briefing, I had the charts, because I was in Headquarters at the time I got tagged with the job—and all these other guys were sitting there. You could see he [Fubini] was getting kind of red in the face, and . . . [he] said, "OK you guys, who's responsible for this? I want an individual name that's responsible for spending this $12 million." And there was silence—all of them had participated. And then Duffy finally got the courage and he stood up and said, "Well, I think we're all responsible for this, we were trying to be responsive to a need that we saw." He made a very eloquent speech, and [Fubini] said, "OK, I'll give you

144. Ibid.
145. Interview with Kenneth Fertig, Cambridge, Mass., November 12, 1984.

the $12 million this time, but I don't want to see any of you guys back here again." And of course we were back again![146]

SABRE was very much a pre-MIRV conceptualization—MIRV's separate warheads made the idea of guided entry unattractive for a long time to come. SABRE without that feature was considered as a guidance option for Minuteman III. Both AC Delco and Autonetics had contracts to build single prototype SABRE systems. But "there was a big price tag ahead for SABRE," and in the face of a budget reduction the Department of Defense opted for the cheaper and safer option (one consequence of the Minuteman II crisis, which "ruined a lot of people's careers" was "conservativism" in new developments).[147] As we have seen, this involved simply improving the by then "debugged" Minuteman II guidance system.

The next stage in nurturing the beryllium baby was played by a program named MPMS [Missile Position Measurement System] particularly fostered during the tenure of the position of Secretary of the Air Force from February 1969 to May 1973 by Robert C. Seamans, Jr. Although Seamans was not a member of the guidance mafia, he had worked at MIT from 1940 to 1955, sometimes along with Draper, and was willing to support the continued development of the technology.[148]

By around 1969, however, deeper problems than securing retroactive authorization for spending had come to be involved. The ideas of assured destruction—even mutual assured destruction—had become well entrenched in Congress. In particular, Edward Brooke, Senator from Massachusetts, and his aide, Alton Frye, had come to view the development of counterforce capability in U.S. missiles as dangerous not just to Soviet security, but also to American, in view of the pressure to preemptive launch they might generate. Brooke occupied a strategic position: not only was he the first black senator, but he was also a Republican, and thus of considerable potential

146. Hepfer interview.
147. Gasper interview. This is my only source on this episode, so the description in the text is put forward tentatively.
148. Interview with Robert C. Seamans, Jr., Cambridge, Mass., October 9, 1984; transcript of interview with C. Stark Draper by Barton Hacker, MIT Oral History Program, January 19, February 2, March and April 5, 1976, 117 (MIT Archives and Special Collections, Oral History Collection, Charles Stark Draper, MC 134, box 1).

political importance to President Nixon. Frye and Brooke were well informed on Department of Defense programs, and determined if not to block, at least to force open debate on programs that seemed to them to go beyond the requirements of assured destruction.[149]

The extent of Brooke's influence can be seen from a letter he sent Nixon on December 5, 1969, seeking reassurances "that the United States will not seek a capability to disarm the Soviet Union." After some delay, and consultations between the White House and Department of Defense, Nixon replied on December 29: "There is no current US program to develop a so-called 'hard-target' MIRV capability."[150] As we shall see in chapter 5, this pressure from Brooke was a major factor leading to the cancellation of a program for highly accurate guidance of Poseidon.

So the beryllium baby had to be put forward not as a missile guidance system, but as a piece of instrumentation to assist the process of missile testing: hence Missile Position Measurement System. It was a thin subterfuge, but it does seem to have assisted in the "stay alive" job that was seen as required in that period, and it was as MPMS that the beryllium baby first flew in a missile, "riding piggyback" on a Minuteman III test flight in 1976.[151]

The inertial instruments to go inside the beryllium baby were also developed over the 1960s and 1970s, with much more active support from Draper: *that* was where his heart was. We have already discussed the decades-long development effort at the Draper Laboratory that led to them. Both the AIRS gyroscope (the TGG, or third generation gyroscope; figure 4.11) and the AIRS accelerometer (the SFIR, or specific force integrating receiver) were paradigmatic Draper floated instruments. The SFIR in particular could be seen as the linear descendant of the Minuteman 16 PIGA-G, leading to the jibe from those exasperated with the expense and production difficulties of PIGAs that the letters stood for same f---ing instrument renamed!

149. Interview with Alton Frye, Washington, D.C., April 2, 1985.
150. Alton Frye, *A Responsible Congress: the Politics of National Security* (New York: McGraw-Hill, 1970), 69–70.
151. Anonymous, *Peacekeeper AIRS*. There was *some* rationale for the use of AIRS in this role: see chapter 7.

Figure 4.11
MX Third Generation Gyroscope
Photograph courtesy Northrop Corporation, Electronics Division.

A Guidance System in Search of a Missile

The development of both the overall beryllium baby configuration, and of the instruments to go into it, had been supported by the guidance mafia with extreme missile accuracy clearly in mind, but with a less clear notion of the particular missile in which it would be realized. The beryllium baby was in a sense a guidance system in search of a missile.

The idea that was eventually to become its missile was far from new. A large-diameter solid-fueled "Improved Capability Minuteman" or "ICBM-X" had been under consideration within the Air Force at least since 1965, but its status was for several years "shrouded in uncertainty largely because of the inability of Air Force planning, developing, and using agencies to convince the Defense Department of the urgency of an advanced ICBM."[152] But the new missile began finally to take shape with the issuing of a Strategic Air Command "require-

152. Irwin Stone, "ICBM Studies Focus on 156-in. Motors," *Aviation Week and Space Technology* (March 15, 1965), 140.

ment" in 1971 and an "advanced development program" in late 1973.[153]

The new missile was to be a big missile: one of the legacies of Minuteman's origins was that it was, relative to Soviet ICBMs, physically small and thus limited in its "throw-weight" and the warheads it could carry. The new missile was also to be a highly accurate one, even if the accuracy specification was not quite as stringent as the Draper Laboratory believed possible and the Strategic Air Command and Air Force Headquarters wanted. General Hepfer, who played the key role in negotiating its specifications, struggled hard, against pressure from all sides, "to keep the numbers at something I was sure we could do."[154]

Large payload and high accuracy added up to a counterforce weapon. Although counterforce was still largely alien to stated national strategic policy in that period, it was by then well entrenched in the Air Force, and there seems to have been little question but that the new missile, MX as it was christened, was going to be designed to have as great a counterforce capability as possible.[155] Nor does there seem to have been much realistic doubt that the beryllium baby would guide it. Now known as AIRS, the Advanced Inertial Reference Sphere, it was passed from the Draper Laboratory to Northrop for "advanced development" in May 1975,[156] although—with AIRS already selected as prime choice—Autonetics was asked by the Air Force to design a conventional gimbaled system, the ASP, or Advanced Stable Platform, for comparative purposes.[157]

153. Lieutenant General Alton D. Slay, "MX, A New Dimension in Strategic Deterrence," *Air Force Magazine* (September 1976), 44.
154. Hepfer letter.
155. It had been Air Force practice since the 1940s to refer to missile development programs as MX followed by a number (the early Convair ICBM was MX-774, for example); if it meant anything it was Missile eXperimental. The "ICBM-X" or "Missile X" version, with its somewhat glamorous and mysterious connotations, may have been a public relations ploy. See, for example, Edgar Ulsamer, "M-X: The Missile System for the Year 2000," *Air Force Magazine* (March 1973), 38–44.
156. Anonymous, *MX/AIRS: Missile X Advanced Inertial Reference System* (Hawthorne, Calif.: Northrop Corporation Electronics Division, n.d., unpaginated).
157. Gasper interview. The conclusion appears to have been that to get the same freedom from fear of gimbal lock and the same accuracy, the system had to be much bigger than AIRS.

The developers of MX faced a deep problem, however. In the climate of the early 1970s, with the popularity of arms control and of the ideas of mutual assured destruction, it was difficult to press for a new missile simply on the grounds of enhanced counterforce capability. Accordingly, the dominant public rationales for MX became either the "bargaining chip" argument (that it should be procured so as to be negotiated away), or, more importantly, the threat allegedly posed by Soviet ICBMs to the Minuteman force. As Colin Gray, one of MX's civilian supporters, put it: "MX advocates have tended to be nervous of making a case for the MX missile, as opposed to "survivable ICBMs," because a focus on the missile has to lead to a discussion of why large payload (for U.S. ICBMs) and small CEP (high accuracy) is deemed to be desirable."[158]

Gray was prepared openly to endorse a counterforce, even a first-strike, strategy:

The long-standing anticipated inability of the United States and its allies to contain a Soviet attack in Europe has meant that unless the Soviet Union preempted strategically, it would be the United States that would need a credible (limited) first-strike capability. It would be the United States that would be compelled, by the logic of impending, if not actually consummated, defeat in the theater, to take the conflict to a higher level in quest for an improved outcome. In short, the role of the U.S. strategic nuclear forces in a conflict around the Eurasian littoral would not, in all likelihood, be confined to that of a "counterdeterrent."[159]

But then Gray was not abashed to put his name to articles entitled "Nuclear Strategy: The Case for a Theory of Victory" and "Victory is Possible."[160]

In politically more sensitive circles, the counterforce capabilities of MX could be advocated only in the more sophisticated form of the theory of "second-strike counterforce." According to this, the United States needed counterforce capabilities not to preempt, but to deter the Soviet Union from launching a strike against U.S. ICBMs. Deprived of its counterforce weapons—by then the earlier conventional wisdom in

158. Colin S. Gray, *The MX ICBM and National Security* (New York: Praeger, 1981), 27.
159. Ibid., 19.
160. *International Security*, Vol. 4, No. 1 (1979), 54–87, and *Foreign Policy*, No. 39 (1980), 14–27, the latter co-authored by Keith Payne.

favor of the bomber had been reversed to the extent that it could be asserted as if it was uncontentious that "the ICBM is the counterforce weapon par excellence"[161]—the only retaliation open to the United States would be to strike at Soviet cities. So an American President would be left with only the terrible choice between surrender and a countercity attack that would amount to national suicide because of the certainty of Soviet retaliation. Thus reasoned Paul Nitze, whose article "Deterring our Deterrent" did much to popularize the "second-strike counterforce" theory.[162]

The "second-strike counterforce" theory did help build a "respectable" case for MX's counterforce capabilities. But, precisely in its emphasis on *second* strike, it too posed the issue of vulnerability as central. There were those in the Air Force, all along, who felt either that ICBM vulnerability did not matter crucially, because the United States ought, on sure intelligence warning of a Soviet strike, to use its nuclear force preemptively, or was a nonissue, since there was no real Soviet threat to Minuteman. An important figure who held elements of both views was General Bruce Holloway, Commander of the Strategic Air Command at the time when the "Required Operating Capability" for MX was written.[163]

The political and strategic climate of the early 1970s meant, however, that this sort of viewpoint could not readily be taken to its natural conclusion that MX should be built as large and accurate as possible, and should simply be placed in silos like those occupied by Minuteman. The public case for MX was hinged, irreversibly, around the "ICBM vulnerability" argument. But the argument proved to be a double-edged sword.

Summarizing a complex and fascinating history, what essentially happened was that "invulnerability" could not be found.[164] What for a long time seemed the most promising scheme was a massive "race track": "transporter-erecter-launcher" vehicles would carry MX missiles around specially

161. Gray, *MX ICBM*, 39.
162. *Foreign Policy*, No. 25 (Winter 1976/77), 195–210.
163. See his remarks to Michael Parfit, quoted in *Boys behind the Bombs*, 250–257. Holloway is also one of those who do not accept the facticity of missile accuracy figures; see chapter 7.
164. See John Edwards, *Superweapon* (New York: Norton, 1982) and Lauren H. Holland and Robert A. Hoover, *The MX Decision: A New Direction in US Weapons Procurement Policy* (Boulder, Colo.: Westview, 1985).

constructed roads linking thousands of shelters in Utah and Nevada, if necessary dashing from one shelter to the next in the event of warning of an attack. But this was defeated by the strength of local opposition, aided by the close links between Nevada Republican Senator Paul Laxalt and President Reagan. In all, dozens of "basing modes" were proposed, and none proved both "technically" and "politically" viable; it was a tale of failed heterogeneous engineering on a grand scale.

By 1983, then, there was no option but to take the possibility that had been open right at the beginning, to place MX in Minuteman silos.[165] But since the case for MX had been built around Minuteman being vulnerable, it was natural that critics of the program should conclude that MX was vulnerable. An attempt was made by the Air Force to argue that this was not necessarily so. "A better understanding of the inherent geological characteristics [at Warren Air Force Base, Wyoming, where MX began deployment late in 1986] has led to the determination that existing silos for the Minuteman 2 and Minuteman 3 ICBMs based there are much harder than originally thought."[166] General Bennie Davis, Commander of the Strategic Air Command, drew the conclusion that "[s]urvivability prospects for U.S. intercontinental ballistic missiles in the case of a nuclear attack are much better than had been predicted earlier and can be enhanced further."[167]

This weighed little in Congressional debate, however, and MX's "vulnerability" was a powerful argument against the missile. The Carter Administration plan had been for 200 MX missiles, and the Air Force wanted more than that.[168] But Congress refused to allow more than fifty MX missiles to be deployed unless an "invulnerable" basing mode could be found. Deploying MX on railroad cars is currently being explored as a possibility, but it remains to be seen whether this basing mode proves any more successful than its predecessors.

So the "external" heterogeneous engineering of MX went, from the point of view of its proponents, horribly wrong. These

---

165. This was recommended by the "Scowcroft Commission" set up by President Reagan. See Brent Scowcroft et al., *Report of the President's Commission on Strategic Forces*, April 1983.
166. Anonymous, "ICBM Survivability Improves," *Aviation Week and Space Technology* (March 11, 1985), 23.
167. Ibid.
168. Edwards, *Superweapon*, 202.

difficulties were then compounded by a failure of "internal" heterogeneous engineering at the missile's core, the Advanced Inertial Reference Sphere. The probable ultimate roots of the failure lay in the whole process of technical development examined in this chapter. While the nonmissile sectors of the inertial market pursued designs that were easier to produce—by the mid-to-late 1980s tending to employ laser gyroscopes in mechanically simple "strapdown" configurations—the beryllium baby, for all its elegance, is an intricate, complex design placing high demands on human skill in its fabrication. In the picture on the cover of this book, for example, the inner sphere of the AIRS is being rocked gently backwards and forwards in a bath of fluid, while the technician makes manual adjustments to ensure its sphericity.

Something went badly wrong with the process of linking together a multitude of human skills and many different subcomponent suppliers to produce the beryllium baby. Part of the problem may have been trying to combine a skill-intensive process with Northrop Electronics Division's "transitioning from a 500–employee research and development firm to a 5,000-employee production firm," though there were also problems with the subcontractors that supply the no fewer than 19,000 separate parts of AIRS.[169] Whatever the immediate cause, the schedule of production began to slip, and corners began to be cut.

Employees of Northrop Electronics Division set up "dummy companies" so that components could be acquired quickly, circumventing the normal procurement procedures. Even with these practices, the schedule could still not be met, and by July 1987 a third of the small number of MX missiles deployed at Warren Air Force base were "not on alert because of the lack of I[nertial] M[easurement] U[nit]s." In August of that year Northrop Electronics Division was charged with fraud by the Justice Department.[170]

169. Michael Mecham, "Air Force defends MX Management as Northrop is Charged with Fraud," *Aviation Week and Space Technology* (August 31, 1987), 18; interview with David N. Ferguson, Los Angeles, September 10, 1986; letter to author from Ferguson, February 3, 1989; Peter Adams, "MX Guidance Electronics Critics Satisfied with Improvements," *Defense News* (October 24, 1988), 11.
170. Subcommittee on Research and Development and Subcommittee on Procurement and Military Nuclear Systems of the Committee on Armed

The episode did not end there, but led to doubt being cast on the results of the test flights of MX that were guided by the operational (as distinct from the developmental) version of AIRS, with a Congressional report reprinting test results (without a scale) and seeking to reanalyze them. An unprecedented public squabble broke out between the House of Representatives Armed Services Committee and the Air Force over the detailed interpretation of MX test results.[171] Though "corporate radical surgery" at Northrop began to set them right soon thereafter, the difficulties of AIRS had allowed the opponents of MX right "inside the black box."[172]

The Small ICBM, Economics, and the Future

Many of the opponents of MX, furthermore, had "their own" ICBM to offer in its place: the Small ICBM, colloquially known as "Midgetman." Proposals for some kind of small ICBM, including a 30,000-pound road-mobile missile, had been around for as long as the ICBM-X.[173] What they had against them, by comparison with their larger rival, was largely economic. Even the most hawkish analyst could be persuaded that "as a counsel of perfection, one might well prefer that the payload of 200 MX missiles be spread over, say, 1,000 launch vehicles."[174] But that would be far more costly, and would undermine one of the arguments for the ICBM that had been made since the days of Hall's proposals for Minuteman; that it is a cheaper weapon system than a submarine-launched missile.[175] The costs per warhead of protecting and maintaining a small ICBM were much larger than those for a large multiple-warhead missile. Furthermore, in a large MIRVed ICBM the

Services, House of Representatives, *The MX Missile Inertial Measurement Unit: A Program Review* (Washington, D.C.: U.S. Government Printing Office, 1987), quote on p.12; Mecham, "Northrop Charged with Fraud."

171. Subcommittee on Research and Development and Subcommittee on Procurement and Military Nuclear Systems, *MX Inertial Measurement Unit*, chart 1, 16; briefing by Brigadier General Charles May (U.S. Air Force), *Federal News Service*, August 24, 1987.

172. Adams, "Critics Satisfied."

173. See Stone, "ICBM Studies," 143.

174. Gray, *MX ICBM*, 44.

175. According to Reed, *Minuteman ICBM*, 91, "The Air Force claimed that Polaris was 100 times more expensive than Minuteman on a per missile basis."

cost of the guidance system was "amortized" over many warheads, rather than over just one or two for the case of the small ICBM.

The troubles of MX raised the small ICBM from dormancy, particularly the inability to find a politically viable basing mode that everyone would agree was invulnerable. These troubles might, it was reckoned, be less with a much smaller missile, since it could be made mobile in a way that was difficult with a large ICBM, and a multitude of small ICBMs might present a less tempting target for Soviet preemption than a few big ones. Furthermore, if the small ICBM shrunk to being a single warhead weapon, then it might represent a way of starting to shut the genie of MIRV back up in the bottle. And having an alternative to propose enabled defense "liberals" to oppose MX while not being open to the accusation of being "soft on the Soviets."

The Small ICBM is thus unique among the missiles considered in this book, in the sense that its origins, and many of the forces shaping it, lay in the formal political system, especially Congress, rather than deep within the armed services, corporations, and technological institutions. The forum that gave birth to it was the Scowcroft Commission, established by President Reagan to resolve the deadlock over MX basing, and the Small ICBM seems to have been part of a compromise with liberal elements in Congress, especially influential Democratic Representative Les Aspin, worked out by Commission member James Woolsey, a "defense insider" who was also a good friend of Aspin's.[176] "In late January [1983], Woolsey and Scowcroft began a series of meetings with Aspin on Sunday afternoons—at Woolsey's house, at Aspin's house—to talk about what the effect would be on the House floor of various possible commission proposals," and Aspin sounded them out with leading figures such as House Speaker Thomas P. O'Neill.[177]

The deal worked out involved limiting the MX deployment to 100, the rationale being that 100 ten-warhead missiles could not be seen as a first-strike threat against the Soviet Union's 1,398 ICBM silos (as we have seen, though, MX's troubles are

176. I am here following the fascinating account by Elizabeth Drew, "A Political Journal," *The New Yorker* (June 20, 1983), 39–72, quote on p. 49.
177. Ibid., p. 49.

so deep that even this compromise figure may not be achieved). But it also involved the Scowcroft Commission's conclusion that:

a single-warhead missile weighing about fifteen tons (rather than the nearly 100 tons of MX) may offer greater flexibility in the long-run effort to obtain an ICBM force that is highly survivable. . . . The Commission thus recommends beginning engineering design of such an ICBM, leading to the initiation of full-scale development in 1987 and an initial operating capacity in the early 1990s.[178]

The Small ICBM was thus the doves' missile. But, outcome of a compromise as it was, it had hawkish characteristics too. The commission went on to recommend that the Small ICBM, "should have sufficient accuracy and yield to put Soviet hardened military targets at risk."[179] As a "mobile political missile"— to borrow the tag attached to a missile of two decades before that we shall discuss in the next chapter—the Small ICBM thus worked rather well, and the Scowcroft Commission's report did indeed lead to Congress unfreezing the funds to produce and deploy MX.

But turning it into a real missile was something else again. The Small ICBM, shaped as it were on the floor of the House of Representatives, lacked any real constituency of support within the Air Force. So by the late 1980s there was the disconcerting spectacle to be seen of the Air Force proposing the cancellation of one of its own major weapons programs, with the system being kept alive by Congress.[180] In large part economics was the root issue. As one opponent of the Small ICBM argued:

Estimates of the total cost of the Midgetman program range from $40 billion to $60 billion for the planned 500 missiles. This breaks down to almost $100 million for each Midgetman warhead. The per warhead cost of the survivable Trident II submarine-launched ballistic missile system will be only $13 million, including the cost of the new Trident submarine. The cost per warhead of the B-1B strategic

178. Brent Scowcroft et al., *Strategic Forces*, 15. Fifteen tons is roughly 30,000 pounds, so there was a strong echo of the 1965 road-mobile ICBM!
179. Ibid.
180. See for example, Brendan M. Greeley, Jr., "Budget Slashing would kill A-6F, Small ICBM and Aquila," *Aviation Week and Space Technology* (December 21, 1987), 31.

bomber is $16 million, and the next 50 MX warheads will cost only about $4 million apiece.[181]

Calculations such as this led to proposals to redesign the Small ICBM to carry more than one warhead. They also led to considerable pressure to reduce guidance system cost. Though AIRS was the prime choice to ensure the Small ICBM had the accuracy to threaten Soviet hard targets, and one of the alternatives also considered was a version of the Mk6 stellar-inertial guidance system of Trident II (see chapter 5), two systems based around Litton and Honeywell laser gyroscopes were also considered. As one of the backers of the last-named put it: "We really believe, we know, that the ring laser gyro system would be much more cost effective, let me say adequate performance, maybe not quite as good performance as the AIRS for MX."[182]

For a moment it looked as if the first strand of inertial technology development considered in this chapter might come to occupy the traditional territory of the second. It was not to be. The laser gyroscope options were eliminated from the competition for Small ICBM guidance by December 1987, and then the stellar-inertial system was eliminated early in 1988.[183] It seems that the key argument for AIRS, and against the laser gyroscope, was the perception that with the latter it was impossible to align the guidance system in a wholly black-box manner while maintaining hard-target accuracy.[184] Precision gyrocompassing to achieve this, if it is done with the inertial measurement unit's own gyros as in AIRS, is held to require gyro random drift rates as much as two orders of magnitude below the requirements for in-flight guidance: it is the major rationale for the Draper third generation gyroscopes on MX.

181. Pete Wilson, "Midgetman: Small at any Cost?" *Issues in Science and Technology*, Vol. 2, No. 4 (Summer 1986), 50.
182. Bailey interview.
183. Anonymous, "GE will compete for Guidance System," *Aviation Week and Space Technology*, December 21, 1987, 31; and Anonymous, "New Briefs," ibid., (February 29, 1988). 34.
184. In 1984 a committee of the Air Force Scientific Advisory Board, chaired by ICBM pioneer General Bernard Schriever, concluded that "strapdown ring laser gyros do not have a gyrocompassing level sufficient for small missile accuracy": James B. Schultz, "En Route to Endgame: Strategic Missile Guidance," *Defense Electronics* (September 1984), 59.

So it seems that the Draper floated gyro, born in the 1940s, has again fought off what has become its most dangerous rival, though the probability that the Small ICBM will never be deployed may make the victory Pyrrhic.[185] There is, too, an increasing sense that AIRS, its third generation gyroscope and specific force integrating receiver, together with the distinctively Draper elements in the Trident II guidance system, represent the "end of the line" of ever-increasing accuracy based on evolutionary improvement of floated inertial instruments.

Draper's fourth generation, even more accurate than the third generation MX and Trident II instruments, may well be stillborn. Apart from at the Draper Laboratory and among a few who, in the words of one of them, are "too much of a Draper disciple to believe AIRS can't be improved on,"[186] the belief has grown that inertial instruments are no longer a reverse salient in the growth of missile accuracy. In the words of General Hepfer:

Now, you can take that accelerometer error on down to zero and [the overall system] error doesn't change because you're in the noise level. The accelerometer is no longer *the* major contributor to the error sources. There are other error sources, and every time we fix one, we find there are others, and so we're sort of at a plateau with probably anywhere from 200 to 500 sources in there of equal magnitude, and trying to push each one of those down, the guys will do it, they'll identify an error source, and they'll say, now we can work on that and we'll get that out. As soon as you do, you find that that was just one of many in there. . . . We're at the point now where we identify new sources rather than fixing things.[187]

The growth of this perception among key decision makers translates into a practical funding difficulty for the Draper Laboratory. With the military unconvinced that more accurate inertial instruments would mean more accurate missiles, and with NASA no longer in the business of handing out contracts like Draper's "third generation instruments" contract, the

185. In July 1989 uncertainty over the Small ICBM's future led to several development contracts not being renewed, at least temporarily. As a result, by August only 400 people were working on Small ICBM development contracts. Anonymous, "Washington Roundup," *Aviation Week and Space Technology* (August 21, 1989), 19.
186. Ferguson interview.
187. Hepfer interview.

Draper Laboratory has had to finance fourth generation work out of its relatively small independent research and development budget.

Related to the "physical" difficulty of achieving inertial system accuracies beyond that of AIRS (which is, as I have indicated, not universally accepted), there is also an "economic" difficulty, which no one denies. David Hoag of the Draper Laboratory, who as noted above argued in 1970 that ICBM accuracies substantially better than even those of today were possible, writes: "A barrier is hard to recognize distinctly. I believe more accuracy of an unassisted inertially guided weapon system is possible . . . but who can afford it?"[188]

The cost of AIRS is already seen as very high, even in the generous world of strategic system economics. Not only are the gyroscopes and accelerometers very expensive in the first place: the latter are currently failing on average within less than a year ("little more than one-third of the reliability planned for them"), and it "takes about six months and $300,000 to rebuild an MX accelerometer."[189] These accelerometers are, in the words of their virtuoso, "labor-sensitive" devices, and, he feels, there has been a lack of "the awareness and commitment of industry management to preserve all the detailed manufacturing disciplines required to build these instruments."[190] With each AIRS containing three accelerometers, and entire industry standard inertial navigators being purchasable for significantly less than $300,000, the economic costs of achieving accuracy by traditional means have indeed become salient.

So if the United States builds another ICBM inertial guidance system after AIRS, which is uncertain, it is unlikely to be a continuation of the path of technical development that led to the beryllium baby. There are perceived ways of increasing ballistic missile accuracy, but these involve radical changes of technology. The most probable is starting to guide missile warheads during reentry: as explained in chapter 1, all current U.S. and Soviet strategic missile reentry vehicles are unguided once they leave the "bus" carrying them.

188. Letter to author, November 17, 1985.
189. David F. Bond, "House cuts Funding for Modernizing USAF's ICBM Launch Control Centers," *Aviation Week and Space Technology* (August 21, 1989), 71.
190. Letter to author from Michele S. Sapuppo, January 23, 1989.

This might[191] involve an inertial guidance system in each reentry vehicle. But such an inertial system would not be an AIRS, or anything like it. It would have to be cheap, since each missile would have to carry several, not just one; it would have to be very small, light, and rugged; and it might not need to be very accurate, since inertial sensor errors have only seconds in which to cumulate during reentry compared with minutes for guidance in the boost phase. In other other words, what might be needed might be more like what the wider inertial industry, rather than the Draper Laboratory, has focused on, and indeed both dry tuned-rotor gyroscopes and laser gyroscopes have been considered for reentry vehicle guidance.

Conclusion

The beryllium baby may thus be indeed an end-point to the particular form of technological development that brought it into being. That form is the "technological trajectory" of ever-increasing missile guidance system accuracy based, above all, on the continuous refinement of the same basic inertial sensors.

The central argument of this chapter has been that, for all its apparent naturalness, this trajectory is a construct, a contingent product. Among the factors sustaining it were Draper's technological vision and the practice of his Laboratory, the Air Force's interest in counterforce, and the acceptance, in circles with the resources to help make them a reality, that the extreme accuracies in inertially guided missiles predicted in the early 1960s were a possibility rather than a fantasy.

No one of these factors was sufficient on its own. Remove the technological vision and practice, and extreme missile accuracy, if it was to be achieved at all, would have had to be achieved by a quite different means. Remove the Air Force's interest in counterforce, and the chief resources sustaining this technological practice would no longer have been there. As we saw, Draper became dependent on that interest. His attempt to create another interest (particularly in inertially guided civil aircraft landing) to tie his technological trajectory to was a failure, but the interest in counterforce cannot be seen as the creation or puppet of technologists wishing support for their

191. A non-black-box system, for example to match radar images of terrain with a stored "map," is an alternative possibility.

favored lines of technological development. We saw, in the development of Minuteman II in particular, that in pursuit of its goals the Air Force was perfectly capable of overriding these technological preferences. We saw, too, some of the organizational and political complexities of counterforce. Nor, finally, is much left once confidence in the prediction of greatly increased accuracies has gone. That, perhaps, is the current situation.

The fact that what was sustained by all these factors was no natural trajectory of technology was highlighted in this chapter by the existence of an alternative form of development of inertial technology, a form whose most distinctive products are the dry, tuned-rotor gyro and, above all, the laser gyro. It is indicative of the hold of the "natural trajectory" way of thinking that even well-informed people often assume that MX must be guided by laser gyros, because they are "the latest thing."[192] But, to date at least, the laser gyroscope has inhabited quite a different institutional world, one where the key goals are reliability and economy, rather than ultimate accuracy.[193]

That different world, while no more "natural," is perhaps more normal: it is the world of large firms competing for market-share and profit, concerned with product improvement but also with not being too reliant on overly labor sensitive production processes. In a word, it is the world of capitalism.

One of the paradoxes of defense, particularly of nuclear weaponry, has been that though there is a sense in which it is defense of capitalism, it has created islands of industry and of technological change where the normal capitalist processes are at many removes.[194] The Draper Laboratory, at least in its central technological effort, has been one such island. It brings one up with a start to read, in the annual report of the Draper

192. A version of the photograph of the MX inertial measurement unit on the cover of this book was published to illustrate an article based on my research in a British newspaper. The newspaper's science and defense correspondents, to my horror, rewrote the caption to include the words "laser gyro guidance system." See *The Independent* (August 7, 1989), 13.

193. Signs of change are the failed attempt to secure the laser gyroscope's acceptance for Small ICBM guidance, and efforts, particularly at Autonetics, to develop highly accurate laser gyroscopes for use in Ships Inertial Navigation Systems.

194. For some of the resulting pathologies, see Gansler, *Defense Industry,* and Mary Kaldor, *The Baroque Arsenal* (London: Deutsch, 1982).

Laboratory, passages skeptical of computerization and in favor of human skill that would not be wholly out of place in Marxist writing about the labor process.[195] Similarly surprising on first contact is just how skill-intensive and labor sensitive the processes of ballistic missile inertial guidance system production are.[196]

Such are the nuances of technical change. I hope that in this chapter I have shown, if nothing else, that matters of technological nuance and detail should not be ignored by the social scientist. Often apparently little things, "mere technicalities"—the nature of the bearings in the accelerometers of the Minuteman II guidance system, for example—are the key to understanding the patterns of social relations in a technology's development.

Such is the case for the technology we shall next consider, the submarine-launched ballistic missile. Though other aspects will be considered too, the central thread will be another "mere technicality"—this time whether the black box of inertial guidance should be opened a fraction to allow a sighting on the stars.

195. The Charles Stark Draper Laboratory, *1977 Annual Report* (Cambridge, Mass.: Draper Laboratory, 1977), 13: "We are concerned over a trend, a national one, that assumes that manual and mechanical skills are no longer necessary in a world populated by computers, which are impersonal, omniscient, and infallible. A result of this sort of thinking is that the people who are skilled at making things that work are considered expendable. . . . No computer ever made a piece of working hardware. No software ever made a measurement, ground a fitting, sealed a vacuum, cast a bearing, wound a core, heat-treated a metal, magnetized a spoon, polished a lens, or etched a plate. . . . The letdown in standards has not yet become a problem in this Laboratory; but if the pool of the highly skilled dries up, we would be seriously affected, working as we do in areas where craftmanship is a principal ingredient." Very similar points are made in the Laboratory's *1978 Annual Report*, 31 and *1979 Annual Report*, 40. Compare Harry Braverman, *Labor and Monopoly Capital: The Degradation of Work under Monopoly Capitalism* (New York: Monthly Review Press, 1974).

196. There of course have been and are efforts to deskill them, but not at the cost of deterioration in missile accuracy, when that would be the perceived result. Thus both the MX and Trident II accelerometers are PIGAs, even though the production difficulties these have brought with them were wholly predictable to anyone knowing the device's history.

5

Transforming the Fleet Ballistic Missile

The U.S. Navy's Polaris submarine-launched ballistic missile was justified, from fairly early on in the Fleet Ballistic Missile program, as the ideal "ultimate deterrent." Claimed to be invulnerable under the ocean, Polaris could strike a devastating blow at Soviet cities in retaliation for a Soviet nuclear strike. Submarine-launched missiles were generally seen as intrinsically less accurate than the Air Force's ICBMs, because of the extra uncertainty about the position, velocity, and orientation of the submarine at the point of launching. But to the proponents of Polaris, given their vision of its strategic role of countercity retaliation, that did not matter.

By the end of the 1980s the Fleet Ballistic Missile had become a quite different weapon. In the design of Trident II (Trident D5 in the Navy's nomenclature), first deployed in March 1990, accuracy was a primary concern. While the problem of rapid communication with submerged submarines still limits the extent to which it can be used flexibly, Trident D5's counterforce capacities rival those of MX. Unlike MX, however, the D5 was designated by Congress a "noncontroversial" program that could therefore receive, instead of the usual annual funding, a five-year authorization.[1]

The subject of this chapter is the transformation undergone by the Fleet Ballistic Missile from Polaris, through the intermediate steps of Poseidon and Trident C4, to Trident D5 (table 5.1). As in chapter 4, we will find that this is no mere natural trajectory of technology. There, the key evidence lay in the simultaneous existence of a quite different form of technical

1. Michael Mecham, "Congress favors Conventional Defense, Production Efficiency," *Aviation Week and Space Technology* (November 23, 1987), 23.

Table 5.1
U.S. Fleet Ballistic Missiles

Year of first deployment	Name	Range	Guidance system	Estimated accuracy
1960	Polaris A1	1,200	Mk1 inertial	2.0
1962	Polaris A2	1,500	Mk1 inertial	2.0
1964	Polaris A3	2,500	Mk2 inertial	0.5
1971	Poseidon C3	2,500+	Mk3 inertial (Mk4 stellar-inertial cancelled)	0.25
1979	Trident C4	4,000	Mk5 stellar-inertial	0.12–0.25
1990	Trident D5	4,000+	Mk6 stellar-inertial	0.06

Poseidon C3, Trident C4, and Trident D5 are in service with the U.S. Navy. Polaris A3, though no longer in service with the U.S. Navy, is still in service with the Royal Navy, but Trident D5 is being acquired by the United Kingdom to replace it. All are launched from submerged submarines.
All guidance systems were designed by the MIT Instrumentation Laboratory, later the Charles Stark Draper Laboratory, Inc., with significant input from the Kearfott Division of the Singer Corporation into the design of stellar-inertial systems.
Ranges and estimated accuracies are expressed in nautical miles.

change. But the two forms mostly coexisted side-by-side, rather than conflicting. And while there was external opposition to increasing ICBM accuracy, within the program there was no dispute about this overall goal, however much disagreement there might have been about the best means to achieve it.

The Fleet Ballistic Missile program was different: the appropriate direction of technical development was contested not externally but internally, and goals as well as means were at issue. There were deep differences of opinion within the Navy and "defense establishment" about the place of the Fleet Ballistic Missile in nuclear strategy. These differences, together with other factors, led the design of each new generation of Fleet Ballistic Missile to be the subject of remarkable controversy.

With one major exception, these remained insider conflicts, not spilling into the public arena or formal political system.[2] If

2. I am here discussing issues concerning the missile. The fraught process of Trident submarine construction became the subject of intense controversy, but while there is a relation to missile developments, it is a subject that cannot

one face of Janus was internal controversy, the other was a public image of a program evolving smoothly. Thus much of the debate in the United Kingdom about whether to purchase Trident D5 missiles from the United States proceeded as if the D5 was a simple modernization of Polaris. The insiders' view that it was a transformed missile, reflecting quite different priorities, surfaced only sporadically.

The exception where conflict became public concerns the technology that its proponents claim permits the Fleet Ballistic Missile to overcome the handicap of submarine launching and rival the ICBM in accuracy. That technology is stellar-inertial guidance, in which the missile's self-contained inertial system is supplemented by an inflight star sight. Tracing the dispute over stellar-inertial guidance takes us into the heart of both the organizational structure of the Fleet Ballistic Missile program and of the contested meaning of the submarine-launched missile for nuclear strategy.

Given the centrality of ballistic missile stellar-inertial guidance to the transformation of the Fleet Ballistic Missile, we need first to examine its origins. Paradoxically, given the ultimate role of stellar-inertial systems in the relations between the two services' missiles, these origins were closer to Air Force programs than to the Navy ones.

Origins of Ballistic Missile Stellar-inertial Guidance

There was, of course, nothing novel in the basic idea of stellar-inertial guidance. As we saw in chapter 2, navigating a bomber or guiding a cruise missile by stellar-inertial means was originally a less radical proposal than navigating or guiding in purely black-box fashion. What was novel and radical, however, was the proposal to guide a ballistic missile by stellar-inertial guidance. Thus a 1958 article by a very well informed technical journalist simply ruled it out of court: "The inability to use stellar . . . references in a ballistic missile, as can be done in an aircraft or winged missile . . . greatly increases the accuracy

be done justice here. The purchase by the United Kingdom of Trident D5 missiles has also been the subject of controversy, but the effects of this on Trident development have been slight.

requirements for ballistic missile acceleration sensors."[3] Even proponents of stellar-inertial guidance omitted the ballistic missile from the list of systems on which it might be used.[4]

The problem was obvious: could usable star-sightings be taken from a ballistic missile in flight, given the short period of time available before rocket motor cut-off, the vibration and acceleration of ballistic missile boost phase, and the difficulty of constructing a "window" strong enough to survive boost but not distorting optically? With a fairly modest definition of "adequate" accuracy for an ICBM, and with first radio and then pure inertial guidance capable of providing it, little motivation seems to have existed to explore these difficulties for fixed-base missiles.

What did provide the context for their exploration were proposals around 1960 to make Air Force missiles mobile, either in the air or on land. Air mobility or land mobility raised the same issue that submarine-basing did: the error introduced by inadequate knowledge of initial conditions, of position, velocity, and orientation prior to firing.

The air-mobile missile was Skybolt. This was an intensely ambitious scheme for a long-range ballistic missile that could be fired from an aircraft. The problem Skybolt was designed to solve was as much bureaucratic as strategic. Both the United States Air Force and the British Royal Air Force (which was also supposed to purchase Skybolt) remained deeply committed to the manned bomber as a weapon system. But with massive Soviet investment in air defenses, the capacity of such an aircraft to reach its target seemed increasingly in doubt. If a ballistic missile could be fired from a bomber, then this would not be necessary. The bomber need not even enter Soviet air space, and so its physical survival (and therefore political survival) could be assured.

Stellar-inertial guidance was proposed for Skybolt, and a system was developed by Northrop, whose earlier stellar-inertial work was described in chapter 2. But Skybolt ran into deeper and deeper trouble, and eventually was canceled. Though it too was unsuccessful, it was the other mobile missile

3. Philip J. Klass, "Titan Guidance Reliable, Accurate," *Aviation Week* (May 12, 1958), 93.
4. R. B. Horsfall, "Stellar-Inertial Guidance Reduces Error," *Aviation Week and Space Technology* (March 17, 1958), 73.

system that is of greater relevance for our story. That was the Mobile Medium-Range Ballistic Missile, or MMRBM.

Like Skybolt, the MMRBM was a technology designed to enter a difficult strategic and political environment. It was to be a "European theater" weapon, and reflected a growing sense in the late 1950s and early 1960s of the need to plan for a war, which, though nuclear, might be in some sense limited. Unlike the existing fixed-base Thor and Jupiter missiles deployed in Great Britain, Italy, and Turkey, it would be mobile, and thus protected against a Soviet preemptive attack. It was also required to be highly accurate, because its "limited" use was anticipated, and ready to be fired at extremely short notice, perhaps as little as five minutes.

But the idea that a nuclear war could be limited was far from universally accepted, and the territory on which the MMRBM would be deployed might not be welcoming. Saboteurs were one worry. More immediately, France was already on the path that was to take it to disengagement from the military structure of NATO, and the French government wished complete control over MMRBMs deployed on French soil, a proposal that would have been unacceptable in Washington.[5]

Nor was Washington unified in what the MMRBM meant to it. It was to be an Air Force missile, but the Army was still smarting from the way the Air Force had defeated it in the battle for control over intermediate range missiles (see chapter 3), and could not be expected to welcome the MMRBM. The Department of Defense, particularly Assistant Secretary John H. Rubel, believed the Air Force to be too ambitious in the range and accuracy requirements being proposed for the MMRBM.[6] There were accusations that the State Department was trying to block the MMRBM program by slowing negotiations with the European members of NATO for deployment rights, the suspicion being that the State Department saw the MMRBM as a rival to its pet project for a multilateral nuclear force.[7] This unstable political environment made the stabiliza-

5. Anonymous, "U.S.A.F. Wins Space Role Support; Overruled on Deterrent," *Aviation Week and Space Technology* (March 12, 1962), 70–71.
6. Philip J. Klass, "MMRBM to use new Development Plan," *Aviation Week and Space Technology* (April 2, 1962), 22–23.
7. Anonymous, "Ford charges State Dept. killed MMRBM," *Aviation Week and Space Technology* (October 12, 1964), 29. For the multilateral force, see J.

tion of the MMRBM as a technical system extremely difficult: "Since the decision was made last October [1961] to go ahead with the development of the missile, international and military politics have so affected attempts to complete a set of specifications on which to base requests for proposals to industry that some Pentagon officers have dubbed it the 'Mobile Political Missile.'"[8]

Designing the guidance system for the MMRBM was a process that could not be wholly insulated from these wider considerations. The mobility of the MMRBM immediately raised the issue of initial conditions, especially when combined with the demand for high accuracy. How was the system to "know" the exact location of its launch point, and how was it to be aligned?

One way would have been to have a range of presurveyed launch locations, each equipped with a marker whose orientation was precisely known and which could be used for optical alignment of the MMRBM's guidance system. "While this approach involves the least complexity for both missile and transporter, it runs the risk that the transporter might be blocked from reaching a launch site by bomb damage or sabotage. Additionally, it would require a permanent continuous guard at launch sites to assure their integrity and prevent sabotage of the azimuth reference benchmark."[9]

Another would be to black box the problem of initial alignment within the transporter but not within the missile. For example, at the order to fire the transporter could move to the nearest presurveyed launch point, but alignment could then be performed using a gyrocompass on the transporter. This would do away with vulnerable benchmarks, but it was argued that it would compromise the requirement for rapid readiness to fire: "the north-seeking gyro would require 10–20 minutes to establish a reasonably accurate azimuth reference, and the transporter still must proceed to the nearest launch site to fire its missile."[10] A further complication was that there were seen

D. Steinbruner, *The Cybernetic Theory of Decision: New Dimensions of Political Analysis* (Princeton, N.J.: Princeton University Press, 1974).
8. L. Booda, "Mobile Mid-Range Missile Delayed Again," *Aviation Week and Space Technology* (March 26, 1962), 16.
9. Philip J. Klass, "MMRBM Guidance Challenges Industry," *Aviation Week and Space Technology* (April 23, 1962), 89.
10. Ibid.

to be military and political advantages in altogether avoiding land-basing:

Some Defense Department officials are inclined to believe that ship-based launch may ultimately prove the more feasible for a variety of reasons. One is the belief that water-based launching platforms would be less vulnerable to sabotage than transports operating over West European highways. Another is that water-based launchers would be less conspicuous and therefore less likely to provide a symbol which Communist or neutralist propagandists could use to create public opposition in some Western European countries.[11]

Such a system could not rely on fixed external references for either position or orientation. Similarly, an entirely free-ranging land-based system would also have to have its own means of ascertaining both of these. But was this possible without radical deterioration of accuracy?

It was into this situation, in which the "goals" to be met by an MMRBM guidance system were not stable and were strongly affected by its being "the mobile political missile," that the proposal to guide the MMRBM by a stellar-inertial system emerged. Where the proposal first came from is not wholly clear. It may have been from a group at the AC Spark Plug Division of General Motors, which was one of the leading members of the nascent guidance industry. That group's leader, Hyman Shulman, and another prominent member of it, Martin Stevenson, had both worked at Northrop, and were well acquainted with the stellar-inertial work there.[12]

The advantage that stellar-inertial guidance offered (if it worked—an issue to which I shall turn shortly) was that the problems of initial position and alignment could be shifted into the missile. This prospect was attractive enough to persuade the Air Force to issue three exploratory contracts under the Stellar Inertial Guidance System (STINGS) program, a key role being played by the same young technical officer, Robert Duffy, who appeared several times in chapter 4.

With neither the technology nor the environment stable, a variety of different approaches were pursued under the STINGS program. The most radical was the approach of the

11. Ibid.
12. Interviews with Martin Stevenson, Goleta, Calif., September 11, 1986, and Hyman Shulman, Santa Monica, Calif., January 13, 1987.

United Aircraft Corporate Systems Center, which involved an effectively complete detachment of the missile guidance system from the political-operational problems of initial conditions. It involved designing the missile guidance system so that it included *two* stellar sensors, which could take fixes on two stars simultaneously as the missile climbed into space: "From measurements of azimuth [orientation in the horizontal plane] and elevation angles of two stars relative to the horizon established by the stable platform, an on-board computer would calculate the missile's geographic position, compare this with the position it should have to hit its intended target and initiate steering signals to alter the missile's course as required."[13] With this system, it was argued, missiles could thus be fired from any point, and no azimuth reference, nor complex navigational equipment in the transporter, would be necessary.

Doubt about the feasibility of this approach to stellar-inertial guidance seems, however, to have been strong, and the two other STINGS contractors, AC Spark Plug and the Kearfott Division of what was then General Precision Equipment Corporation (it later became part of the Singer Corporation), took more modest approaches. Unlike the United Aircraft version of stellar-inertial guidance, the point of these was to use the stellar information to correct accumulated errors rather than directly to work out missile position. It was assumed that the system would have some information about both initial position and azimuth. Any errors in the system's best estimate of these would translate into errors in angular orientation vis-à-vis the stars once the missile was in flight. (If one thinks of a frame of reference with its origin at the center of the earth, then latitude and longitude are angles with respect to this frame; that is why they, as well as azimuth, translate into errors of angular orientation.) So differences between the actual and predicted positions of star images in the star sensor's field of view could be used to deduce errors in angular orientation and thus to correct for errors in knowledge of initial position and azimuth.

It was taken for granted that, in general, the star sensor had to sight on two stars. As a 1962 textbook put it, "Since a star tracker cannot detect angular errors about the line-of-sight, in order to correct all three axes of the platform, at least two stars

13. Klass, "MMRBM Guidance Challenges Industry," 90.

must be tracked."[14] One star would be adequate under some circumtances, but these circumstances were as much "politico-military" as "technical." If presurveyed launch sites could be used, then errors in latitude and longitude could be discounted, leaving only azimuth error, which could be corrected by sighting on a single star.[15] If this were not so, however, sightings on two stars, with the added complexity that brought, seemed inevitable.

This kind of issue, one star versus two, never got resolved for the MMRBM. Faced with the difficulties outlined above, the MMRBM was to remain merely a political missile, not a real one. After various fluctuations in its fortunes, it was finally canceled with effect from August 31, 1964.

Stellar inertial guidance's proponents at Kearfott were, however, not prepared to sit back and let the death of the program kill the technology. Kearfott had hired a Washington "insider" as its president, Dr. Marvin Stern, a veteran of the Air Force ballistic missile program who had served as Assistant Director for Strategic Systems in the Office of the Director of Defense Research and Engineering:

Practically the day I walked in there [Kearfott] the MMRBM was cancelled. That's life. But that stellar acquisition thing . . . could potentially have some future payoff . . . So . . . I went down to Washington. Harold Brown was DDR&E, Director of Defense Research and Engineering. I literally arrived at the river entrance [to the Pentagon], Harold Brown was coming walking down the steps to go to Congress to testify amongst other things on the cancellation of the MMRBM. So I said, "Whoops, Harold, MMRBM's been cancelled. You know Kearfott had the program, it's gone, that's life. However that stellar acquisition feasibility . . . that's interesting for its own sake, you ought to keep that alive." It was forty million bucks. Harold had . . . confidence in me . . . I usually didn't lead him astray. He was going to Congress and he kept that forty million alive.[16]

So funding for stellar-inertial guidance survived the cancellation of the MMRBM, in the form of a program to test whether the technology worked; an issue that was still regarded as not

14. R. L. Doty and R. F. Nease, "Augmented Inertial Cruise Systems," in G. R. Pitman, Jr., ed., *Inertial Guidance* (New York: Wiley, 1962), 216.
15. Klass, "MMRBM Guidance Challenges Industry," 89.
16. Interview with Marvin Stern by Graham Spinardi, Washington, D.C., July 13, 1987.

settled. Though as we shall see there were more sophisticated levels of doubt to be dealt with, the basic one, referred to above, concerned the very possibility of taking a star-fix from a ballistic missile in flight, especially in daytime firings.

According to one interviewee, a whole series of experiments had been attempted, using Polaris A2 and A3 missiles, Minuteman missiles, and high-altitude balloons.[17] These experiments proved wholly ambiguous. There was a general problem of knowing whether or not a star-sighting had successfully been made. If a star-sighting was in error, then the problem could lie in the star-sensor system, but it could be that the inertial platform, on which the sensor was mounted, was not in the correct orientation. Even more seriously, if no star could be "seen" at all, what did this mean? That apparently is what happened in the early experiments, and it led those centrally responsible for one evaluation (reportedly contracted out to the Astronomy Department of MIT) to conclude that the experiments showed the infeasibility of stellar-inertial guidance. The brightness of the daytime sky was too great for the missile-mounted sensor to "acquire" a star. An alternative interpretation of this "failure" was, however, pushed by the proponents of stellar-inertial guidance. They argued that residue from the rocket engine might have become deposited on the "window," and sunlight reflecting on this layer was what was preventing the stars being seen.

The program that survived the MMRBM's cancellation—the Stellar Acquisition Flight Feasibility (STAFF) program—was an attempt to settle issues such as this. It used Polaris A1 missiles being retired from active service and modified to carry a Kearfott-designed stellar-inertial system employing a Northrop star sensor.

The first such test was on April 14, 1965. Although range safety command aborted the flight when the missile veered off course just over a minute after takeoff, telemetry from the missile was accepted as showing that the star-tracker had successfully locked onto the North Star, Polaris.[18] This did not

17. Interview with John Brett, Wayne, N.J., January 15, 1987. Though the STAFF tests were widely reported, I have found no documentary traces of these earlier tests, and so am relying on Mr. Brett's account of them.
18. "Air Force Cites Good Test Data in First STAFF System Flight," *Aviation Week and Space Technology* (April 19, 1965), 36.

end doubts. The scientist responsible for the earlier evaluation noted that it was a night flight, and reportedly told a proponent of stellar-inertial guidance that he "would eat his hat" if the system succeeded in acquiring a star in daylight.[19] But the telemetry from the later daytime STAFF flights was taken as showing this. Ballistic missile stellar-inertial guidance had finally passed its first feasibility test.

So the technology might work. But who needed it? The Air Force looked increasingly unlikely to deploy a mobile ballistic missile, and the case that a fixed-base ICBM needed stellar-inertial guidance was hard to make.[20] The hopes of its proponents therefore increasingly turned to the Navy, which, after all, had deployed and would continue to deploy mobile ballistic missiles. But Polaris's combination of a submarine inertial navigator plus inertially guided missiles worked perfectly well according to Charles Stark Draper's MIT Instrumentation Laboratory, the Polaris guidance system's influential designers, and the U.S. Navy's Special Projects Office. The latter had built up an enviable reputation for developing technology on time and on budget. Why should it risk that reputation on an unproven technology that would add both weight and complexity? To get down to the most basic level of "office politics," why should those within Special Projects responsible for missile guidance (the branch known as SP 23) take on a task that could easily go wrong to compensate for errors in the information provided by the submarine's navigator, the province of another branch (SP 24)?[21]

19. Brett interview.
20. The initial conditions problem was obviously much reduced for a fixed-base missile, though this does not foreclose all avenues for arguing that there would be advantages to stellar-inertial guidance of a silo-based ICBM: see Stephen F. Rounds and George Marmar, "Stellar-Inertial Guidance Capabilities for Advanced ICBM," paper 83–2297 read to American Institute of Aeronautics and Astronautics, 1983 Guidance and Control Conference (I am grateful to Matt Bunn for a copy of this). Some would say that the influence of Draper counted against all forms of ICBM guidance that were not wholly black box: "The Air Force had fallen in love with Stark Draper, who had become a guru, and the Air Force was so persuaded by Stark Draper about his ability to drop a r[eentry] v[ehicle] into a pickle barrel at 6,500 miles . . . that they rejected anything that was ancillary to pure inertial guidance" (interview with James Schlesinger, Washington, D.C., September 22, 1986).
21. One early proponent of stellar-inertial guidance concluded that this was

To answer those questions involved redefining both what stellar-inertial guidance was and what the strategic and bureaucratic roles of the Fleet Ballistic Missile were. The proponents of stellar-inertial guidance succeeded in the first, but they could not, on their own, succeed in the second.

Unistar

It was originally assumed that, except under special conditions, a ballistic missile stellar-inertial system would have to take sightings on two stars. A crucial stage in the development of stellar-inertial guidance in the United States was the discarding of that assumption. The reduction in weight and complexity of the system made possible by the shift from two stars to one greatly helped the acceptance of stellar-inertial guidance by the U.S. Navy.

Indeed, the exigencies of securing that acceptance may have been the crucial factor in ensuring that the argument for the shift from two to one was found compelling. As we shall see, it was *not* found compelling in the Soviet Union, and there are those in the United States who still do not find it so. Certainly, when Kearfott, who had become the central proponents of stellar-inertial guidance, attempted to persuade Special Projects of the virtues of the original two-star version of stellar-inertial guidance they found resistance. David Gold, the influential chief civilian engineer of Special Projects' guidance and fire control branch, SP-23, "was against stellar-inertial guidance because of the complications of having two additional gimbals for the stellar sightings."[22]

This may have helped create a climate at Kearfott receptive to the implications of the findings of two Kearfott engineers with strong statistical backgrounds.[23] In 1965 Irwin Citron and Ted Mehlig were doing mathematical analyses of error behavior in stellar-inertial systems and came to see a relationship in their models between the angular position of the stars on which

one reason why he "never would get the Navy interested in it." Interview with Hyman Shulman, Santa Monica, Calif., January 14, 1986.

22. "If you got through Dave Gold it was all over" (Brett interview). Mr. Gold confirmed his opposition to stellar-inertial prior to Unistar (interview with David Gold, Arlington, Virginia, April 2, 1985).

23. "Some of us sat around a room thinking 'is there a [simpler] way?'" (Brett interview).

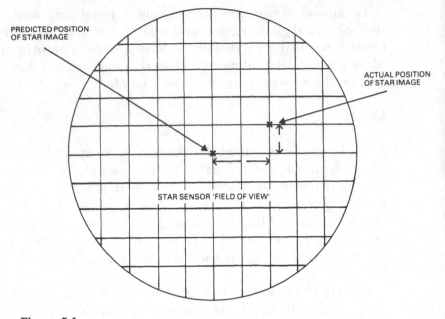

PREDICTED POSITION
OF STAR IMAGE

ACTUAL POSITION
OF STAR IMAGE

STAR SENSOR 'FIELD OF VIEW'

Figure 5.1
The Two Pieces of Information from a Star-Sighting

fixes were taken and the final miss-distance of the missile. Gradually, they came to formulate the principle that undermined the previous assumption that two star-sightings were necessary. If the first star were properly chosen, they suggested, then the second sighting was redundant. As a later statement was to put it, the "unistar principle" implied that "a single fix from an optimally located star will yield accuracy equivalent to a two-star fix."[24]

This was, patently, a counterintuitive argument, so it is worth following how it is justified in this later paper, one of the few publicly available detailed discussions of ballistic missile stellar-inertial guidance. Assume that the three errors in initial conditions for which the star-fix has to correct are in latitude, longitude, and azimuth (orientation in the horizontal plane). On the face of it, it seems hard to imagine how the two pieces of information (see figure 5.1) from a single star-fix could correct three errors. Surely a fix on a second star would help. But consider the situation of firing from the equator to a target located at the North Pole, with the star-sensor prealigned on

24. Rounds and Marmar, "Stellar-Inertial Guidance," 849.

the star Polaris, located vertically over the pole. Any error in initial latitude can be corrected for, because it will lead to a determinate deviation in elevation between the predicted and actual images of Polaris. An error in azimuth will likewise be correctable from the deviation in azimuth of the position of Polaris. An error in longitude cannot be corrected for, however, since it has no effect on the location of the image of Polaris in the star-sensor's field of view, as all lines of longitude converge at the pole. But precisely because all lines of longitude converge there, an error in initial longitude does not cause a target miss. So a second star-fix that could be used to determine error in initial longitude would not enhance final accuracy.

This can be generalized to any target, argue the paper's authors, Stephen Rounds and George Marmar of Kearfott:

Generally it is true that there is a direction in space about which attitude errors are uncorrelated with target miss errors. Therefore, if the single star to be sighted is the optimum star, the only axis about which the stellar sensor cannot measure attitude errors is exactly that axis for which it cannot make any correction. It follows that a second sighting to measure this third component of attitude error can provide no useful information. This is the Unistar Principle: multiple sightings of non-colinear stars cannot provide better accuracy than a single sighting of an optimum star.[25]

Three kinds of doubts had to be overcome before the unistar argument was accepted. The first was the basic matter of intuitions. Even though the proponents of unistar could argue that the two cases were not analogous, the widespread understanding that in ordinary manual stellar navigation a single star-fix was not adequate to yield position was clearly a barrier that proponents of unistar had to overcome. In "insider" circles—guidance engineers, analysts, and the like—unistar in fact seems to have won complete acceptance at the abstract level at which I have just presented it. But not all decision makers were in this sense "insiders," and their contrary intuitions had to be reckoned with.

The second issue was the more practical one that there would not in general be a star of sufficient brightness in the optimum position. This was accepted by Kearfott, who went on to argue—and to devote considerable analytical energy to seeking

25. Ibid., 853–854.

to prove—that the degradation in accuracy caused by use of a not quite optimally placed star would not be too serious. This seems to have been widely agreed, though it is worth noting that this kind of consideration led to what is to my knowledge the first published statement of the unistar principle to be significantly qualified: "A single star-direction sighting will give only two components of data. Sighting on a second star at a substantial angle from the first star can give added data, but in a practical situation only one star properly chosen should provide nearly the full value that star sightings can give."[26] Furthermore, to ensure that even a near-optimum star would always be available required access to a large, accurate computerized star map.

A third issue was that what was envisaged by Kearfott was not merely use of a single star, but a single sighting of that star. The reason why this was of significance is that errors in platform orientation would not simply be caused by errors in knowledge of initial conditions. "Drift" in the gyros that were used to maintain the stable platform in known orientation would obviously also generate errors in platform orientation. Since errors with different causes necessitated different corrections, it was important to be able to separate them out.

Prior to unistar, the systems envisaged at both Kearfott and Northrop envisaged taking repeated fixes on the same star. As Northrop manager Leonard Baker argued "this would make it possible to determine drift of the platform gyros in dynamic conditions of acceleration."[27] Thus gyro drift and initial conditions errors could be separated out, and the proper correction applied.

With a single star-fix, however, this could only be done with an a priori model. As Rounds and Marmar put it: "the attitude error observed by the stellar sensor results from a number of error sources and the flight computer software uses a priori knowledge of system statistics in order to make an optimum correction."[28] Confidence in this had to rest on a general con-

26. David G. Hoag, "Ballistic-Missile Guidance," in B. T. Feld et al., *Impact of New Technologies on the Arms Race* (Cambridge, Mass: MIT Press, 1971), 93.

27. Philip J. Klass, "MMRBM Guidance Techniques Described," *Aviation Week and Space Technology* (November 19, 1962), 83–91; Miller, "Star Tracker," 103.

28. Round and Marmar, "Stellar-Inertial Guidance."

fidence in this kind of statistical algorithm. More specifically, there had to be confidence in the a priori knowledge of system statistics, a point to which we shall return.

In the United States, none of these three forms of doubt was sufficient to prevent widespread acceptance of the unistar argument. What this meant, and what it contributed to the stabilization of stellar-inertial guidance, was a considerable simplification of the mechanical design of the system. With only one star to point at once, there was no need to give the star-sensor freedom to move with respect to the stable platform. It could be mounted integrally to it, and prealigned before launch so that it would point to the predicted position of the chosen star at the appropriate point in time. Not only was this an overall reduction in complexity, it also made it easier to design the system so that its "line of sight" would not be distorted by the acceleration and vibrations of missile flight. The stable, "paradigm" American design of a ballistic missile stellar-inertial system, shown in figure 5.2, was coming into being.

Not everyone was convinced. One of the original, pre-uni-star, proponents of stellar-inertial guidance told me in interview that he still regarded two stars as better. It was the best choice in terms of accuracy, and gave important flexibility in picking the chosen stars. For example, it obviously is important to have one and only one star of the proper magnitude in the star-sensor's field of view when taking the star-sight. With uni-star this was difficult to achieve always, since one was restricted to a star close to the optimum position, while in a two-star system a much wider range of choice was available. "The added complexity" of a two-star system is "peanuts," he went on: "all it is is that you've added a gimbal."[29]

Such possible counter-arguments were ineffective, however. Kearfott successfully convinced senior members of the "defense establishment" of the virtues of unistar. Two key insiders they persuaded were Rand Corporation physicist Albert Latter, one of the "inventors" of MIRV, and John Foster, a veteran of the Livermore nuclear weapons laboratory who served as Director of Defense Research and Engineering from 1965 to 1973.[30] Both were procounterforce, and both quickly

29. Shulman interview.
30. Ted Greenwood, *Making the MIRV: A Study of Defense Decision-Making* (Cambridge, Mass.: Ballinger, 1975).

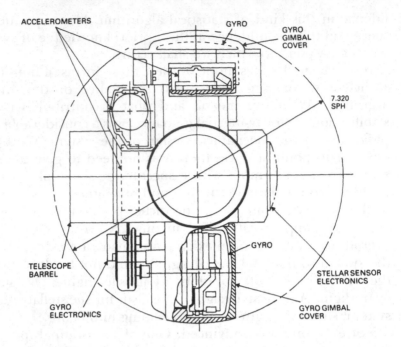

Figure 5.2
American "Paradigm" Design for Ballistic Missile Stellar-inertial Guidance
System
Source: Redrawn from figure in Stephen F. Rounds and George Marmar
"Stellar-Inertial Capabilities for Advanced ICBM," paper 83–2297 read to
American Institute of Aeronautics and Astronautics 1983 Guidance and
Control Conference.

saw the potential of unistar stellar-inertial guidance in boosting
missile accuracy.

Marvin Stern, President of Kearfott, made the connections
and played an important role of his own in presenting the
unistar argument. He sent stellar-inertial advocate John Brett
to see Latter: "he [Latter] said, 'that's it, I understand it [Unis-
tar],' so we had an ally." Latter was "all for getting missile
systems more accurate. . . . The Latter brothers [Richard Latter
was also a Rand physicist] were convinced the Russians were
going for accuracy, and had to be countered by accuracy." Stern
also arranged a meeting for Brett with Foster, whose senior
position meant that it was "hard to get in to see him. Foster
looked at it, and said, 'It's an elegant solution.' He called in
[Special Projects' Director] Levering Smith, and said, 'Levering, what do you think?'" When Smith failed to share his im-

mediate enthusiasm, Foster asked, "Levering, what's the problem?"[31]

Polaris, Poseidon, and the Politics of Accuracy

Special Projects was not to be won over as easily as that.[32] The simplification of a stellar-inertial system that the acceptance of unistar permitted certainly helped build a sense that the technology could be adopted without endangering the Fleet Ballistic Missile program's reputation. The leading guidance expert in Special Projects, David Gold, was now convinced, and, as he put it in interview, began to feel that "I could provide it, so I should provide it."[33]

But Special Projects' Director Vice Admiral Levering Smith did not take this attitude: "Smith's strength was to know that if you could do it, it didn't necessarily mean you ought to."[34] Most importantly, Smith was unconvinced of the need to increase the accuracy of the Fleet Ballistic Missile, and to increase accuracy was the rationale of stellar-inertial guidance.

Missiles of modest accuracy had worked. They had been built on schedule and without a cost overrun; they had performed satisfactorily in tests; they had avoided overt conflict with the Air Force by occupying a differentiated "ultimate deterrent" niche and not competing for the counterforce mission; they had, to the extent anyone could tell, deterred the Soviet Union.

As Special Projects moved from the early A1 and A2 versions of Polaris, guided by the Mk 1 guidance system described in chapter 2, to the A3 version (first deployed in 1964, and also supplied to the Royal Navy), they had not only increased the missile's range but tightened its accuracy considerably.[35] The original accuracy goal for the A3 was to maintain at least the same accuracy at the greater range, but that was consid-

31. Brett interview; John S. Foster, personal communication.
32. During this period the Special Projects Office was renamed Strategic Systems Project Office, but for simplicity I shall use the original, and still familiar, name throughout.
33. Gold interview.
34. Ibid.
35. Aside from having a longer range than the A1, the A2 was designed to be a more robust and reliable system, without the delicate aspects of the A1 referred to in chapter 3.

erably surpassed, with circular error probable reduced to half a nautical mile from the one to two miles of the A1.[36]

The approach taken in designing the Mk 2 guidance system for Polaris A3 was incremental improvement of the Mk 1 system, both to increase accuracy and to make the system lighter.[37] The unceasing inertial instrument development work at Draper's Instrumentation Laboratory was drawn upon to provide smaller and more sensitive instruments, the physical structure of the system was redesigned, and welded rather than soldered electronics used. In similar fashion, the Autonetics Mk 2 Ships Inertial Navigation System (SINS) was progressively modified during this period to increase considerably its accuracy (see figure 5.3).[38]

Taken by itself, the increase in accuracy might have indicated a shift to a counterforce role for the A3. But the design of its warhead indicates that a countercity, assured destruction, role remained the priority. The original goal for Polaris, as with all the early American ballistic missiles, was a one megaton warhead, chosen, according to a participant, not for reasons of strategic calculation but because it was a round number.[39] Although there were Navy fears that if they could not achieve a megaton yield, "LeMay [General Curtis LeMay, commander of the Strategic Air Command] would laugh us out of this business," in order to meet their schedule Polaris A1 and A2 had to be deployed with warheads of lesser yield.[40] By the time Polaris A3 was being designed, the United States and Soviet Union were observing a moratorium on nuclear testing in the atmosphere. "The Atomic Energy Commission proposed scaling up a 200 kt warhead that had already been tested. But Special Projects did not want to rely on a weapon that had not

36. Levering Smith, Robert H. Wertheim, and Robert A. Duffy, "Innovative Engineering in the Trident Missile Development," *The Bridge* (National Academy of Engineering) Vol. 10 (1980), 11; Graham Spinardi's interviews.

37. Benedict O. Olson, "History of FBM Guidance at CSDL," March 10, 1975.

38. For the development of SINS technology, see B. McKelvie and H. Galt, Jr., "The Evolution of the Ship's Inertial Navigation System for the Fleet Ballistic Missile Program," *Navigation: Journal of the Institute of Navigation*, Vol. 25 (1978), 310–322.

39. Herbert York, *Race to Oblivion: A Participant's View of the Arms Race* (New York: Simon and Schuster, 1970), 89.

40. Chuck Hansen, *U.S. Nuclear Weapons: The Secret History* (Arlington, Texas: Aerofax, 1988), 204.

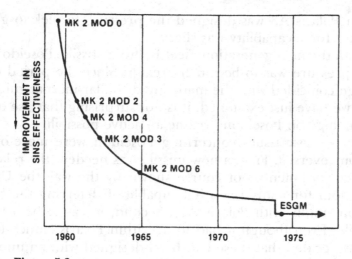

Figure 5.3
Development of Ships Inertial Navigation Systems (SINS) for U.S. Ballistic Missile Submarines
Scale omitted in original. ESGM is the Electrostatically Supported Gyro Monitor. Source: Rockwell International, *The Ships Inertial Navigation System Story* (Anaheim, Calif.: Rockwell International, 1979), figure 6.

been tested in its operational configuration," and may also have preferred to avoid one designed for Air Force use.[41]

An ingenious solution was found, one that also eased fears about a possible Soviet defensive missile that would destroy American warheads as they came down through the atmosphere. Three of the smaller tested warheads were placed on each missile, and a mechanical device released them into slightly different trajectories, so that they fell in a triangular "claw" about a mile wide, tripling the targets for a defensive missile but causing damage to about as great an area as would a single large warhead.[42] The "claw" arrangement, ideal for attacking cities, was, of course, patently not optimal for attacks on hard "point" targets. But that was of little concern, since

41. Greenwood, *Making the MIRV*, 161. Captain Steven R. Cohen, then of Special Projects' successor, the Strategic Systems Projects Office (SSPO), commented: "Many in this organization indeed might resign rather than be involved in an effort which would deploy an untestable or untested system. Some might also prefer not to use an Air Force product, but to give this and the first point equal emphasis, I believe, is insulting to SSPO" (letter to author, November 20, 1986).
42. Greenwood, *Making the MIRV*, 4, 161. CLAW was an acronym for CLustered Atomic Warheads: see Hansen, *U.S. Nuclear Weapons*, 205.

when Polaris A3 was designed the pressure to seek to give it counterforce capability was slight.

With the next generation Fleet Ballistic Missile, Poseidon C3, that pressure was to be much greater. Since the period of its design coincided with the maturing of stellar-inertial guidance that we have just examined, it is not surprising that use of the technology on Poseidon became an active possibility.

Three basic issues concerning Poseidon were more-or-less uncontroversial. First, a new missile was needed (the relativity of that judgment is of course shown by the way the United Kingdom for many years felt capable of deterring the Soviet Union simply with Polaris A3). Second, it was to be a larger missile. Even though it had to fit within Polaris launch-tubes, it was decided that these had been designed with an unnecessarily conservative allowance of space for shock protection.[43] Third, the new missile should have not simply multiple reentry vehicles, like the A3, but multiple independently targetable reentry vehicles, MIRVs.

But what was a new missile needed *for?* The basic technology of MIRV was open to a range of different interpretations and concrete realizations: indeed, as mentioned in chapter 4, its particularly evident "interpretative flexibility" seems to have been what caused its remarkably ready acceptance. There were at root two different, clashing rationales for Poseidon that translated into different priorities for its design.

One rationale was that of Special Projects. Special Projects feared that the Soviets might deploy a defensive missile effective enough to blunt the Fleet Ballistic Missile's retaliatory city-destroying capacity, a missile that would operate not inside the atmosphere but outside it, where the separation of warheads in Polaris A3's "claw" would not be enough to prevent a single explosion destroying them all. With MIRV, however, Poseidon's warheads would follow much more widely separated trajectories, greatly complicating the defense's task.

---

43. The diameter of Polaris was 54 inches. Special Projects proposed an increase to 66 inches (Greenwood, *Making the MIRV*, 34), but Director of Defense Research and Engineering Harold Brown insisted it be increased to the maximum regarded as possible (74 inches). "Harold Brown said, 'I want to remove all of it [shock protection] that you don't absolutely have to have.' In other words, 'I don't want still a larger missile coming round in a few years from now'" (interview with Rear Admiral Robert Wertheim, U.S. Navy, rtd., Burbank, Calif., March 4, 1985).

On this rationale, Poseidon's warheads did not individually have to be particularly big; indeed a larger number of small warheads was superior to a small number of large warheads. Accuracy, too, was a secondary concern. At most, it might have to be increased somewhat to make sure that "assured destruction" capacity did not deteriorate with the move to smaller warheads. To Special Projects, this approach seemed to maintain their by now traditional posture both vis-à-vis the Soviet Union and vis-à-vis the Air Force: Poseidon was to be a "modernized" ultimate deterrent, not a counterforce weapon.

But by the mid-1960s there was an active lobby inside the Navy that pushed an alternative vision of Poseidon. It was centered in the Office of the Chief of Naval Operations, and particularly in the "Great Circle" planning group set up in 1964 by Secretary of the Navy Paul Nitze to study strategic issues. These planners felt that the Navy should adopt a more aggressive posture, both against the Soviet Union (they favored both counterforce and active defense against ballistic missiles) and against the Air Force: "There were advocates in the Office of the Chief of Naval Operations . . . who [felt that] anything the Air Force could do the Navy also needed to do." They were impatient with Special Projects, and at least one of them suspected that Special Projects, "had made a deal with the Air Force not to try to gain counterforce capability."[44] The vision of Poseidon put forward by the Great Circle group reversed that of Special Projects: they wanted a counterforce missile, and, in particular, preferred a smaller number of large warheads.

The Great Circle group wanted Poseidon equipped with the large Mk 17 reentry vehicle being designed by the Avco Corporation, and associated Los Alamos multimegaton W67 hydrogen bomb, a combination that was being developed to enhance the counterforce capability of Minuteman.[45] Special Projects wanted the smaller Mk 3 reentry vehicle, being designed by the Lockheed Corporation, who were and are

44. Wertheim interview. The formation and position of the Great Circle group are described in Greenwood, *Making the MIRV*, quote on p. 55.
45. For the Mk 17, see ibid. Development of the W 67 began at the Los Alamos National Laboratory in 1965, and was cancelled (see below) in December 1967, according to Chuck Hansen, *U.S. Nuclear Weapons*, 107, from whom I am taking the description of its intended yield as "multimegaton."

system integrators for the Fleet Ballistic Missile and who had also designed and built the Mk 1 and Mk 2 reentry vehicles for Polaris.[46] The Mk 3 would carry the 40- to 50-kiloton W68 warhead under development from 1965 onward at the Lawrence Livermore Laboratory.[47]

The advocates of counterforce in the Office of the Chief of Naval Operations had to be taken seriously. While originally Special Projects was set up outside the normal Navy hierarchy, in 1963 it was formally subordinated to the Chief of Naval Material, and in 1966 the latter was in turn subordinated to the Chief of Naval Operations.[48] Furthermore, Secretary of Defense Robert McNamara and Director of Defense Research and Engineering Harold Brown seemed to have some sympathy with the view that counterforce capability should be built into Poseidon. McNamara was at least prepared to allow the Mk 17 heavy reentry vehicle to be developed as an option with counterforce explicitly in mind, and Brown—quoting McNamara as his authority—personally pressed Special Projects' Director Levering Smith to increase Poseidon's accuracy.[49]

In this potential minefield, Admiral Smith and the Special Projects Office trod carefully, avoiding the risk of further of loss of autonomy (which might easily have happened if they

46. John Brett suggested to me that Lockheed made most of its profit from the Fleet Ballistic Missile program not out of its "system" role but out of its reentry vehicle work, so the latter was of special significance for it.

47. Hansen, *U.S. Nuclear Weapons*, 107.

48. Harvey M. Sapolsky, *The Polaris System Development: Bureaucratic and Programmatic Success in Government* (Cambridge, Mass.: Harvard University Press, 1972), 198–200.

49. Thus McNamara's Fiscal Year 1966 "Posture Statement," quoted by Greenwood, *Making the MIRV*, 7, 45, stated: "Alternatively, [Poseidon] could be used to attack a hardened point target with a greater accuracy and a heavy warhead." For the pressure on Smith, the source is my interview with Vice Admiral Levering Smith (U.S. Navy, rtd.), San Diego, Calif., February 23, 1985. See also James McCormack, "Memorandum to the Files, subject: the Poseidon Contract," November 10, 1968, MIT Archives and Special Collections, Albert Hill papers, 83–40, box 4. MIT Vice-President McCormack spoke to both Admiral Smith and Harold Brown to investigate the propriety of the Instrumentation Laboratory, rather than a private firm, taking on the Poseidon guidance contract. He concluded: "With Mr. McNamara's insistence on getting all the accuracy possible . . . the services of the Instrumentation Laboratory definitely came to be required." I raised this episode in my interview with Robert McNamara (Washington, D.C., March 29, 1985), but, unfortunately, he was unable to recall it.

openly defied those formally above them) while, in the end, successfully implementing their vision of Poseidon. In response to pressure for greater accuracy, Smith agreed to seek to make Poseidon 50 per cent more accurate than Polaris A3, but only as a "goal," not as a "requirement"; thus avoiding the danger to the Fleet Ballistic Missile program and his office of failing to produce "what you commit to produce."[50]

The Special Projects Office commissioned Draper's MIT Instrumentation Laboratory, developers of the Mk 1 and Mk 2 guidance systems for Polaris, to develop Mk 3 for Poseidon. Some marked changes were made, especially to the computer, which became a full general purpose digital computer, using integrated circuits with nine logic gates per chip.[51] But the Mk 3 was still a by now traditional Draper pure inertial system, employing gyroscopes and accelerometers improved on the previous versions by the familiar means of evolutionary refinement.[52] The attitude to the achievement of the accuracy goal was sufficiently relaxed to allow the final stage of a shift that had begun tentatively with the Mk 2 system: the replacement of the expensive and difficult to produce PIGAs (pendulous integrating gyro accelerometers) with simpler, even if somewhat less accurate, PIPAs (pulsed integrating pendulous accelerometers), which contained a pendulous mass rather than a gyro. In the Mk 2 a PIGA had been retained for the crucial

50. Sapolsky, *Polaris System Development*, 221–222. The 50 percent figure is from Smith, Wertheim, and Duffy, "Trident Missile Development," 11. In my interview with him Admiral Smith emphasized his view that "it's the general reputation [for 'producing what you commit to produce'] that stands you in best stead, much better than skillful arguments."

51. Interview with Special Projects staff member. This more powerful computer made it possible to move away from Q-guidance, which seemed to engineers at the Draper Laboratory ill-suited to MIRV: "in the bus mode [i.e. MIRV] . . . you had very complicated Q terms" (interview with Benedict Olson, Cambridge, Mass., October 5, 1984). Though Special Projects was initially reluctant to make the move, it was thus decided to shift to a form of "explicit" guidance, in which the missile constantly, "explicitly," "knows" its position and velocity, and recomputes the change in velocity needed to achieve values of position and velocity at which successive reentry vehicles can be released on trajectories to their targets.

52. One should be careful with this formulation. Thus the Poseidon accelerometer was the first Instrumentation Laboratory instrument to use a permanent magnet (rather than electromagnet) torquer (ibid.). To an outsider this sounds like a small change, but inside the Laboratory, where Draper was suspicious of permanent magnet torquers, it was in its way a little revolution.

thrust axis; with the Mk 3 all three accelerometers became PIPAs, albeit improved on those of the previous generation.[53] Similarly, incremental improvements continued to be made to the Ships Inertial Navigation System (SINS) and to the means of updating it.[54]

The Mk 3 guidance system, navigation improvements and the formulation of the accuracy specification as a "goal" rather than a "requirement" solved some of Special Projects' problems, notably the pressure from the Department of Defense to increase accuracy. But they could not solve the problem of the desire for the heavy Mk 17 reentry vehicle. Special Projects strongly opposed this, and even opposed the obvious bureaucratic compromise of a "mixed force," carrying both heavy and light reentry vehicles: the mixed force would be a "logistical nightmare . . . a specially configured missile assigned to special targets as opposed to submarines which could go on patrol with the flexibility to be targeted from one kind of target to another without having to worry about what you had loaded in the tubes."[55]

Here stellar-inertial guidance came into play. If the proponents of that technology were correct, then a stellar-inertial Poseidon could be made substantially more accurate than an all-inertial Poseidon. While the latter plus a light warhead did not add up to a counterforce weapon, the former plus a light warhead perhaps did. A 1966 study at Lockheed, designers of the small Mk 3 reentry vehicle, concluded that with stellar-inertial guidance the Mk 3 could indeed achieve counterforce capability. As one person involved recalled, "the reason that study was done was to kill the Mk 17."[56] It succeeded: "The Mk 17 [heavy warhead] program . . . was made to disappear by the prospect of still further accuracy improvement which

53. Ibid.

54. See McKelvie and Galt, "Ships Inertial Navigation System," and Owen Wilkes and Nils Petter Gleditsch, *Loran-C and Omega: A Study of the Military Importance of Radio Navigation Aids* (Oslo: Norwegian University Press, 1987), 91–100.

55. Wertheim interview. Again note the relativity of this argument. Two decades later, faced with the inevitability of a heavy warhead for Trident D5, Special Projects welcomed a mixed force as a means of ensuring the deployment on Trident D5 of some of Trident C4's "small" 100-kiloton warheads.

56. Interview with Lockheed engineer by Graham Spinardi, May 8, 1987.

made it possible to show that even with the small yield warheads . . . you could, potentially at least, threaten damage to moderately hard targets."[57] Lockheed, indeed, became active supporters of stellar-inertial guidance: "we . . . sold it with Singer [Kearfott]."[58]

Was the future of stellar-inertial guidance secure? For several reasons the answer was "no." First, Special Projects' argument that counterforce capability could be achieved with a light warhead plus stellar-inertial guidance had something of a tactical nature. It solved the problem of the Mk 17, but once that was canceled, there was little real push for counterforce capability within the office:

> most of us saw the role of Poseidon as not different from the role of its predecessors, namely providing an absolutely dependable, reliable deterrent, and most of us were skeptical about the need to dig out hard targets as an essential element of deterrence. We went along with it to the degree necessary to keep the program. The nature of democracy . . . is that you're constantly making compromises with conflicting constituencies, and we had to serve the reigning constituency even if we, sometimes we felt they were a little nutty.[59]

Second, the Instrumentation Laboratory was not immediately enamored with the new technology. One engineer there recalls: "I must admit it [stellar-inertial guidance] wasn't pushed by us hard enough, it was pushed by Singer [Kearfott]. I remember having tunnel vision myself talking with my boss at the time about a star-tracker: 'Well, the system has been pretty good so far. We don't need to improve it with a star-tracker, necessarily.'"[60] As well as the belief that the development of all-inertial systems was the proven route to increased accuracy, there were worries at the Instrumentation Laboratory about even the restricted "opening of the black box" involved in taking a star-sight: "There was a concern here at the Lab. that it was an unnecessary complication . . . then the other concern that I always had and I still have is that . . . there's a

---

57. Wertheim interview.
58. Spinardi Lockheed interview.
59. Wertheim interview.
60. Olson interview. Olson believes that a particular source of Instrumentation Laboratory skepticism about stellar-inertial guidance had to do with the second star-sighting believed, prior to the acceptance of unistar, to be necessary.

possibility of a nuclear explosion in the atmosphere making the stellar system inoperable."[61]

Proponents of stellar-inertial guidance recalled actual opposition and a "battle" with the Instrumentation Laboratory, "a year worth of arguing and screaming:"[62]

Dr. Draper took the gyro out of his pocket and said if you have a good enough gyro you wouldn't need a [star-tracker] . . . [Draper and the Instrumentation Laboratory] didn't think one should gimmick it by adding these crazy things called stellar sensors, that's like putting a band-aid on an inertial system.[63]

The Instrumentation Laboratory produced various counterproposals to Kearfott's unistar stellar-inertial design. While these were unsuccessful, they did delay the start of serious design work on a stellar-inertial option for Poseidon.[64]

Once stellar-inertial guidance became firmly accepted as part of the Fleet Ballistic Missile program, the Instrumentation Laboratory, by then the Charles Stark Draper Laboratory, loyally supported the technology: any residual skepticism surfaces now only in the form of the occasional less-than-enthusiastic comment.[65] Explicit doubt about stellar-inertial guidance was, however, expressed to me by a senior member of the guidance community not connected to the Draper Laboratory. His argument was that the efficacy of a unistar correction depended on the validity of the statistical error model used. He fears this model may be seriously wrong, and, especially if this is combined with possible errors in the computerized star map from which the predicted position of the chosen star is derived, the correction prompted by the star-fix may increase, rather than reduce, target miss.[66]

61. I am quoting here from a 1987 interview of a Draper Laboratory member by Graham Spinardi.
62. Spinardi Lockheed interview.
63. Interview.
64. Interview by Spinardi.
65. Some proponents of stellar-inertial guidance, it is interesting to note, still doubt the depth of commitment to it at the Draper Laboratory (interview).
66. Interview. Whether missile testing can compel resolution of doubts of this kind is an issue we must postpone to chapter 7. The skeptic I am quoting has had access to the test results for stellar-inertial guided Trident C4 missiles.

A third difficulty for stellar-inertial guidance was that it did not map well onto the organizational structure of the Special Projects Office and its system of corporate contractors. As we saw in chapter 2, the two phases of the task of achieving accuracy, submarine navigation and missile guidance, had become the provinces of two separate sets of contractors. Two distinct branches of Special Projects, SP-23 (Guidance and Fire Control) and SP-24 (Navigation) coordinated and directed the activities of these two contractors. Prior to stellar-inertial guidance, the contribution of the errors of navigation and guidance to overall target miss were separable: the mathematical model of system accuracy, the "error budget", could be partitioned in a clear-cut way corresponding to the organization of the Special Projects Office and its contractors. But the whole point of stellar-inertial guidance was that it was a guidance technology designed to correct retrospectively for navigational errors. Wholly separable guidance and navigation components in the mathematical model could no longer be identified, but an error budget had to be agreed to let detailed system design proceed.

Constructing the mathematical model thus became a task of organizational negotiation of unprecedented difficulty. The careful words of one participant describe the very considerable problems involved:

[With stellar-inertial guidance,] partitioning of the task became more difficult in that accuracy, unlike reliability, did not partition linearly. Even with exhaustive coordination, the accuracy performance of one branch could not be divorced from that of the other, and management visibility into subsystem activity to a level beyond that desired by the branches [SP-23 and SP-24] became essential. Additionally, the process was affected by intra-organizational politics, budget realities and the capability and ambitions of the contractors.[67]

On some accounts (participants are not in agreement in their memories of this episode) what had to be negotiated was not only the details of the error budget but crucial technical characteristics of stellar-inertial guidance. For what, exactly, could a stellar fix correct? No one doubted that, in principle, errors in initial azimuth, latitude, and longitude could be corrected (though there were, as we have seen, doubts about the necessity

67. Interview with Captain Steven R. Cohen (U.S. Navy), Arlington, Virginia, March 29, 1985.

and practical efficacy of a stellar system for doing so). But the submarine's inertial system provided not just these inputs but also data on the submarine's velocity at the moment of firing. If errors in this could not be corrected for retrospectively, then a very good Ships Inertial Navigation System, or possibly additional velocity-sensing navigational equipment, were still needed. The leading technical figure in SP-24, the navigation branch of Special Projects, recalls a long struggle to establish this: "[The star sensor] . . . alleviated performance problems on navigation. [But] some of them it doesn't address at all, like velocity. . . . It took us five years to convince [industry proponents of stellar-inertial guidance] that $1 + 1 = 2$, and when it didn't address it, it didn't address it."[68] These industry proponents deny, however, that any responsible advocate of stellar-inertial guidance claimed that it could correct for errors in initial velocity. The difference in recollection perhaps hinges on the fact that correcting for errors in initial velocity turns out to be a less straightforward notion that appears at first sight, the "problem of the vertical" reappearing here.[69]

The fourth problem for stellar-inertial guidance was that, unique among guidance technologies up to this point (the late 1960s), it sparked controversy in the formal political system. This was controversy over it as a black box: the "insider" opposition to it never became public, and stellar-inertial guidance's opponents in the formal political system did not doubt that it

68. Interview with Thomas A. J. King, Arlington, Virginia, April 2, 1985.
69. One guidance specialist not directly involved suggested to me in a letter of March 9, 1987, "One of the facts that causes some confusion . . . is that velocity errors and the direction of the local vertical are coupled. That is, the velocity error that builds up in an inertial navigation system due to initial conditions and errors in gravity modeling will result in a Schuler frequency [i.e. 84–minute; see chapter 2] oscillation in position error. This has both an initial velocity component and also a tilt of the vertical. 'Velocity error' is sometimes the generic term for this, since the instrumentation errors are integrated to give velocity errors before another integration results in position, and since the contributions of the Schuler oscillation cause more on-target error due to velocity than their equivalent position error. The fact that the position error is related to the orientation leads some people to say that the star sighting corrects the initial orientation which was caused by the velocity error. That argument is probably logically okay, except that at the instant of launch, it is not the velocity error itself that is being corrected, but rather what it has propagated into as a position error through the Schuler oscillation."

would increase missile accuracy, instead opposing it precisely because it would. Nevertheless, the extent of insider disagreement about stellar-inertial guidance was probably what enabled its public critics to succeed in achieving the cancellation of the Mk 4 stellar-inertial guidance system for Poseidon.[70]

Paradoxically, it was an apparent success for top-level lobbying efforts on behalf of stellar-inertial guidance that led to its demise as an option for Poseidon. In January 1969 the Nixon Administration assumed office, and the new Secretary of Defense, Melvin Laird, was charged with both cutting the defense budget and presenting a hawkish profile. Stellar-inertial proponent John Brett of Kearfott suggested that a good way of achieving a cheap toughening of the defense posture would be to speed up the development and deployment of the Mk 4 stellar-inertial guidance system.[71] Laird liked the idea, and in March 1969 proposed it to Congress: "The increase of $12.4 million for the development of an improved guidance system for the Poseidon missile will advance the initial operating capability (IOC) of that system by about six months . . . This is an important program since it promises to improve significantly the accuracy of the Poseidon missile, thus enhancing its effectiveness against hard targets."[72]

This imaginative enrollment of the interests of the Nixon Administration backfired because it made stellar-inertial guidance an inevitable target of the campaign against counterforce mounted by Senator Brooke and his supporters.[73] As we saw, by December 1969 Brooke had pushed Nixon into stating that "there is no current U.S. program to develop a so-called 'hard-target MIRV capability.'"[74] While other hard-target programs

70. See Greenwood, *Making the MIRV*, 136–137.
71. Brett interview. Brett told me that he believed Democratic administrations talked like doves while spending like hawks, and Republican ones talked like hawks while spending like doves, a correlation that does not hold for the Reagan Administration.
72. Cited in Stockholm International Peace Research Institute, *SIPRI Yearbook of World Armaments and Disarmament*, 1968/69 (Stockholm: Almqvist and Wiksell, 1970), 109 (I owe this reference to Graham Spinardi).
73. For the use of the term "enrollment" in the sociology of science and technology, see, for example, Bruno Latour, *Science in Action* (Milton Keynes, Bucks England: Open University Press, 1986), 108.
74. Quoted by Alton Frye, *A Responsible Congress: the Politics of National Security* (New York: McGraw-Hill, 1975), 70.

could still be pursued quietly, the Mk 4, because of its prominence, had to be cancelled so that reality could be brought into line with this assertion. As Director of Defense Research and Engineering John Foster put it the following summer: "[w]e had a program of investigation along these lines [hard-target MIRV] and last year I cancelled it. My purpose was to make it absolutely clear to the Congress and hopefully to the Soviet Union, that it is not the policy of the United States to deny the Soviet Union their deterrent capability."[75] Foster's office had pushed for stellar-inertial guidance, and he himself tended to favor counterforce. The tenor of his statement shows the influence of Congressional opposition, and Foster recalls that he had also been swayed personally by the argument of Admiral Levering Smith of the Special Projects Office that too much accuracy in the submarine-launched ballistic-missile force was strategically destabilizing.[76]

With "insiders" divided, and active opposition in the formal political system, stellar-inertial guidance for Poseidon was dead. Its only physical trace in deployed Poseidon missiles is a "trap door" through which the star sighting was to be taken.[77] But for the proponents of the technology, and those of counterforce capability in the Fleet Ballistic Missile, it was the loss of a battle, not the war. For by the time the Mk 4 was finally canceled a new missile was already on the horizon.

Trident C4: Compromise

The dominant rationale for the new missile was no longer a possible Soviet anti-ballistic missile system, but possible Soviet developments in anti-submarine warfare. With the relatively restricted range of Poseidon, submarines had to approach moderately close to the Eurasian landmass to strike at targets in the heart of the Soviet Union. Although greater range could be achieved by "offloading" some of the maximum of fourteen warheads in the Poseidon MIRV system, Special Projects felt that the invulnerability of the submarine-launched deterrent

75. Quoted ibid. The quotation does not refer explicitly to the Mk 4, but it seems to be the most likely program. Even if it was not, the statement indicates the influence of the opposition to counterforce.
76. Interview with John Foster, Cleveland, Ohio, September 24, 1986.
77. I owe this information to Spinardi's interviews.

force could best be preserved by developing a new missile with a much greater range than Poseidon.[78]

So, as with the step from Polaris to Poseidon, Special Projects's notion of the logical "follow-on" from Poseidon to Trident was one that preserved the assured destruction, ultimate deterrent, conception of the Fleet Ballistic Missile.[79] Again they had to contend with those who had different goals and interests. While they again managed to keep their vision largely intact, this time the compromises made would have effects that were permanent.

One troublesome interest to be dealt with was those whose concern lay not with the missile per se, but with the submarine, particularly its nuclear reactor. The Trident program reopened the issue that Special Projects had successfully closed with the choice of a modified existing attack submarine to carry Polaris—the potentially deeply problematic relationship with Admiral Rickover, "father" of the nuclear-propelled submarine.

The issue around which conflict broke out was the top speed of the Trident submarine. Admiral Smith and many others in Special Projects saw little virtue in a high top speed—the submarine's task was to lurk unnoticed, waiting to deliver its retaliatory blow. But most submarine commanders favored increasing speed. Rickover agreed with them; he also knew well that a high top speed for the submarine would necessitate a new nuclear reactor and, thus, a central role for him and his Naval Reactors Branch in the Fleet Ballistic Missile program.[80]

The Navy hierarchy backed Rickover. Over the years Rickover had carefully cultivated Congressional contacts and Chief of Naval Operations Admiral Zumwalt was concerned that he

78. The flexibility of MIRV in this respect was one of the arguments for it in Navy circles, albeit not a dominant one.

79. Originally Trident was ULMS, the Undersea Long-range Missile System. ULMS had its origins in a study begun in 1966 to ascertain the most cost-effective means of ensuring the delivery of sufficient equivalent megatons to deter the Soviet Union, taking into account possible Soviet developments such as ultra-accurate ICBMs and anti-ballistic missile defenses. Volume 1 of Strat-X's 20 volumes is unclassified: *The Strat-X Report* (Arlington, Virginia: Institute for Defense Analysis, August 1967).

80. See J. Steinbruner and B. Carter, "Organizational and Political Dimensions of the Strategic Posture: the Problems of Reform," *Daedalus*, Vol. 104 (Summer 1975), 137.

needed Rickover's backing "because you couldn't get Congressional support without it." "We badly needed to get the Trident missile to sea. The only way we could do it was to buy Adm. Rickover's reactors."[81]

The resolution of this dispute reflected Rickover's preferences rather than those of Special Projects. The new class of submarines would be very big—the consequences of which may only now be apparent, as the invulnerability of ballistic missile submarines is being questioned—and faster than Special Projects felt necessary.[82] Organizationally, Special Projects lost responsibility for the missile submarine construction program, which it had had for the forty one Polaris/Poseidon submarines. Through the late 1970s the Trident submarine was dogged by delay, cost overruns, and litigation, with the first Trident submarine, the USS *Ohio*, due to be delivered in December 1977, beginning sea trials only in June 1981.[83]

Special Projects, however, had successfully advocated a strategy that at least temporarily divorced the Trident missile from the Trident submarine. This strategy involved designing not one Trident missile but two: Trident I (C4) would be small enough to fit inside the launch tubes of existing missile submarines, while only Trident II (D5) would take advantage of the greater space available in the new, large submarines.[84] So despite the delays in the submarine program, Special Projects managed to keep the missile program on schedule, with the

81. M. Mintz, "Depth Charge: Cost Overruns on New Trident Sub Leave a Muddied Wake," *Washington Post* (October 4, 1981), reprinted in Dina Rasor, ed., *More Bucks, Less Bang: How the Pentagon Buys Ineffective Weapons* (Washington, D.C.: Fund for Constitutional Government, 1983), 213.

82. During 1989 there were reports that ultrasensitive radars could detect the faint "wake" on the ocean surface left by a submerged submarine. If this is true, the concealment of the ballistic missile submarine would be threatened, and having large numbers of missiles in each of a relatively small number of big submarines might be to have "too many eggs in one basket." U.S. Navy circles are, however, skeptical of the radar experiments, and suspect that the Air Force is seizing upon them to bolster threatened Air Force programs like the B-2 "Stealth" bomber. See Martin Walker, "New Radar Threatens Trident," *The Guardian* [London] (August 25, 1989), 1.

83. N. Polmar and T. Allen, *Rickover* (New York: Simon and Schuster 1982), 579, 576. On the whole episode, see Patrick Tyler, *Running Critical: The Silent War, Rickover and General Dynamics* (New York: Harper and Row, 1986).

84. These missiles were originally known as ULMS 1 and ULMS 2.

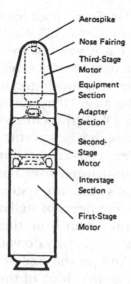

Figure 5.4
Trident C4 Missile
Source: Levering Smith, Robert H. Wertheim, and Robert A. Duffy,
"Innovative Engineering in the Trident Missile Development," *The Bridge*
(National Academy of Engineering), Vol. 10, No. 2 (Summer 1980), 12.

C4 being deployed in existing submarines until the new Trident boats were ready.

The design of the Trident C4 missile raised the same issues of warhead yield and accuracy, of counterforce versus assured destruction, as had the design of Poseidon C3. The pressure for a high-yield warhead was more easily defeated with the C4. All were agreed that the new missile should have greater range. To fit it into existing launch tubes, Special Projects designed a squat missile whose third stage rocket motor went almost to the top of the missile (see figure 5.4). So the guidance system and reentry vehicles had to be arranged around the rocket motor. Thus the decisions about range and compatibility with existing launch tubes translated into a strong, and from Special Projects' point of view, not necessarily undesirable, physical constraint on the size of reentry vehicle that could be used.

Although their room to maneuver was thus (literally) limited, the proponents of counterforce in the Offices of the Chief of Naval Operations and Secretary of Defense pressed for as large-yield a weapon as possible, and were reportedly joined in their efforts by Navy officers from the Joint Strategic Target

Planning Staffs who were tired of Air Force jibes about the Navy's "firecrackers," that is, Poseidon's "small" 40–kiloton warheads.[85] Special Projects argued that it was not cost-effective to spend a lot of money to build a new reentry vehicle and weapon with a yield that could not be very much greater, but in the end succumbed, reportedly because "they recognized the political benefit of agreeing with OSD [the Office of the Secretary of Defense]."[86] So Lockheed built a new reentry vehicle for the Trident C4, containing a new, 100–kiloton warhead.

After the sometimes acrimonious controversy over Poseidon guidance, the design of the Trident C4's guidance system went remarkably smoothly. Crucially, the proponents of stellar-inertial guidance had learned an important lesson from their failure on Poseidon: "the question was asked . . . why do you need this stellar-inertial guidance system? And [in the case of Poseidon] the wrong answer was given; we didn't think of the right answer."[87]

The "wrong" answer had of course been "to enhance counterforce capability." The "right" answer, the one compatible with Special Projects' views and with liberal sentiments in Congress, was "to enhance survivability." One of the key proponents of stellar-inertial guidance, John Brett of Kearfott, was at hand in the Department of Defense to make sure the right answer was given for the C4. When the Nixon Administration caved in to Congressional opposition to Poseidon stellar-inertial guidance, Brett decided that the "only way to get this was to become part of the power structure."[88] Already well connected in Washington, he became Under-Secretary of Defense for Strategic Systems and had the responsibility for turning into detailed specificiations a vague direction from Deputy Secretary David Packard to build a new missile.

Though "what we really wanted was a more accurate missile," Brett "didn't want to fly in the face of all those M[utual]

85. Information from interviews conducted by Graham Spinardi. "Small" is of course a relative term applied to nuclear warheads: the yield of the weapon that destroyed Hiroshima is reckoned at 15 kilotons (Hansen, U.S. Nuclear Weapons, 121).

86. This quotation is from an interview with a participant conducted by Graham Spinardi.

87. Spinardi Lockheed interview.

88. Brett interview.

A[ssured] D[estruction] advocates."[89] The specifications he helped set—the extent of his influence on them was not agreed on by interviewees—cleverly avoided doing so. Trident C4 was to have an accuracy "equivalent at 4000 nm [nautical miles] to that of Poseidon C3 at 2000 nm."[90] This would roughly double the precision of guidance and navigation needed, but would not raise the spectre of counterforce. The increased range was uncontroversial, because the argument for it was to increase submarine survivability. Similarly, the new system was to have the capability of maintaining accuracy after much longer periods of submerged cruise without access to external sources of navigational information—a specification that again was justified by submarine survivability, but also increased the demands on guidance and navigation.

In Brett's opinion there was "only one system," stellar-inertial guidance, that could meet these specifications.[91] Special Projects agreed; the Draper Laboratory did not demur; and this time Congress—care having been taken not to offend the predilictions of its liberal wing—did not oppose the shift to stellar-inertial guidance.[92] Even the internal issues between Special Projects' guidance and navigation branches, SP 23 and SP 24, were on the way to solution.

The Mk 5 guidance system for Trident C4 incorporated not just stellar-inertial guidance, but also some other noteworthy changes. For the first time in the Fleet Ballistic Missile program, Draper floated gyros were not used. Instead, the "dry," tuned-rotor design pioneered at Kearfott was selected. It was a "tearful" decision, recalls one Draper engineer.[93] Volume was important, especially in the context of trying to built a much longer-range missile to fit into the same launch-tube. Since the tuned-rotor gyro was a two-degree-of-freedom gyro, only two of them, not three, were needed, so making "room" for the

89. Ibid.
90. Smith, Wertheim, and Duffy, "Trident Missile Development," 12.
91. Brett interview.
92. Although "in 1974 and 1975 . . . Congress cut back Navy work on high-precision stellar inertial guidance for the traditionally less accurate submarine-launched ballistic missiles (SLBMs), lest the Soviets perceive the SLBMs as becoming a first-strike force": Deborah Shapley, "Arms Control as a Regulator of Military Technology," *Daedalus*, Vol. 109, No. 2 (Winter 1980), 149.
93. Olson interview.

star sensor.[94] Gyro drift could be argued to be less important, since the star sight ought to help correct for it. So the Draper Laboratory could go along with the selection of dry gyros without abandoning the claim that floated gyros were more accurate, and had to accept that the dry gyro was almost certainly cheaper. But all of these factors might not have been conclusive, according to one participant, were it not for Kearfott's "plain flat-out aggressive salesmanship."[95]

The accelerometer choice for Trident C4 was by comparison undramatic. The accelerometer chosen was essentially the PIPA used in Poseidon's Mk-3 system with a few modifications, "largely things that made it more producible."[96] Even though a PIGA might be more accurate, the PIPA was considered by Special Projects to be good enough to meet the accuracy goal and light enough to meet the range goal, and so the extra cost of developing a new accelerometer was judged to be not worthwhile.

[W]e chose to stay with the accelerometer because we didn't have to go out and re-invent the thing. . . . Staying with the accelerometer certainly simplified the job. . . . Inertial components . . . are always difficult to do whenever you start to design some new ones. Not just the design and development, but also getting the production system up to speed. Start-up costs, start-up problems—they're always tremendous.[97]

As with the choice of accelerometers (the lighter PIPAs rather than more accurate PIGAs), the design of the "bus" that carries the reentry vehicles also involved sacrificing accuracy for range:

In designing for range, the "bus" structure was designed to be of minimum weight for structural integrity with adequate margin. The optimized graphite cone structure, as an outcome, had vibrational modes which added a statistically bounded, but not exactly predictable on a body-by-body basis, increment to deployment velocity. This increment of course translates to an addition to the C[ircular] E[rror] P[robable]. . . . While neither large nor affecting performance relative

94. Gold interview.
95. The quote is from an interview by Graham Spinardi.
96. Interview.
97. Ibid.

to the goal, this deployment inaccuracy was nevertheless identifiable and could have been traded for less range.[98]

In submarine navigation technology a major step was made that, like stellar-inertial guidance, could have been made earlier for Poseidon but was not. That was the introduction of electrostatically suspended gyroscopes. As we saw in chapter 4, during the 1960s this technology was touted as providing the ultimate in gyroscope accuracy. Draper seems to have been unconvinced, and it was generally agreed that the harsh accelerations of rocket boost made the relatively delicate electrostatic gyro unsuitable for missile guidance. But it looked potentially ideal for the protected, benign environment of submarine navigation. Indeed, the man who first thought of the device, Professor Arnold Nordsieck of the University of Illinois, originally justified it in 1954 on the grounds that the advantage of nuclear propulsion (which was then becoming a reality)—that the submarine need not surface for air—would be lost if it had to surface for navigational fixes.[99]

By 1961 Honeywell, who as we saw were the electrostatic gyro's main corporate proponents, were confidently predicting that "first production models and deliveries to the Navy for installation in its Polaris submarine fleet will begin in mid-1963."[100] It was not to be. The electrostatic gyro was not adopted for Polaris or Poseidon. One factor certainly was the enormously sensitive production process resulting from the need to keep "the sphericity of the ball [the gyro rotor] to

98. Letter to author from Captain Steven R. Cohen (U.S. Navy), December 2, 1986.

99. I am paraphrasing what appears to be the earliest extant document on the electrostatic gyro, a handwritten report by Nordsieck, "Feasibility of a Free Gyro Navigation Device," dated May 1954, a copy of which was kindly given to me by Nordsieck's collaborator, Professor Howard Knoebel. The report cites an earlier paper, dated March 16, 1954, sent to Office of Naval Research, but Professor Knoebel has been unable to trace this document. There is a lively popular history of the electrostatic gyro's origins in R. A. Kingery, R. D. Berg, and E. H. Schillinger, *Men and Ideas in Engineering: Twelve Histories from Illinois* (Urbana, Illinois: University of Illinois Press, 1967), 129–139.

100. Charles Lafond, "Honeywell ESG Slated for Duty in Polaris Subs," *Missiles and Rockets* (November 13, 1961), 30, quoting an unnamed "Honeywell spokesman."

better than five millionths of an inch during manufacture."[101] Another was the steady increase in accuracy of the conventional Ships Inertial Navigation System (SINS) technology (see figure 5.3). Special Projects came to feel that electrostatic gyro peformance was only "modestly better" than the projected performance of improved conventional SINS, and it "would cost a lot of bucks" to get that modest improvement.[102] With no enormous pressure to increase accuracy, and with a limit on accuracy in any case being placed by the systems used to "reset" the submarine's inertial navigator, the step did not seem worth it.

The electrostatic gyro finally got its opportunity with Trident, though by then Autonetics, the traditional supplier of ballistic missile submarine inertial navigators, had developed its own electrostatic gyro and defeated Honeywell's bid to enter its market.[103] As with the introduction of stellar-inertial guidance on Trident C4, the rationale was not accuracy per se, but survivability, and the new technology was introduced with caution. It was not to replace the conventional SINS, but to "monitor" it, to "reset" the SINS, interrupting the build-up of errors, as the external systems such as Loran-C did, but in a black-box fashion.[104] So external resets would be necessary less often (see figure 5.5), perhaps once every two weeks as compared to once a day.[105]

Overall, then, Trident C4 and its associated navigation technology was a subtle compromise, partially explicit, partially

101. Philip J. Klass, "Navy to Test Electrically Suspended Gyro," *Aviation Week* (February 6, 1961), 87.
102. King interview.
103. Autonetics had developed a different, easier to produce, electrostatic gyro design, involving a small solid ball rather than Honeywell's larger hollow ball. Joseph Boltinghouse, John Slater's successor as Autonetics's "gyro genius," played the key role in the development of the former. The Autonetics work began only in 1959 (interview with Joseph Boltinghouse, Anaheim, Calif., September 9, 1986), after Honeywell's, and, at least in its ballistic missile submarine application, was largely company-funded, not Navy-funded (King interview), presumably in response to the threatened loss of market.
104. That indeed was the role that seems originally to have been planned for it for Polaris. See Lafond, "Honeywell ESG."
105. Wilkes and Gleditsch, *Loran-C and Omega*, 122, quoting Department of Defense and Navy testimony. The two weeks figure may be conservative: interviewees quoted higher figures.

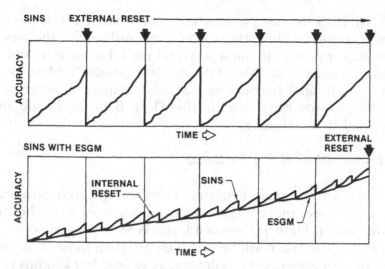

Figure 5.5
Use of Electrostatically Supported Gyro Monitor (ESGM) to Extend
the Ships Inertial Navigation System (SINS) Reset Interval
Source: Rockwell International, *The Ships Inertial Navigation System Story*
(Anaheim, Calif.: Rockwell International, 1979), figure 16.

implicit, that was acceptable to all the interests centrally
involved. But by the same token it was not ideal from the point
of view of any one of them. Thus the expensive move to a
smaller number of somewhat higher-yield warheads was, from
the point of view of assured destruction, silly. On the other
hand, even though the C4 met its accuracy goal easily, its
counterforce capability, though not non-existent, was marginal:
a single C4 warhead, fired at 4,000 nautical mile range, would
have a probability of only about one in four of destroying a
Soviet silo hardened to 3,000 pounds per square inch.[106] In
Brett's words, the C4 "would have been good enough for the
silos back then [when it was being designed in the early 1970s],
but by the time it was deployed [1979] it wasn't good enough"—
the targets had become harder.[107]

106. See the mathematics in appendix B. The figure for the accuracy
achieved for C4 (750 feet or roughly 0.125 nautical miles) is taken from a
secret Department of Defense report to Congress, leaked to William Arkin.
See Arkin, "Sleight of Hand with Trident II," *Bulletin of the Atomic Scientists*,
Vol. 40, No. 11 (December 1984), 6.
107. Brett interview.

By 1979 the compromise that was Trident I (C4) was on its way to unravelling. Despite all their difficulties, the new, big submarines were on their way, and the C4 could thus look only like an interim missile. With the new missile, Trident II (D5), the traditional barriers to the prioritization of accuracy and counterforce capability in the Fleet Ballistic Missile finally ceased to be effective.

Trident D5—"Going for Broke"

There are two main markers of this decisive shift, although as we shall see, there are a host of minor ones too. The first is the new Trident D5 warhead, the W88 hydrogen bomb. The yield of the latter will be much bigger than those of its imme- diate predecessors (475 kilotons, as against 100 kilotons for the C4's W76 and 40 to 50 kilotons for Poseidon's W68). It is also be the largest warhead of any U.S. MIRVed missile, outstrip- ping in particular the 300 kilotons of MX's W87.[108]

For the first time, then, a Navy missile will carry larger warheads than the contemporary Air Force missile. This is no accident of technical design, but a deliberate decision on the allocation of a scarce resource: oralloy (Oak Ridge alloy, the original Manhattan Project name for highly enriched ura- nium). In the 1960s U.S. oralloy production facilities were closed down, the assumption being that oralloy needed for new warheads could be obtained from stocks or from old warheads being retired from service. The magnitude of the Reagan Administration nuclear plans undermined that assumption, and it was concluded that while the yield of either the MX or Trident D5 warheads could be increased from around 300 to 475 kilotons, supplies were insufficient to increase both.[109]

The priority given to the Navy warhead rankled with the Air Force: "Questions are being raised by the [Air Force] over why the Navy will be allowed to deploy the higher yield device

108. I am taking yield figures from Hansen, *U.S. Nuclear Weapons*, 203, 206.
109. On the oralloy problem, see Thomas B. Cochran, William M. Arkin, Robert S. Norris, and Milton M. Hoenig, *Nuclear Weapons Databook*, Vol. 2: *U.S. Nuclear Warhead Production* (Cambridge, Mass.: Ballinger, 1987), 78–86; on the upgradability of the MX warhead, see Thomas B. Cochran, William M. Arkin, and Milton M. Hoenig, *Nuclear Weapons Databook*, Vol. 1: *U.S. Nuclear Forces and Capabilities* (Cambridge, Mass.: Ballinger, 1984), 126.

requiring more oralloy in short supply in the inventory."[110] But given a choice between enlarging the counterforce capacities of MX and of Trident, the Department of Defense chose the latter. The probable reason will emerge below.

The second marker is the changed attitude to accuracy. No longer was accuracy to be a mere "goal" or "objective." With Trident D5 it became a "requirement," and very considerable and quite explicit effort has been made to fulfill it. The requirement is to make Trident D5 "at least twice as accurate as the Trident I [C4] . . . [and] in the same ballpark [as MX]."[111]

Again, this was not how Special Projects saw things at first. Their original view of an appropriate objective was "to achieve at 6000 nm the C[ircular] E[rror] P[robable] of POSEIDON at 2000 nm"—not an undemanding task, but not such a major undertaking as the above.[112] In the words of one person responsible for the latter, "we have . . . an [accuracy] objective, a requirement in this case . . . such that I'm doing just about everything I know how to do with that technology [stellar-inertial guidance]."[113] Furthermore, this requirement is now explicitly justified in terms of hard-target kill capability, while the original formulation pointed to a continuation of the way increased guidance and navigation performance in the C4 had been justified by "survivability."

Why was there a twin shift in warhead yield and attitudes to accuracy? The first factor to consider, one that bears especially on the question of warhead yield, is simply the growth in size made possible when the large Trident submarines finally became available in the 1980s. The possibility of a big, new missile certainly did away with the space constraints of the C4, leaving plenty of room for large warheads. But, as we have seen, those earlier space constraints were in a sense chosen— they were the result of the decision to design the C4 to fit in

110. Anonymous, "Navy to develop new Trident Warhead," *Aviation Week and Space Technology* (January 17, 1983), 26.

111. Testimony by Vice Admiral Glenwood Clark, then Director of the Strategic Systems Projects Office, to the Senate Armed Services Committee, *Department of Defense Authorization Hearings, Fiscal Year 1985* (Washington, D.C.: U.S. Government Printing Office, 1984), 3426, 3428 (I am grateful to Graham Spinardi for a copy of Admiral Clark's testimony).

112. Strategic Systems Program Office, *A Programmatic History of Trident II* (Washington, D.C.: Strategic Systems Program Office, November 1982), 2.

113. Interview with Special Projects officer.

existing launch tubes, a decision that in part reflected the fraught relationship between Special Projects and Admiral Rickover. And the greater throw-weight and volume permitted by the D5 did not dictate that big, heavy warheads be used.

Thus one possibility considered by Special Projects, and presented by them to Congress as an option, was to use the extra space simply to build a "long C4." This would have been a modest, but therefore relatively cheap, modification of the C4, essentially involving a longer and thus more powerful first-stage rocket motor.[114] The "long C4" would carry the same 100-kiloton Mk 4 reentry vehicles as the C4, but would go much further, 6,000 nautical miles rather than 4,000, thus increasing "sea room" and "survivability" even more. Accuracy improvements could have been justified in the same way as they had been for the C4: "to hold the same 750–ft. C[ircular] E[rror] P[robable] at the extended ranges."[115] The "long C4" would thus have remained within the assured destruction tradition of the Fleet Ballistic Missile.

True, the "long C4" would not have made use of the possibility of increasing the missile's diameter as well as length. But even with the decision to do this, to increase diameter from the C4's 74 inches to 83 inches for the D5, no particular warhead choice was compelled.[116] The very large 475-kiloton warhead appears to have been the result of Department of Defense pressure on the Navy, which would have been content to use a modified version of the Mk 12A reentry vehicle and 335-kiloton W78 warhead deployed on "improved" Minuteman III missiles. The Department, "added $88 million to the Navy's Fiscal 1984 budget request to develop a new ballistic reentry vehicle" for the D5.[117]

Even with the new heavy Mk 5 reentry vehicle definitely to be deployed, Special Projects succeeded in retaining the flexi-

114. Testimony of Rear Admiral Robert H. Wertheim, Director, Strategic Systems Programs, to Senate Armed Services Committee, *Department of Defense Authorization Hearings, Fiscal Year 1979, Part 9—Research and Development* (Washington, D.C.: U.S. Government Printing Office, 1978), 6681.
115. Anonymous, "Trident Missile Capabilities Advance," *Aviation Week and Space Technology* (June 16, 1980), 91.
116. The diameter figures are taken from Strategic Systems Project Office, *FBM Facts/Chronology—Polaris, Poseidon, Trident* (Washington, D.C.: Navy Department, 1982), 4.
117. Anonymous, "New Trident Warhead."

bility also to carry the C4's light Mk 4 on Trident D5, thus swallowing their earlier objections to a "mixed force." One reason was the desire to be able to accommodate any severe Congressional antagonism to the blatantly counterforce combination of high accuracy and large warhead by retreating on the latter. Another reason may be to preserve for the United Kingdom—also assumed to be ambivalent about counterforce—the capacity to combine the D5 missile with the older, lighter warhead.[118]

So the growth in missile size did not in itself dictate a counterforce missile. It did, however, reduce the possibility of arguing, as it had been possible to argue for the C4, that large-yield warheads were "physically impossible."

A second factor important to the shift was one that bears upon the changed attitude to accuracy. It was the weakening, for Trident D5, of the traditional argument against hard-and-fast accuracy requirements. That argument, put forward particularly by Special Projects Director Vice Admiral Levering Smith, was based upon technical uncertainty. The error processes in FBM guidance and navigation were not well enough understood, Smith argued, for firm commitments wisely to be entered into. Special Projects could know how accurate its missiles were in test situations, essentially by measuring how far from their targets they landed. But they lacked clear knowledge of the causes of such errors, so had no straightforward way of improving an inaccurate system. Nor, lacking knowledge of the processes leading to test-range performance, could they be completely confident in extrapolating from accuracy in testing to operational accuracy.[119]

Particularly impatient with this situation was James Schlesinger, who became the Nixon Administration's new Secretary of Defense in 1973. Schlesinger, like Robert McNamara in the previous decade, was an activist secretary. An economist by training, Schlesinger had headed the strategic studies division of the Rand Corporation, and there had come to favor "limited nuclear options"—relatively small-scale, selective nuclear targeting, designed to exert political leverage. This was also the main thrust of a Pentagon review of nuclear strategy that had

118. I am here drawing on an unpublished paper by Graham Spinardi on U.S. Fleet Ballistic Missile evolution.
119. Letter to author from Vice Admiral Levering Smith, October 13, 1986.

been conducted during 1972 and 1973, which Schlesinger then adopted and promoted. Known as National Security Decision Memorandum 242, and signed by President Nixon in January 1974, the resultant new policy marked a radical departure from the previous declared policy of assured destruction. As we saw in chapter 4, the 1960s Single Integrated Operational Plan for targeting nuclear forces provided for counterforce attacks, but National Security Decision Memorandum 242 went further in providing preplanned options for small-scale nuclear strikes against military targets.[120]

Although Schlesinger argued that the flexibility of National Security Decision Memorandum 242 could be achieved with the existing arsenal, he considered greater accuracy, and greater confidence in accuracy figures, to be desirable. He did not accept that submarine-launched ballistic missile error processes could not be explained. Admiral Smith recalls: "I remember a couple of sessions with him [Schlesinger] personally when I was trying to show that we were unable to explain the fall of shot. He rolled up his sleeves and said, 'OK, I'll explain it for you.' And we sat down with the raw data a couple of hours each time."[121]

From these discussions, and an earlier exchange with the Chief of Naval Operations, emerged the Improved Accuracy Program. While Special Projects avoided any strict requirement for accuracy improvement in Trident C4, then under development, it committed itself to this program, the goals of which were to identify the sources of inaccuracy in submarine-launched ballistic missile, to assess means of eliminating them, and to develop the most promising solutions.[122]

Although other possibilities (notably the use of the new satellite-based radio navigation Global Positioning System and homing reentry vehicles) were at least nominally considered, the Improved Accuracy Program concentrated on understanding and improving the stellar-inertial system being developed

---

120. Desmond Ball, *Déjà Vu: The Return to Counterforce in the Nixon Administration* (California Seminar on Arms Control and Foreign Policy, 1974); Schlesinger interview.
121. Smith interview.
122. Senate Armed Services Committee, *Fiscal Year 1977 Authorization for Military Procurement, Part 12—Research and Development,* (Washington, D.C.: Government Printing Office, 1976), 6640.

for C4. The program ran from 1974 to 1982, and cost around $600 million dollars.[123]

Though it did lead to "hardware" alterations that were adopted for Trident D5, its most important outcome was consensual knowledge. A central activity was constructing and "verifying" a mathematical model of error processes in submarine navigation and stellar-inertial missile guidance.[124] Accordingly, much of the Improved Accuracy Program expenditure was on greatly increasing the sophistication of the instrumentation monitoring Navy ballistic missile tests. Discussion of the overall significance of this must be postponed to chapter 7. But one important particular consequence was that the involvement of outside agencies in the construction of the error model gave senior management in Special Projects the "leverage" necessary to end the troublesome internal dispute between the guidance and navigation branches, SP-23 and SP-24, referred to above, concerning the Fleet Ballistic Missile "error budget" and its apportioning to navigation and guidance components. A Draper Laboratory team, under the leadership of David Hoag, "openly illuminated the total program in an even handed manner which was then resolvable by SP's [Special Projects'] management."[125]

The more general consequence of the Improved Accuracy Program, however, was to make "uncertainty" no longer a credible argument against taking on an increased accuracy specification, even as a definite "requirement." Simultaneously, too, the more implicit grounds for resistance—that it was not in the interests of the nation, nor in those of the Navy—were also undermined in the late 1970s and early 1980s. Opposition to building a counterforce Fleet Ballistic Missile persisted in Special Projects, but it was personal opposition, not organizational opposition, and ineffective.[126]

123. Interview with senior officer in Special Projects.
124. See Captain Robert L. Topping (U.S. Navy), "Submarine Launched Ballistic Missile Improved Accuracy," paper AIAA-81–0935 presented to American Institute of Aeronautics and Astronautics, Annual Meeting and Technical Display, May 12–14, 1981, Long Beach, Calif.
125. Letter to author from Robert Duffy, July 6, 1987.
126. One member of Special Projects told me in interview that he had advised opponents of counterforce in the Office "for their own sake to get out." No one was fired, but Spinardi's interviews confirm that resignations took place over the issue. In general, both serving and retired members of Special

The "interests of the nation" and "interests of the Navy" questions were interrelated. In general, of course, the overall strategic climate in the United States shifted towards counterforce during the later 1970s and 1980s. Arms control agreements and détente with the Soviet Union seemed to many to have produced a world in which American power simply eroded, both in the face of Third World revolution and vis-à-vis the Soviet Union.[127] The "new right" gradually grew in strength and associated the strategy of assured destruction with this decline in American power.

In a series of steps, American stated nuclear policy moved away from assured destruction. Schlesinger's 1974 National Security Decision Memorandum 242 was the first clear one. The election of Jimmy Carter to the presidency in 1976 seemed initially a step in reverse, as the former nuclear submariner began by affirming a position close to the old Navy "finite deterrence" strategy. But faced with Third World setbacks and the perception of an increasing Soviet threat, Carter came under increasing pressure as a result of the image of "softness" that was being exploited by his Republican challenger, Ronald Reagan. Carter's summer 1980 Presidential Directive 59 was leaked to the press to counter this image (and possibly written in part to counter it). It was a "strategy for nuclear war [which] gives priority to attacking military targets in the Soviet Union rather than destroying cities and industrial complexes."[128] Finally, of course, the election later that year of the Reagan Administration brought with it, initially, a strongly "war-fighting" nuclear rhetoric.[129]

Insiders could of course afford to view this process with some skepticism. Changes in "stated posture," as we saw in chapter 4, did not automatically translate into changes in the practice of nuclear targeting, much less changes in weapons acquisition policy. Nevertheless, they did strengthen the hand of the pro-

Projects that I interviewed did not express the unalloyed enthusiasm for counterforce I tended to find in Air Force interviewees.

127. See, for example, Fred Halliday, *The Making of the Second Cold War* (London: Verso, 1983).

128. Quoted from one of the newspaper articles describing it in Fred Kaplan's useful analysis of the episode: "Going Native without a Field Map: The Press Plunges into Limited Nuclear War," *Columbia Journalism Review* (January/February 1981), 24.

129. See Scheer, *With Enough Shovels*.

ponents of counterforce. One of them was Seymour Zeiberg, Carter's Undersecretary of Defense for Strategic and Space Systems, and a man pressing for as much accuracy as possible from the D5. As he put it, with reference to Carter's Presidential Directive 59; "With the PD-59 story having been pulled together that provided everybody with the theological framework around which you could rally and say: 'we need it [counterforce] because of the following. It's national policy, it's no longer intuitive judgment and whimsy.'"[130]

These high-level policy changes did have direct effects on the Fleet Ballistic Missile program. We have examined what is probably the most immediate one: the role of Secretary Schlesinger in the creation of the Improved Accuracy Program. But, at least in the U.S. political system, higher decision makers like the Secretary of Defense in practice lack the capacity simply to command the armed services in matters like this. The term chosen independently by Admiral Smith and Secretary Schlesinger to describe their interaction over accuracy is symptomatic: to "push." Schlesinger "just kept pushing for improved accuracy," said Smith. "We had to push the Navy at that time," said Schlesinger.[131]

"Pushing" can of course be resisted, and the example of the Air Force, discussed in chapter 4, shows how an armed service could for years hold to a doctrine at odds with stated national strategy. Nor was the shift towards counterforce uncontested. Congress was far from fully persuaded, and during the 1980s the influence of a burgeoning peace movement particularly concerned with isssues of counterforce and "first strike" grew.[132] President Reagan had to distance himself publicly from the "nuclear warfighting" statements of members of his administration.

So the change in overall strategic climate was not, on its own, sufficient to explain the shift in priorities manifest in the design of the D5, and particularly not sufficient to explain why a counterforce D5 proved relatively uncontroversial. The final

130. I am quoting from the interview with him by Graham Spinardi.
131. Smith interview; Schlesinger interview.
132. A person who played a major role in the growth of this concern was a former Lockheed engineer who had worked on Fleet Ballistic Missile reentry vehicles, Robert C. Aldridge. See his *First Strike! The Pentagon's Strategy for Nuclear War* (London: Pluto, 1983).

factor we need to examine, therefore, is the changed relationship between Navy and Air Force strategic programs.

The strategy of differentiating Navy programs from Air Force ones by emphasizing their countercity retaliatory role, and disavowing the pursuit of counterforce, emerged at a time when the Air Force dominated the strategic mission. But, gradually, the relative strength and confidence of the Navy grew, in part because of the troubles of the Air Force programs discussed in chapter 4. In the mid 1970s the Navy's budget for strategic program acquisition costs surpassed that of the Air Force, and in the late 1970s, with the B-1 bomber canceled and MX floundering, the Navy budget was more than twice as big.[133]

"Differentiation," seen as unnecessary by the Navy's "Great Circle" group even in the mid-1960s, became even less so in this changed situation: if any leg of the "triad" of submarine-launched ballistic missile, bomber, and ICBM was under threat, it certainly was not the submarine-launched ballistic missile. Though Navy representatives were still careful not to present Trident D5 as a complete alternative to MX,[134] the hard-target role began openly to be claimed and the two missiles directly compared on that criterion. Testifying before the House of Representatives' Appropriations Committee in 1983, Chief of Naval Operations Admiral James Watkins spoke in terms inconceivable in the 1960s. "By 1991, we believe you could have four to five D5 equipped Trident submarines, which is more than the equivalent of an MX field in terms of hard target kill capability."[135]

More was involved, of course, than a growth in organizational confidence. During the 1960s and early 1970s, differentiation between the Navy's Fleet Ballistic Missile and Air Force ICBMs did not simply suit the two services. It also allowed practical compromise between the two competing strategies of counterforce and assured destruction. Even an advocate of counterforce such as Director of Defense Research and

133. Pauli Olavi Järvenpää, *Weapons Development and Arms Control: A Case Study of the Flexible Options Doctrine* (Ph.D. dissertation, Cornell University, 1978), table 8, p. 164.
134. The problem of rapid communications with submerged submarines remained one barrier.
135. Quoted in Arkin, "Sleight of Hand," 5.

Engineering John Foster could be persuaded against counter-force capability in the Fleet Ballistic Missile. That capability existed and could be fostered in Air Force ICBMs.

The troubles of MX, however, meant that this was no longer certain. Advocates of increasing U.S. hard-target kill capability realized the importance of Trident II as a hedge against non-deployment of MX. Seymour Zeiberg, who as noted above was pressing for greater accuracy in the D5, spelled out the relationship:

> If we move out with a vigorous MPS [Multiple Protective Shelter basing mode for MX] program and we buy a new strategic capability which has high accuracy and has the potential to cope with counter-force missions, certainly the urgency to move out with the Trident II for that reason diminishes. . . . If we don't have an accelerated MX program of that sort, we would endorse the very accelerated Trident II program.[136]

Congressional "doves," on the other hand, saw MX as the main enemy. Although many opposed it because of its increased counterforce capability, they made a tactical decision to fight it on the basing issue, where opposition was greatest. But Trident's basing mode was unproblematically invulnerable, or so it was widely believed. Because of this, and the Fleet Ballistic Missile's enduring image as a retaliatory deterrent, many defense liberals saw Trident as the lesser of two evils at a time when it was politically difficult to oppose both outright. Even for them, therefore, counterforce capability in the D5 might be no bad thing. If Trident could "do anything MX could do," then the case for the vulnerable MX was further undercut.

The relationship between Trident D5 and MX, therefore, built support for a hard-target D5 both within the Navy and outside it, and meant that counterforce capability in the D5 was positively welcomed by "hawks," while generally not opposed, at least not too vigorously, by Congressional defense liberals.[137] This remarkable consensus over the fundamental

136. Quoted in Joel S. Wit, "American SLBM: Counterforce Options and Strategic Implications," *Survival*, Vol. 24 (1982), 168.
137. The main opponent has been in the House, not the Senate: Representative Edward Markey.

strategic shift represented by the D5 was thus the most important factor leading to the program's official noncontroversial status referred to at the start of this chapter.

The detailed design of the D5's Mk 6 guidance system, and associated navigation systems, reflects the high priority given to counterforce accuracy, the by now established status in the United States of unistar stellar-inertial guidance, and the detailed outcome of the Improved Accuracy Program. Perhaps the single most telling shift is that after the simpler and cheaper PIPA accelerometers had become accepted for Polaris A3, Poseidon, and Trident C4, the PIGA returned with the D5. "They went from 3 PIGAs to 1 PIGA to 0 PIGAs as they went through the early generations, and then of course, now they decided, 'Hey, we're going to go for broke,' and now they're back talking PIGAs again."[138] Although a Kearfott-designed vibrating beam accelerometer was also considered, a small Draper-designed PIGA (the 10-PIGA, with its one-inch diameter gyro) was selected as the D5 accelerometer.[139]

The gyroscope remained a Kearfott tuned-rotor design, with improvements on the version used on the C4, but a major change was made in the stellar sensor. Here trade-offs continued longest into Trident II development. Some argued that the "television-camera" vidicon technology used on the C4 was obsolete and that better performance could be achieved by moving to a solid-state sensor. Others felt these new technologies too immature for incorporation into the baseline system, and one interviewee felt that the reason for shifting away from the vidicon, where Kearfott had formidable expertise, was that it would allow the Draper Laboratory to regain design authority lost with the decision to move from inertial to stellar-inertial.[140] In the end a solid-state "charge-coupled device" was selected as the sensor, but its development proved troublesome.[141]

138. Interview with Draper engineer.
139. See anonymous, "Trident Capabilities Advance," 99, for the accelerometer choice.
140. Brett interview.
141. Interview data. For a description of the charge-coupled device, and a discussion of its introduction in another area, see Robert W. Smith and Joseph N. Tatarewicz, "Replacing a Telescope: The Large Space Telescope and CCDs," *Proceedings of the IEEE*, Vol. 73 (1985), 1221–1135.

Guidance changes alone, were, however, understood as not enough to meet the accuracy requirement without improvements in navigation. With the stellar sensor believed capable of correcting for errors in initial position and orientation in the horizontal plane, attention switched to other aspects of launch condition. As we saw, errors in knowledge of initial velocity came to be agreed not correctable by star sighting, and so measuring the submarine's velocity was seen as critical. A Doppler sonar system was chosen to measure velocity from ocean bottom reflections. It will "be operated only moments before missile launching and [will] provide a very accurate initial velocity determination for the guidance initializations."[142]

The increased accuracy demanded of Trident D5 brought back into prominence, at least amongst "insiders," the "problem of the vertical." Anomalies in the gravity field (caused for example by underwater mountains) are believed to lead to what are at this level of accuracy significant errors both in knowledge of submarine velocity and of the local vertical. Like errors in the former, errors in the latter are held to be uncorrectable by the star-sighting.[143]

Two means of solving the revived "problem of the vertical" were pursued. One was an extensive program of mapping the gravity field, using both gravity measurements from surface ships and satellite altimetry. In the latter, signals from satellites are "bounced" off the surface of the sea to determine its exact shape, and hence to discover the precise geometry of the gravity field's "equipotential surface." The 1978 Seasat satellite was used for this purpose, and though its data was expected to contribute to the Trident system, it was found to be insufficient. A new Geosat satellite was developed for 1983 launching with Trident program funding. "The Navy believes the improved Earth gravity models expected from the Geosat spacecraft will provide up to a 10% improvement in circular error target accuracy for certain Trident 2 launch areas. The Geosat data will be most useful for Trident submarine patrol areas in the

142. Topping, "Improved Accuracy," 7.
143. See the entry on "Underwater Navigation," by T. A. King and H. Strell, in the *McGraw-Hill Encyclopedia of Science and Technology*. This is a well-informed piece; King, as noted above, was the leading technical figure in SP-24.

southern hemisphere and parts of the Northern Pacific where gravitational survey data are limited."[144]

The satellite method has limitations, however. Oceanographic effects—variations in the salinity and temperature of the sea, and the fact that its surface is not "flat," but disturbed by ocean currents—limit what it is believed can be learnt from it.[145] Survey ships proceed expensively and slowly, and though the "politicization of the gravity field" that has taken place as a result of the increased military significance of gravitational data has not to date spread from land to sea to delay the process, Special Projects felt it impossible within a reasonable timespan and budget to survey all potential patrol areas for a missile with the physical range of the D5. Accordingly, although the D5 is capable of considerably longer range than the C4, the former's accuracy specification was set for the same nominal range of 4,000 nautical miles.[146]

The other approach taken to the "problem of the vertical" was to black box it: to provide the submarine with an internal means of measuring the gravity field. The emergence of the problem with the D5 was no surprise: it had been argued for some time that, in the words of Milton Trageser of the Draper Laboratory, "gravity system disturbances [rather than inaccuracies in gyroscopes or accelerometers] would soon limit inertial navigation system performance."[147] Again in prima facie violation of the general theory of relativity, a device had been developed for black-box measurement of the gravity field even from a possibly accelerating base. Some of the interest in it arose because of the navigational problems involved in one of the MX basing proposals, to launch the missile from aircraft.

The key to the device was that it was not a gravity meter but a gravity gradiometer. If acceleration was measured at two points in the same "stabilized coordinate frame" (say a gyroscopically stabilized platform in an aircraft or submarine), then it was in principle possible to measure changes in the gravity

144. Anonymous, "Geosat Data to Aid Trident 2 Accuracy," *Aviation Week and Space Technology* (July 19, 1982), 26.
145. Waves are not seen as a problem, since averaging techniques remove their effects. I am grateful to Dr. Roger Hipkin of the Department of Geophysics, University of Edinburgh, for discussion of these points.
146. Data from Spinardi's interviews.
147. Milton B. Trageser, "Floated Gravity Gradiometer," *IEEE Transactions on Aerospace and Electronic Systems*, Vol. 20 (1984), 417.

field between the points—that is, to calculate the gravity gradient. Since the "linear inertial acceleration" at the two points would be the same, the difference in output between sufficiently accurate accelerometers at the two points would yield the gravity gradient.[148]

So this remarkable new device, the gravity gradiometer, was to be added to Trident's suite of navigational equipment, and supplemented by a more orthodox gravity meter.[149] In 1988, however, the combination of gradiometer and gravity meter was canceled. Poor performance, good quality in the gravity maps obtained by ship and satellite, and better-than-specification accuracy in other system components were cited.[150] Again, Special Projects was being ruthless with a "sweet technology" that it felt was not contributing to program goals.

Other changes in Trident navigation have been less dramatic. The traditional SINS is finally being done away with, and the electrostatic gyro system will cease to be a mere monitor and instead will be the primary inertial navigator. Receivers for the traditional external navigation updates, Loran-C and Transit, have been retained, and a receiver for signals from the new Global Positioning System will be added "after GPS has demonstrated continous, worldwide capability equal to or better than Transit."[151]

This expensive, complex interconnected system of guidance and navigation technologies is, as we have seen, designed to ensure unprecedent accuracy in the Fleet Ballistic Missile. As with MX's all-inertial beryllium baby, to at least some of those involved, this system also may represent an end point. One interviewee, though admitting to being "ambivalent" about counterforce capability, saw the D5's Mk 6 guidance system as the "crown jewel" of his career, the "last word, in the sense of

148. Ibid., 419
149. The gradiometer selected was one developed by Bell Aerospace Textron at Buffalo. The competing devices are described in M. A. Gerber, "Gravity Gradiometer: Something new in Inertial Navigation," *Astronautics and Aeronautics* (May 1978), 18–26.
150. General Accounting Office, *Trident II Acquisition* (Washington, D.C.: General Accounting Office, 1988; GAO/NSLAD-89–40), 20. I owe the reference to Graham Spinardi.
151. Anonymous, "Navstar offers 'Little Improvement' in SLBM Accuracy," *Aerospace Daily* (October 19, 1983), 257.

the ultimate in accuracy."[152] Another saw no obvious way to go further without a radical break in technologies:

DMcK: So . . . if somebody came along and said, "Well, here's 50% more money" . . . ?
Interviewee: With the basic technology right now, I'm not sure I'd know what else to do.[153]

End point or not, with MX deployed in limited numbers and the Small ICBM program in deep trouble, Trident D5 will be the centerpiece of U.S. counterforce capability in the 1990s and perhaps beyond, barring quite radical disarmament measures. The Fleet Ballistic Missile has indeed been transformed.

Conclusion

The transformation of the Fleet Ballistic Missile was a labyrinthine process. The central thread of stellar-inertial guidance is clear enough, but it was tugged this way and that by a range of priorities and actors: different views of national nuclear strategy, different views of the appropriate strategy of the Navy with respect to the Air Force, the conflicting attitudes of different organizations to the opening of the black box of inertial guidance represented by the star sight, and the difficulty of negotiating a technology that disturbed an established organizational division of labor.

Within the labyrinth, technology was politics pursued by other, harder, means, and was understood to be so by insiders.[154] The moves sometimes resemble chess in their cunning: support stellar-inertial guidance so as to destroy the chances of a reentry vehicle one does not like, then withdraw support for it because it does not wholly fit one's interests, and so on. I impute no insincerity. One of the most remarkable aspects of the tale of the transformation of the Fleet Ballistic Missile is the tenacity with which an austere vision of its role in national strategy was preserved in the face of both political pressure and the lure of "sweet technology." But those who held to this

152. Interview.
153. Interview.
154. On this general point, the work of Bruno Latour is particularly enlightening. See, for example, Latour, "The Prince for Machines as well as for Machinations," in Brian Elliott, ed., *Technology and Social Process* (Edinburgh: Edinburgh University Press, 1988), 20–43.

vision knew that the most effective way of maintaining it was to imprint it on the system's metal. Those who opposed them also discovered that the formal right of command, which they often possessed, meant little unless they found means of etching their preferences into the technology.

From outside the labyrinth, little of this could be seen. We ended the last chapter with one missile, the Small ICBM, having been in effect designed within Congress, while the political opponents of the other, MX, were able to contest even the details of what was, viewed from a different perspective, a remarkably smooth and successful test program. We end this chapter quite differently, with a program seen as uncontroversial and with a missile that insiders saw as transformed seen in public as a simple "modernization." Even the most dramatic test failure imaginable, the televised cartwheel of the first underwater launch of Trident D5 on March 21, 1989, proved survivable, even though that and subsequent failures brought an opening of the Trident program's black box unprecedently close.[155]

Within the Fleet Ballistic Missile program, in other words, a sphere of technology insulated from politics of the formal, public kind had been created. Given the degree of insider politics within the sphere, and the potential for the two kinds of politics to link up, as they did over Poseidon stellar-inertial guidance, that was a remarkable achievement in boundary maintenance. It was assisted by the breadth of support in Congress and elsewhere for the Fleet Ballistic Missile, support that was generated at the time of Polaris. It was also assisted by the sheer fact that the Fleet Ballistic Missile was a sea-based system. Land-basing, especially if it involves movement outside military reservations, always threatens to raise the visibility of a missile, literally and figuratively. "Out at sea, doesn't bother me," is

155. See Edward H. Kolcum, "Navy Assesses Failure of First Trident 2 Underwater Launch," *Aviation Week and Space Technology* (March 27, 1989), 18–19. The third underwater test launch on August 15, 1989, was also a widely publicized failure, and by then doubts about the program's black box were becoming more serious: John D. Morrocco, "Second Trident 2 Test Failure points to Missile Design Flaw," ibid., August 21, 1989, 26. Even then these doubts did not seem to resonate with wider opposition to Trident D5, as they would have if similar mishaps had occurred in the MX test program, and early in 1990 it appeared that the problems had been solved without lasting damage to the Trident program.

how insiders, with good reason, sum up the public relations advantages of sea-basing.

The insulation of the Fleet Ballistic Missile was, however, also the result of the skill of its managers in the Special Projects Office. As we have seen, they trod a difficult tightrope with care. One risk was to bend too much in response to the pressures on them, taking on requirements that it turned out could not be met, or shaping a system to satisfy powerful insiders that would be too much at variance with Congress's view of a deterrent. The opposite risk was to bend too little and to insist on their visions and preferences when circumstances made this strategy untenable.

By and large, they have succeeded in this balancing act. The result is a paradox. To the outsider, Fleet Ballistic Missile development seems like a sphere of rational technical decision making, uninfluenced by politics. That appearance, however, is the result not of the autonomy of technology, but of its careful shaping so as to maintain that very separation of technology and politics.

The Soviet Union and Strategic Missile Guidance

If technological development had a logic of its own, then one would expect it to follow the same course independent of circumstances such as the country in which the development took place. If, on the other hand, there is no such inherent logic to technological change, then national circumstances might shape technological development. I write "might" shape it, rather than "will" shape it, because a host of factors other than an inherent logic of technology could lead to similar development in different countries. If one country is in the lead in a particular technology, others may take what is done in that country as their model. Cross-national institutions such as the world market, multinational corporations, and technological communities may be more important in shaping a given technology than local circumstances. Different countries may turn out to have similar priorities and similar organizational structures for technological development.

How different has strategic missile guidance development been in different countries? Sensible comparison is possible between only four cases: the United States, the Soviet Union, France, and the People's Republic of China. The German developments discussed in chapter 2 were so much earlier than those in other countries that the factor of time distorts any comparison. Indigenous British strategic ballistic missile guidance development ceased without any system being deployed. Guidance systems—indeed entire missiles, apart from the nuclear warheads—together with the associated submarine navigation and fire control systems are purchased by the United Kingdom from the United States, with no British influence on guidance design that I have been able to detect. Information is not yet available on the guidance systems of the growing number of Third World ballistic missile projects.

The cases of France and China can be summarized briefly. The difference between developments in those countries and that in the United States is stark. In neither France nor China has counterforce accuracy been pursued. Strategic doctrine in each country has emphasized countercity deterrence, resources have been limited, and the relatively small size of their ballistic missile forces has made a counterforce attack by them on a superpower opponent an unconvincing scenario. French missile guidance work initially relied heavily on technology from the U.S. firm Kearfott,[1] though a sophisticated gyro culture firm existed that was able to absorb this technology and make it its own: SAGEM (la Société d'Applications Générales d'Electricité et de Méchanique). Covert collaboration between the United States and France seems to have continued even after the original link was publicly broken. Guidance improvements to the French ballistic missile force have meant increased accuracy and these have proceded in tandem with some refinement of French nuclear targeting, but all within a countercity framework in which the pursuit of accuracy has not been a major priority.[2]

The development of Chinese missile guidance technology took place without a gyro culture tradition analogous to that of France, and, it appears, also with considerably less help from abroad. Although details are sparse, early Chinese nuclear missiles seem to have been equipped with guidance systems analogous to that of the V-2, and only at the end of the 1970s did stable platform designs appear.[3] As with France, strikes against cities, and perhaps large "soft" military targets, seem to have

1. Interview data.
2. Interview data and David S. Yost, *France's Deterrent Posture and Security in Europe.* Part I: *Capabilities and Doctrine,* Adelphi Paper No. 194 (London: International Institute for Strategic Studies, 1984/85), 13–29. General Pierre Gallois, a central figure in the establishment of the French nuclear force and a leading French strategic theorist, became a proponent, after his retirement, of counterforce, but to my knowledge his personal conversion was not widely shared by his serving colleagues. His advocacy seems to have been of a counterforce arsenal for NATO, including the United States, not France in particular. See Pierre Gallois and John Train, "When a Nuclear Strike is Thinkable," *The Wall Street Journal,* March 22, 1984, editorial page.
3. See the remarkable study by John Wilson Lewis and Xue Litai, *China Builds the Bomb* (Stanford, Calif.: Stanford University Press, 1988), 214.

been anticipated. One source remarks explicitly: "It is unnecessary for us to achieve tremendous accuracy. If a nuclear war breaks out between China and the Soviet Union, I don't think there is too much difference between the results, provided China's ICBM misses its predetermined target, the Kremlin, and instead hits the Bolshoi Theater."[4]

By far the most interesting comparison, of course, is between U.S. and Soviet guidance technology. The interest lies not just in the crossnational comparison of technological development, but in what Soviet guidance can tell us about Soviet nuclear strategy and military priorities. Often, the Soviet Union has been seen as having been committed to building the capability to fight and win a nuclear war, a strategy contrasted with the United States's adherence to deterrence by the threat of assured, but mutual, destruction. As we have seen in the previous three chapters, the latter image is too simple by far; but what of the former? The priority awarded accuracy in Soviet missile guidance system design may indicate something of the role of first strike in Soviet nuclear strategy, at least prior to the fundamental rethinking undertaken during the Gorbachev era.

Another issue concerning Soviet guidance technology is the relative weight of different sources of it. A persistent Western suspicion has been that much Soviet technological advance, especially in key military technologies, has been the result of smuggled or stolen Western artifacts and knowledge. How true is this for guidance? Have Soviets developments been generated indigenously, or has legal or illegal technology transfer from the West been crucial?

How We Know about Soviet Guidance

The reader might be beginning to suspect that these questions are unanswerable. The internal workings of Soviet guidance systems are hidden from even the most high resolution photoreconnaissance satellite, and to my knowledge no demonstrably reliable high-level human intelligence has been received from an agent or émigré privy to the technical secrets of Soviet guidance.[5] How, then, can anyone in the West know anything of Soviet guidance technology?

4. Zhang Aiping, quoted in ibid.
5. In Peter Wright's celebrated *Spycatcher: The Candid Autobiography of a Senior*

The key source of information is telemetry, a by-product of the process of Soviet missile development and testing. For a missile test to be of maximum use to the missile's designers and its future users, as much information as possible needs to be gathered on the in-flight performance of key components such as the guidance system. A continuous record of the performance of these components is therefore transmitted from the missile to the ground. From relatively early on in the Soviet missile program, this telemetry has been intercepted more-or-less comprehensively by Western intelligence.[6] Telemetry can be coded before transmission, but the Soviet Union has started to do this only relatively recently, and even now this encryption does not completely prevent the making of inferences.[7]

The various radio signals constituting the telemetry having been intercepted, the next task obviously is to discover what each "channel" of telemetry represents. In early years this posed a fascinating puzzle to analysts fresh to intelligence. The job demanded direct technical experience of American missile programs, so the necessary expertise initially was not to be found within the existing intelligence service. Gradually a picture was built up, partly by the recognition of patterns similar to those in the telemetry from American tests and partly through the use of physics and engineering knowledge.

Intelligence Officer (New York: Viking, 1987), he discusses the intelligence received from the British agent Oleg Penkovsky and two later defectors "Top Hat" and "Fedora." Fedora in particular gave Western intelligence detailed information about guidance, but this information, says Wright (p. 211) later came to be regarded as deliberately misleading. Wright also suggests that the Soviets "doctored" test telemetry, and may have "introduced a fake third gyro on their missiles to make them appear less accurate than they in fact were" (ibid.). But no direct evidence for these latter claims is cited, and the passage leaves one in doubt as to Wright's grasp of the technical issues involved.

6. As early as 1948 a team of British intelligence officers, posing as archaeologists in Iran, reportedly intercepted telemetry from Soviet testing of V-2s. See John Prados, *The Soviet Estimate: U.S. Intelligence Analysis and Soviet Strategic Forces* (Princeton: Princeton University Press, 1986), 57.

7. In recent years, however, the United States has charged that Soviet encryption of telemetry has increased to the point where it impedes verification of Soviet compliance with the unratified SALT II treaty. For a discussion of this issue, see James A. Schear, "Arms Control Treaty Compliance: Buildup to a Breakdown?" *International Security*, Vol. 10, No. 2 (Fall 1985), 162–164.

Particularly important is telemetry of on-board computation. This reveals the nature of the guidance formulation, the "mathematics" of the guidance system, which, in turn, allows inferences about the nature of the technology being used. If stellar-inertial guidance is being employed, for example, an on-board calculation has to be made to turn the information from the star fix or fixes into the requisite guidance correction. In modern guidance systems the on-board computer is programmed to correct for predictable instrument errors, such as drift in the gyroscopes or bias in the accelerometers, and if telemetry representing this can be intercepted, then something of the nature of Soviet beliefs about the errors of their guidance systems can be recovered.

Identifying and interpreting telemetry channels are not simple matters. But although the analysts' experience and intuition are indispensable, their inferences can be checked, both against the radar record of actual missile trajectories (and sometimes satellite photographs of impact craters), and in the internal debates of the intelligence community. The question of encryption aside, experienced analysts reckon that 95 to 98 percent of the telemetry channels from Soviet missile testing can be identified with confidence.

Of course, this does not answer all questions. It does not suffice for production of an unequivocal accuracy figure for a Soviet missile of a given type, though it is at least as important an input into that process as identifying final impact points. Nor does telemetry permit analysts to see as far into the black box as they might like. The output of the accelerometers appears directly in the telemetry, allowing what are taken to be reasonably secure conclusions as to their nature. Gyroscope output, on the other hand, does not, so inference has to be indirect.

The records of intercepted telemetry, radar tracking, and impact point analysis naturally remain classified. So there is a double indirectness in our knowledge of Soviet missile guidance. Nevertheless, over the years some of the conclusions reached by the intelligence community about Soviet missile guidance—chiefly estimates of the accuracy of Soviet missiles, but also very occasional comments about the nature of Soviet guidance technology—have appeared in the specialist press and literature, most frequently *Aviation Week and Space Tech-*

nology.[8] In addition, some early accuracy estimates have now been declassified.[9]

Soviet technical writings should not wholly be dismissed as a source of information, even though inference from them to design details of deployed systems is normally (and, obviously, deliberately) impossible. As we have seen, ballistic missile guidance is part of a wider field of guidance and navigation technology that has space, submarine, military and civil aircraft, and land as well as missile applications. The history of this overall field in the Soviet Union, in at least its theoretical aspects, can to some extent be traced from the technical literature.[10]

Aside from these published sources, in researching Soviet missile guidance, and, more generally, the history of inertial navigation in the Soviet Union, I was able to make use of interview data. Only one interview was conducted in the Soviet Union, but others included people who had some direct contact with missile guidance and inertial navigation in the Soviet Union.[11] Most helpful of all in regard to missile guidance were discussions I was able to have with members of the U.S. in-

8. See, especially, Anonymous, "Soviets' Nuclear Arsenal Continues to Proliferate," *Aviation Week and Space Technology* (June 16, 1980), 67–76, a remarkable compendium of Soviet missile accuracies and yields. The correspondence of the figures in this article to more recently declassified official estimates strongly suggests an authoritative leak, possibly from someone with access to Defense Intelligence Agency (DIA) estimates, given the closeness of the article's estimate of the CEP of the SS-19 (see appendix A) to that of the DIA.

9. See Barton Wright, *World Weapon Database*, Vol. 1: *Soviet Missiles* (Lexington, Mass.: Lexington Books, 1986). This is a most useful compilation, which seeks to reproduce and compare the full range of Western published estimates of the technical characteristics of Soviet missiles.

10. There is a helpful historical survey of this literature in V. D. Andreev, I. D. Blyumin, E. A. Devyanin, and D. M. Klimov, "Obzor Razvitiya Teorii Giroskopicheskikh i Inertsial'nykh Navigatsionnykh Sistem," (Survey of the Development of the Theory of Gyroscopic and Inertial Navigation Systems) in *Razvitie Mekhaniki Giroskopicheskikh i Inertsial'nykh Sistem* (The Development of the Mechanics of Gyroscopic and Inertial Systems) (Moscow: Nauka, 1973), 33–72.

11. These were an émigré computer engineer, a German gyroscope specialist who worked in the Soviet Union after 1945 (interviewed by my colleague, Dr. Wolfgang Rüdig), and several American specialists who were in contact with their Soviet counterparts.

telligence community, especially guidance system analysts. Though they could not stray onto classified terrain, these discussions considerably supplemented what I was able to glean from the published record and other sources.[12]

All this evidence must be viewed with caution. Western intelligence does not know as much as it would like about Soviet guidance.[13] The intelligence community is also notoriously divided in the conclusions it draws from the available evidence, so in interviewing or use of published "leaks" one must take into account the source's likely position in these divides.[14] In what follows, therefore, the methodological issues referred to in chapter 1 are compounded by a whole extra layer of indirectness in much of the data I am drawing on. That layer— the processes by which the U.S. intelligence community turns intercepted radio signals into knowledge of Soviet guidance systems—is worthy of a sociology of knowledge study in its own right. I discuss this briefly in the following chapter. Here, though, I shall concentrate on what we can learn about the development of Soviet guidance technology and about the contrasts with and similarities to what happened in the United States.

The German Inheritance

Like the Americans, the Soviets seem to have learned a lot from pre-1945 German work. The exemplar of the V-2 was central to post-1945 Soviet rocketry, perhaps even more than in the United States, where more "aviation-minded" engineers applied ideas drawn from airframe design to the Atlas ICBM.[15]

With no powerful tradition of manned strategic bombers to compete with, ballistic missile development received much

12. The latter interviews were conducted with the promise of anonymity. Where no reference is given for claims about Soviet guidance, these interviews are the source.

13. One intelligence analyst told me of a repeated dream he had, in which he was shown around the Soviet missile test center at Tyuratam and allowed to look at whatever he wanted. Alas, in the morning he could never quite remember what he had seen!

14. Prados, *The Soviet Estimate,* and Lawrence Freedman, *US Intelligence and the Soviet Strategic Threat* (London: Macmillan, 1977).

15. See Edmund Beard, *Developing the ICBM: A Study in Bureaucratic Politics* (New York: Columbia University Press, 1976).

higher priority in the Soviet Union than in the United States.[16] Perhaps, too, the fact that the Soviet Union had potential enemies closer to hand than did the United States eased acceptance of the ballistic missile. A missile could have much less than intercontinental range, and thus be a less daunting project, while still counting as a "strategic" weapon.

While the Americans obtained the bulk of the technical expertise from the V-2 program, a certain number, led by Helmut Gröttrup, Assistant to the Director of the Guidance, Control and Telemetry Laboratory, chose to work for the Soviets, apparently for personal and career reasons rather than out of ideological preference. They reopened the V-2 production line, and by autumn 1946, when they were forcibly transferred to the Soviet Union, had produced some 30 V-2s.[17]

Although aside from Gröttrup this group contained several guidance specialists and gyroscope expert Kurt Magnus, it probably would have been insufficient to sustain a major guidance development effort, even if the Soviets had trusted it more. But this group was not the only source of Soviet access to German navigation and gyroscope technology.

Also of long-term significance were the links between the Soviet Union and the Navy gyro culture firm Kreiselgeräte. As we saw in chapter 2, Kreiselgeräte had lent its expertise to the V-2 program, and its technical director, Johannes Maria Boykow, had been an early enthusiast for the idea of fully self-contained inertial navigation. Under Boykow's successor, Johannes Gievers, inertial navigation was put on the back burner while sophisticated gyro culture work continued.

Most significant in the present context was Kreiselgeräte's systematic effort to improve the accuracy of gyroscopes and accelerometers. A major source of error was believed to be bearing friction. A gyroscope or gyroscopic accelerometer

16. For the history of the Soviet ICBM program, see David Holloway, "Military Technology," in Ronald Amann, Julian Cooper, and R. W. Davies, eds., *The Technological Level of Soviet Industry* (New Haven, Conn.: Yale University Press, 1977), 407–498; David Holloway, "Innovation in the Defense Sector: Battle Tanks and ICBMs," in Ronald Amann and Julian Cooper, *Industrial Innovation in the Soviet Union* (New Haven, Conn.: Yale University Press, 1982), 368–414; and David Holloway, *The Soviet Union and the Arms Race* (New Haven, Conn: Yale University Press, 1983), 150–154.

17. Frederick I. Ordway III and Mitchell R. Sharpe, *The Rocket Team* (Cambridge, Mass.: MIT Press, 1982), 318–319, 321.

could not precess smoothly in response to rotation or acceleration unless it could do so on a frictionless bearing. Gievers, along with other Kreiselgeräte specialists F. Mueller and H. Rothe, came to the conclusion that the way to solve this problem was to replace the mechanical bearings on which the gyroscope precessed by a film of gas supplied by an external gas-bottle or pump.[18] By the end of the war Kreiselgeräte was working on gas-supported gyroscopes to improve V-2 accuracy, and had built a miniature submarine gyrocompass, designed by Gievers, incorporating a gas-supported accelerometer.[19]

After the war, Mueller and Rothe moved to the United States to become part of the German rocket team at Fort Bliss and Huntsville. The externally pressurized gas-bearing gyroscope (figure 6.1) and analogous gas-bearing gyro accelerometer were widely used in their systems.[20] But it remained a minority approach in the United States, its last major use being in the guidance system of NASA's Saturn booster in the 1960s.

In the Soviet Union, however, the externally pressurized gas bearing became the dominant approach to inertial instrument design. Soviet gas bearing designs almost certainly derive ultimately from Kreiselgeräte. Links between the latter and the Soviet Union predated World War II. During the time of the Hitler-Stalin pact, Kreiselgeräte contracted to provide gyroscopic instruments and fire control systems for German warships to be supplied to the Soviet Union. From 1939 to 1941 Soviet "acceptance officials," versed in gyro technology and fluent in German, were present in the Kreiselgeräte plant in Berlin. These officials returned with the occupying forces

18. Johannes G. Gievers, "Erinnerungen an Kreiselgeräte" (Recollections of Kreiselgeräte), *Jahrbuch der Deutschen Gesellschaft fur Luft- und Raumfahrt*, 1971, 263–91; Cdr. W. E. May, W. G. Heatly, and H. C. Wassell, "Draft Report of Visit to B.N.G.M., Minden, June 12th, 13th, 1946," in the file on Gievers at the U.K. Admiralty Compass Observatory, Slough; Walter Haeussermann, "Developments in the Field of Automatic Guidance and Control of Rockets," *Journal of Guidance and Control*, Vol. 4 (1981), 232–233.

19. G. Klein and B. Stieler, "Contributions of the late Dr. Johannes Gievers to Inertial Technology—Some Aspects on [sic] the History of Inertial Navigation," *Ortnung und Navigation*, Vol. 3 (1979), 436–443.

20. It should not be confused with the self-activated gas bearing described earlier. The source of pressure in the latter comes from the spin of the gyro rotor, not a pump or gas-bottle, and it is used to substitute for the ball-bearings on which the gyro wheel spins. In the German design these ball-bearings were retained.

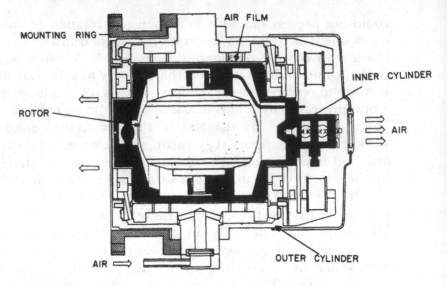

Figure 6.1
Cross-section through Externally Pressurized Gas-Bearing Gyroscope
Source: Fritz K. Mueller, *A History of Inertial Guidance* (Redstone Arsenal,
Ala.: Army Ballistic Missile Agency, n.d.), 27.

in 1945 to obtain navigational devices, gain the cooperation of
Kreiselgeräte employees, and find out more about German
developments. Among those who worked temporarily for the
Soviets was Gievers, who provided them with a detailed
description of the miniature gyrocompass. The Soviets also
obtained a working prototype of it, including the gas-bearing
accelerometer.[21]

It would be a mistake to see the Soviets as simply "acquiring"
advanced German technology. A high level of indigenous technical expertise must have existed to assimilate and make use
of it. Gievers, too, testifies that the Soviet "acceptance officials"
were "more knowledgeable of certain aspects of gyroscope
technology than we were: they could tell us a lot about gyro
development in American firms that we had no knowledge
of."[22] Certainly, the Soviet Union was careful not to become
overreliant on captured German personnel. The Gröttrup

21. Gievers, "Erinnerungen an Kreiselgeräte."
22. Ibid.; I am quoting from the Defense Intelligence Agency translation of
this article (DIA Translation No. LN 040–82, entitled "History of the Gyroscope"), 27.

group became aware that their efforts were being paralleled by Soviet teams: the technical questions they were asked, as their work proceeded in the late 1940s, could only have emanated from people directly involved.[23]

So the Soviet Union had both the capacity and the desire to learn from German work, not merely to copy it. German work was thus an input into post-1945 indigenous Soviet development rather than a straightforward determinant of it. Its importance as an input, however, can be seen by the fact that as late as 1953, when the bulk of the Gröttrup group was allowed to return home, twelve guidance specialists were retained for a further five years "with good salaries and excellent accommodation."[24] By then, though, the Soviets' own research and development effort in missile guidance must have been well on the way to maturity.

The Social Organization of Soviet Guidance System Research, Development, and Production

The indigenous Soviet effort that emerged in missile guidance has three vertical layers: one visible and two hidden from view. The visible face of Soviet guidance and navigation technology is an active and sophisticated theoretical effort, openly published. The best-known source of it is the Institute for Problems in Mechanics of the Academy of Sciences, the head of which is the leading Soviet theoretician, Academician A. Yu. Ishlinskii. Born in Moscow in 1913, Ishlinskii worked on the theory of elasticity, before he moved to work on the applied mathematics of gyroscopes and inertial systems.[25] Ishlinskii has since the 1950s been in close touch with developments in the United States, and he maintained frequent contact with Charles Stark Draper. A congratulatory telegram from Ishlinskii is, for example, to be found in the volume published in 1963 to celebrate

23. Interview with member of Gröttrup group. See also Irmgard Gröttrup, *Rocket Wife* (London: Deutsch, 1959).
24. Ordway and Sharpe, *The Rocket Team*, 342.
25. See, for example, A. Yu. Ishlinskii, *Mechanics of Gyroscopic Systems* (Jerusalem:Israel Program for Scientific Translations, 1965), which is a translation of Ishlinskii's *Mekhanika Giroskopicheskikh Sistem* (Moscow: Izdatel'stvo Akademii Nauk SSSR, 1965); Ishlinskii, *Inertsial'noe Upravlenie ballisticheskimi Raketami* (Inertial Guidance of Ballistic Missiles) (Moscow: Nauka, 1968).

Draper's sixtieth birthday.[26] Overseas specialists can visit Ishlinskii's institute, and freely discuss with Ishlinskii and his staff
the work done there.

From the 1940s to 1965 Ishlinskii had considerable practical
experience of missile and space guidance.[27] His Academy of
Sciences institute is, however, not responsible for missile guidance systems. Development and design work for these is the
business, in missile guidance as elsewhere in the Soviet Union,
of design bureaus. In the case of guidance these are three
"closed" (i.e. secret) bureaus. Two specialize in work for the
land-based missiles of the Strategic Rocket Forces, and the
third in Soviet navy missiles. My intelligence sources did not
wish to identify these bureaus, but the first two are presumably
those established by Nikolai Alekseyevich Pilyugin and Boris
N. Petrov; the third is headed by Viktor Kuznetsov.[28] These
guidance bureaus are separate from the missile bureaus
headed by S. P. Korolev (until 1966), M. K. Yangel (until 1971),
V. N. Nadiraze and V. N. Chelomei.[29] One Soviet source notes
a problem that flows from this separation. "During the course
of experimental development the problem of individual determination of the probability of reliable missile operation $P*_m$
and dispersion of impact points $(sigma*)^2$ is often encountered.
Sometimes this is caused by the fact that the guidance system,
which determines $(sigma*)^2$, and the missile design are pro-

26. Sidney Lees, ed., *Air, Space and Instruments: Draper Anniversary Volume*
(New York: McGraw-Hill, 1963), vi.
27. Interview with Academician A. Yu. Ishlinskii, Moscow, May 23, 1988.
28. Letter from Steven J. Zaloga, September 24, 1988; Alexander Radin,
Miniature Bearing Technology in the USSR (Falls Church, Va.: Delphic Associ-
ates, 1986), 23.
29. For these bureaus, see Holloway, *The Soviet Union and the Arms Race*,
p. 151; Holloway, "Innovation in the Defense Sector," 399; and Robert P.
Berman and John C. Baker, *Soviet Strategic Forces: Requirements and Responses*
(Washington, D.C.: Brookings, 1982), 53–55, 102–107. According to the
latter, Korolev's bureau, which seems to have stopped strategic missile work
at his death, designed the SS-6 and SS-8. Yangel's bureau designed the SS-
7, SS-9, SS-17, and SS-18, as well as the early medium range missiles, SS-4
and SS-5, and submarine-launched missiles SS-N-4 and SS-N-5. The Nadiraze bureau designed the SS-13, the unsuccessful solid-fueled SS-16, and the
SS-20. The Chelomei bureau designed the SS-11 and SS-19, and seems to
have taken over responsibility for the later submarine-launched ballistic missile program.

duced by different organizations."[30] The guidance bureaus are not tied to particular missile bureaus, although as the latter have also tended to specialize (into "heavy" ICBMs, "light" ICBMs, solid-fuel systems, etc.), the pattern of interconnections is not wholly random.

The third level of the Soviet guidance effort, production, is separate yet again: the plants that produce guidance components and systems are distinct from both the design bureaus and research institutes. So the social organization of guidance system technology in the Soviet Union, like that of other military technology there, appears highly compartmentalized vertically.[31]

It is also compartmentalized horizontally. Each of the guidance bureaus has developed its own style, and there appears to be little communication or crossover of technology between them; common features seem to have their roots long in the past. Further, there appear to be barriers between missile guidance system design and work in the related technologies of submarine and aircraft inertial navigation. Only at the level of theory and basic research does "inertial guidance and navigation" seem to exist as a unified entity in the Soviet Union.

This situation of vertical and horizontal compartmentalization contrasts with that prevalent in the West, though perhaps not quite as strongly as it first appears. Inertial guidance and navigation is conceived of in the West as a largely unified field dominated by commercial firms rather than by government plants or laboratories. These firms perform basic research, design, and production, and mostly seek to cover more than

30. V. I. Varfolomeyev and M. I. Kopytov, *Proyektirovaniye i Ispytaniya Ballisticheskikh Raket* (Design and Testing of Ballistic Missiles) (Moscow: Military Publishing House of the Ministry of Defense, 1970). I am quoting from the Joint Publications Research Service translation (no. 51810), 305.
31. For interesting general descriptions of Soviet military research and development, and the place of design bureaus in this, see Holloway, *The Soviet Union and the Arms Race;* David Holloway, "The Soviet Style of Military R & D," in Franklin A. Long and Judith Reppy, eds., *The Genesis of New Weapons: Decision Making for Military R & D* (New York: Pergamon Press, 1980), 137–157; Arthur J. Alexander, *Weapons Acquisition in the Soviet Union, United States and France* (Santa Monica, Calif.: Rand Corporation, 1973, P-4989); John A. McDonnell, "The Soviet Weapons Acquisition System," *Soviet Armed Forces Review Annual*, Vol. 3 (1979), 175–203; and the eyewitness account of Mikhail Agursky, *The Soviet Military-Industrial Complex*, Jerusalem Papers on Peace Problems No. 31 (Jerusalem: Hebrew University, 1980).

one application area. But, as we have seen, Western work on strategic ballistic missile guidance became been increasingly dominated by the Draper Laboratory. This is a not-for-profit organization, not a commercial firm; it does only basic research and design work, not volume production; and it became heavily specialized in ballistic missile guidance rather than any other area of inertial technology. So American missile guidance is predominantly the product of an institution specializing in that one field, largely insulated from the commercial market-place, and vertically separated from production—in many ways more like a Soviet design bureau than a commercial firm. This we will need to bear in mind when we consider the effects on Soviet guidance technology of the social organization of its research, design, and production.

Soviet Land-based Missile Guidance Technology

Land-based ballistic missiles—first medium range, later inter-continental range—were and still are of primary importance in the Soviet nuclear arsenal, so it is appropriate to begin to review Soviet ballistic missile guidance by looking at land-based systems (table 6.1).

Radio versus Inertial Guidance

Two guidance technologies could be inherited from the V-2: radio and inertial. Radio appeared to be more accurate than inertial, but suffered from an obvious military disadvantage because enemies could perhaps jam or interfere with it. Never-theless, given the very great difficulties involved in developing inertial components accurate, reliable and light enough to guide missiles of substantially longer range than the V-2, radio guidance appeared to many in the United States to be the more immediately feasible proposition.

That also seems to have been the situation in the Soviet Union. From the 1940s on, both radio and inertial guidance were pursued in the Soviet Union. Irmgard Gröttrup, who accompanied the German technologists who went to the Soviet Union, describes the use of radio guidance in 1948.[32] Radio guidance, or inertial guidance supplemented by radio, domi-nated in the early Soviet ballistic missiles. It was gradually

32. Gröttrup, *Rocket Wife*, 87–88.

Table 6.1
Main Soviet Land-based Ballistic Missiles of Intermediate and
Intercontinental Range

Year first deployed	Missile	Guidance
IRBMs		
1959	SS-4	radio, then inertial
1961	SS-5	inertial
1977	SS-20	inertial
"First generation" ICBM		
1961	SS-6	radio
"Second generation" ICBMs		
1962	SS-7	radio, then inertial
1964	SS-8	radio
"Third generation" ICBMs		
1966	SS-9	inertial
1966	SS-11	inertial
1969	SS-13	inertial
"Fourth generation" ICBMs		
1974	SS-18	inertial
1975	SS-17	inertial
1975	SS-19	inertial
"Fifth generation" ICBMs		
1986	SS-25	inertial
1987	SS-24	inertial
1989(?)	"SS-18 Follow On"	inertial

IRBM = intermediate-range ballistic missile
ICBM = intercontinental ballistic missile
These are Western intelligence's designations of the missiles. "SS" stands
for surface-to-surface. Variants of many exist, such as the SS-18 mod 2
(1977), the SS-18 mod 3 (1977), and the SS-18 mod 4 (1979).

replaced by all-inertial systems, first in shorter-range missiles and then eventually in ICBMs.

The first seriously operational Soviet nuclear missile was the medium-range SS-3 "Shyster."[33] It first entered service around 1955, and though Soviet-designed it incorporated much V-2 technology, some of it improved. It appears to have been radio guided. Its successor, the SS-4, deployed from 1959 onward, was also radio guided. But its radio guidance was replaced, possibly as early as by 1962, with inertial guidance. The 1961 SS-5 may well have been inertially guided from deployment onward.[34] All subsequent Soviet medium and intermediate range missiles have been inertially guided.

A similar pattern, though slightly later, is to be found in ICBM development. Khrushchev's memoirs indicate use of radio guidance for the first experimental Soviet ICBM, the SS-6, tested in the late 1950s and deployed in small numbers around 1961.[35] The widely deployed second generation ICBM, the SS-7, operational from 1962/1963 onward, may, like the SS-4, have had early radio guidance replaced by inertial.[36] Radio guidance may also have been used on the less successful SS-8.[37] Only with the third generation Soviet ICBMs, the SS-9, SS-11, and SS-13, first deployed in the latter half of the 1960s, is there evidence of an irreversible shift to inertial.[38] With the fourth (SS-17, SS-18, and SS-19) and fifth generation ICBMs (SS-24, SS-25, and "SS-18 Follow On"), pure inertial guidance has been universal.

This pattern of initial use of radio guidance and its gradual replacement by inertial guidance, more slowly in the more demanding intercontinental missiles, is perhaps significant. If

33. Earlier, two versions of the V-2 had been built: one by the Gröttrup group, the other, known as the R-1, by a Soviet group under designer Sergei Korolev. Then a longer range, improved version of the R-1, the R-2, was built, followed by the SS-3. See Holloway, "Military Technology," 455–458.
34. Wright, *Soviet Missiles,* 309, 314, 323.
35. N. Khrushchev, *Khrushchev Remembers: The Last Testament* (London: Deutsch, 1974), 48. See also Wright, *Soviet Missiles,* 109.
36. Wright, *Soviet Missiles,* 114.
37. Berman and Baker, *Soviet Strategic Forces,* 104.
38. Though one of the sources quoted by Wright, *Soviet Missiles,* 152, suggests radio guidance of (presumably early versions of) the SS-11, which entered service in 1966, and Prados, *Soviet Estimate,* 206, suggests radio guidance of the SS-9.

the Soviets shared the belief, common to both the German and early American ballistic missile programs, that radio guidance was the more accurate technique, then the decision to replace radio with inertial represented accepting less than maximum possible accuracy, presumably in return for lesser perceived vulnerability. Since accuracy matters a great deal to a first-strike nuclear strategy, and vulnerability considerably less, this is of some interest.

Soviet Inertial Instrument Technology

Having focused on inertial rather than radio missile guidance, the Soviets have developed it by means similar to those adopted in the United States: painstaking incremental improvement rather than radical change. Novel forms of gyroscope and accelerometer have been absent from the Soviet ICBM program just as they have from the American. Instead, in both countries, we have seen continuous improvement to traditional designs.

But the traditional designs differ. In the United States, they have been the Draper floated gyro and accelerometers derived from it. In the Soviet Union, they appear to have been the externally pressurized gas-bearing gyroscope and accelerometer, most likely inherited from Kreiselgeräte.[39]

39. The evidence for this conclusion is not conclusive, even though it appears to be the dominant opinion in the U.S. intelligence community. The difficulty is that telemetry does not directly reveal the nature of the bearings used in inertial sensors. My guidance analysis interviewees told me, however, that they were in possession of indirect evidence from the telemetry that points to the conclusion that gas-bearing sensors were being employed.

Alexander Radin's émigré account of *Miniature Bearing Technology in the USSR* suggests that the use of gas bearings cannot be universal in Soviet gyroscopes. He notes considerable Soviet use of sleeve (or "pivot-and-jewel") bearings, originally developed in watch-making, for the output axes of Soviet gyroscopes. Such sleeve bearings have been used in the West, notably in early Draper floated gyroscopes, but are not necessary in gas-bearing gyroscopes, since gas pressure keeps the inner float centered.

Radin's account is, however, consistent with the conclusion in the text. He notes (p. 24) a relative lack of interest in sleeve bearings, at least up until around 1980, in Pilyugin's and Kuznetsov's "classified scientific research institutes" (i.e., missile guidance design bureaus). In the early 1980s interest in sleeve bearings grew, but apparently as a replacement for the ball-bearings traditionally used on the gyro spin axes, rather than for the gyro output axes (p. 35). The chief fields of application of sleeve bearings in gyro output

That course of incremental improvement of sensors different from those dominant in the United States has been followed at all three guidance bureaus and continues to this day. It cannot be a selection continued with in ignorance. Fluid-floated gyroscopes and accelerometers have been described in detail in open Western literature from the mid-1950s onward, their use in U.S. strategic systems has not been hard to detect, and through Ishlinskii the Soviets have had close contact with their most enthusiastic proponent, Draper. The Soviet Union would have had little difficulty in obtaining actual examples of at least medium-accuracy fluid-floated gyros 15 years ago: they are to be found in the standard AC Delco inertial navigator of the widely sold Boeing 747. Earlier floated gyros could have been obtained from U.S. aircraft shot down over Vietnam. Nor indeed is the Soviet Union ignorant of the floated gyroscope's Western competitors. Soviet theoretical work exists on dry tuned-rotor gyroscopes, laser gyroscopes, and even the hemispherical resonator gyro being developed by Delco in defiance of the laser gyro revolution. From a visitor to the Soviet Union, there is evidence that dry tuned-rotor gyroscopes are produced in quantity, but not for missile guidance (presumably for aircraft navigation). There is also apparently at least development work (if not production) on laser gyroscopes and fiber-optic gyroscopes.[40]

Are Soviet strategic missile inertial instruments then inherently inferior to Western instruments, given the dominance of instrument types abandoned in the West over twenty years ago? Apart from in one important area, the answer given by Western intelligence's guidance system analysts is "no." In the sensing of acceleration and in-flight rotation, the best Soviet instruments appear to those I interviewed to be in the same league as the best Western ones.

Despite this overall adherence to gas-bearing instruments, Soviet strategic missile inertial instrument design is not completely uniform. One important difference among the different guidance bureaus is in accelerometer design. This is of

axes must, therefore, presumably be areas other than strategic missiles, such as aircraft navigation.
40. The fiber-optic gyroscope is based on a principle similar to that of the laser gyroscope, using optical fiber rather than gas-filled cavities in a solid block.

considerable interest, because, as we have seen, in ballistic missile use (with high accelerations to be measured accurately, but only a short period of time in which the stable platform need be kept in known orientation) it is the performance of the accelerometer rather than that of the gyroscope that is commonly understood to be the more crucial to overall accuracy.

From the V-2 onward, missile guidance designers, Soviet and American, normally have chosen between two different kinds of accelerometers.[41] One is the pendulous integrating gyro accelerometer (PIGA), in which a gyroscope is supported in an unsymmetrical way so as to make it sensitive to acceleration, not just rotation (see figure 2.6). The other is the restrained pendulum, where the sensitive element is not a gyroscope but simply a mass. When the device is subject to acceleration, a restraining force has to be applied to the mass to keep it from moving from its equilibrium position. The size of this force is then the measure of acceleration.

Although in the V-2 program the restrained pendulum design appears to have been understood to be the more accurate, since then there has been practical consensus that the PIGA is the more accurate accelerometer.[42] On the other hand, it is also seen as inherently more complex, and so much more difficult to make and more expensive. In U.S. programs the PIGA has dominated, except in Navy missiles of the 1960s and 1970s, a period during which Navy missile designers did not give accuracy the highest priority (see chapter 5).

Soviet ICBM programs use both PIGAs and restrained pendula; the two types of devices are characteristic of different bureaus. This divergence is an important contributor to a vital aspect of Soviet ICBMs to which I now turn: within the same generation of ICBM, "high-accuracy" missiles, using PIGAs, can be differentiated from "medium-accuracy" missiles using restrained pendula.

41. Thomas M. Moore, "German Missile Accelerometers," *Electrical Engineering*, November 1949, 996–999.
42. Ibid., 996–97. DMcK: "But the PIGA is intrinsically more accurate than the [restrained pendulum]?" Draper Laboratory interviewee: "Yes, I guess that is the statement that 99 percent of the community would agree with. To make it a horse race there's always a couple of guys that say 'we think we can do the job for you with a [restrained pendulum],' and that's attractive in terms of being a whole lot cheaper."

Accuracy, Institutional Interests, and Strategic Roles

This differentiation in accuracy can be seen in Soviet ICBMs of the first all-inertial generation, the SS-9, SS-11 mod 1, and SS-13 mod 1 of the late 1960s. The SS-9, although the earliest missile of the generation, seems to have been the most accurate (see appendix A). The estimates in the classified versions of the Secretary of Defense's reports to Congress, for example, gave an SS-9 accuracy of 0.5 to 1.0 nautical miles, but an SS-11 accuracy of only 1 to 1.5 nautical miles.[43]

A similar differentiation, albeit less clearly seen among a welter of leaked and guessed accuracy figures, may be found in the next (fourth) generation of Soviet ICBMs—the SS-17, SS-18, and SS-19—although by then the meanings of "high" and "medium" accuracy had changed. One interviewee suggested to me that, in round terms, the SS-18 is a "tenth of a nautical mile system," while the SS-17 and SS-19 are "quarter of a nautical mile systems." Taken literally this would probably be seen by most intelligence analysts as slightly overestimating the accuracy of the SS-18 and underestimating that of the others (see appendix A). But there is, I believe, good reason to think that the fourth generation of ICBMs, like the third generation, includes both high- and medium-accuracy missiles.[44]

One possible explanation lies in institutional interests, beliefs, and competences. Perhaps the two guidance bureaus specializing in ICBM guidance have simply evolved different technical styles, one prioritizing accuracy, the other "producibility." Perhaps they have different beliefs about how best to achieve accuracy. Or perhaps one is simply better at its job than the other, but the inferior one is too well entrenched politically to be done without.

Without the detailed interview evidence that can be gathered on equivalent U.S. technical decisions, we cannot directly compare these possible explanations. It is interesting, however, to correlate the nature of guidance design with overall missile design in third generation ICBMs. The SS-9 was much larger physically than the other ICBMs of its generation, and it car-

43. Wright, *Soviet Missiles*, 60, 61.
44. See appendix A, though it will be noted that this conclusion is dependent on the judgment there that the SS-19 is a "medium-" rather than "high-" accuracy missile.

ried a warhead with a yield estimated officially in the United States at around 18 megatons, while the yield of the SS-11 mod 1 was around one megaton and that of the SS-13 mod 1 probably less than that.[45]

So the high-accuracy missile was also high-yield, while the medium-accuracy missiles were medium-yield. This suggests a deliberate design policy: to optimize the SS-9, even at considerable expense, for hard-target kill capability, while designing the SS-11 and SS-13 more cheaply for attacks on softer targets. As it was put at the time: "Savage [SS-11] is intended as a city buster while Scarp [SS-9] is directed towards neutralization of specific hardened targets such as Minuteman silos."[46] In other words, the medium accuracy of the SS-11 and SS-13, and high accuracy of the SS-9, might have been the result of a deliberate policy decision.

The pattern in the fourth generation ICBMs is not so clear. The high-accuracy missile is again the "heavy ICBM," the SS-18, but the distinction in yields is not clear-cut.[47] In the most recent fifth generation, however, the pattern has remerged

45. Wright, *Soviet Missiles*, 60.
46. Donald C. Winston, "SS-9 seen spurring Nixon ABM Effort," *Aviation Week and Space Technology*, March 31, 1969, 18. It was later believed that the probable prime targets of the SS-9 force were the Minuteman launch command centers, rather than the more numerous silos.
47. The single warhead versions of the SS-18 (mod 1 and mod 3) still had enormous yields; they may simply have used the SS-9s thermonuclear device. But the single warhead versions of the SS-17 and SS-19 are also large yield, if not as large: perhaps 2 and 3.5 megatons respectively. There is some evidence that in the first MIRVed subgeneration (SS-17 mod 1, SS-18 mod 2, and SS-19 mod 1), the SS-18 carried both more and heavier warheads. But in the second, current subgeneration (SS-17 mod 3, SS-18 mod 4, and SS-19 mod 3) the evidence suggests very similar yields, possibly even the same approximately 500-kiloton warhead, with the SS-18 simply carrying more of them. As that later generation was being developed, a new factor may have been important: the haste resulting from the need to test any warhead of more than 150 kilotons, if it was to be test-fired at full yield, before the Threshold Test Ban Treaty came into effect on March 31, 1976. See Lynn R. Sykes and Dan M. Davis, "The Yields of Soviet Strategic Weapons," *Scientific American*, Vol. 256, No. 1 (January 1987), 29–37, on this latter point; and Sykes and Davis, "Yields," together with Wright, *Soviet Missiles*, 62–68, on the above yield estimates. So we have to be much more tentative in a differentiation of intended strategic roles for fourth generation Soviet ICBMs than for third generation. While the SS-18 was clearly intended for a counterforce role, it is harder to infer an unambiguous role for the SS-17 and, especially, SS-19.

clearly. That generation consists of three ICBMs. One is a missile similar to the SS-18, the "SS-18 Follow-On," which was flight-tested during 1987.[48] With greater accuracy than the SS-18 mod 4 but a similar formidable load of warheads, this is a clear counterforce weapon.

The other two fifth-generation missiles are mobile ICBMs: the multiple warhead, rail-mobile SS-24 and the single-warhead, road-mobile SS-25. Neither, it appears, have accuracies comparable to the SS-18 Follow-On. Secretary of Defense Frank Carlucci told Congress in September 1988: "the SS-24 and SS-25 are presently suited most appropriately to soft and medium hard targets—rather than hard targets—due to lower accuracies and reliability."[49] So there is continuing evidence of differentiation of strategic roles, with some of the ICBM force being given the greater protection of mobility, at a cost in accuracy, while the best silo-based counterforce missile is improved even further in its accuracy.

The question of instrument design and the relative priorities apparently awarded to accuracy in different systems are not, however, the only interesting features of Soviet ICBM guidance. The guidance "mathematics" used by the Soviets differs from American, both overall and in the lesser extent to which the Soviets seek to compensate mathematically for instrument errors. Prelaunch alignment of ICBM guidance systems is achieved differently; and Soviet and American systems differ also in certain significant physical respects. It is to these differences that I now turn.

The Mathematics of Guidance

The most basic question in the mathematics of guidance is how the missile is to be guided so that the warhead(s) will end on or near target. As we have seen, two polar possibilities can be envisaged. In the first, the requisite trajectory is calculated in advance, on the ground, and the guidance and control system is then required simply to return the missile to the preplanned

48. Brendan M. Greeley, Jr., "Soviets Increase Deployment of Mobile Ballistic, Cruise Missiles," *Aviation Week and Space Technology*, (March 30, 1987), 24–25; "Washington Roundup," *Aviation Week and Space Technology* (October 5, 1987), 17; Department of Defense, *Soviet Military Power: An Assessment of the Threat, 1988* (Washington, D.C.: U.S. Government Printing Office, 1988), 47.
49. Quoted in *Defense Monitor*, Vol. 18, No. 1 (1989), 5.

trajectory when it appears to deviate: "fly-the-wire" guidance, this is sometimes called. In the second, known as "explicit" guidance, the missile guidance system is given only the coordinates of the launch point and target, continuously calculates its instantaneous position and velocity, and constantly recalculates the changes in velocity required to bring the missile to the target.

The questions of mathematics involved are closely connected to other aspects of system design. Fly-the-wire guidance minimizes, and explicit guidances maximizes, the demands on onboard computer capacity. Guidance formulations are also affected by the type of propulsion used. The size of the thrust of solid-fuel rocket engines is very difficult to control or even predict closely (it is affected by weather conditions, notably temperature). A common metaphor is that being solid-fueled is like having to drive with the gas pedal stuck to the floor: a wholly preplanned trajectory is hard to achieve. At least in principle, however, liquid-fuel rocket engines can be throttled, and the greater flexibility makes following a preplanned trajectory easier.

It is therefore not surprising that the Soviets have remained much closer to the "fly-the-wire" end of the continuum than the Americans. On-board digital computers were introduced to the Soviet ICBM program only with the fourth generation ICBMs of the 1970s; while the greater flexibility of the liquid motors that dominated (and still dominate) the Soviet ICBM program made a "fly-the-wire" system not too difficult to implement.[50] This was particularly evident in the testing of early Soviet ICBMs. The SS-6s and SS-9s would shut off their rocket motors at the same place and same time in one test after another.

As we have seen, even early American systems moved some way away from fly-the-wire guidance. Thor (which had an on-board analog computer), and Polaris (which had a simple special-purpose digital computer) used Q-guidance, which permitted most of the calculations to be done on the ground while avoiding the worst of the inflexibility of "fly-the-wire."[51] Titan, some versions of Atlas, and Minuteman I and II used "delta

50. Berman and Baker, *Soviet Strategic Forces*, 104.
51. See chapter 3.

guidance," where the missile's position and velocity at the end
of powered flight are preplanned, but where there is flexibility
in how those conditions are achieved.[52] With Minuteman III
and MX the Americans have moved much closer to explicit
guidance, although the impossibility of storing on board an
accurate model of the earth's gravitational field comprehensive
enough to cover all possible trajectories means that fully
explicit guidance still cannot be pursued without compromising
accuracy.

Soviet guidance formulations have remained similar to delta
guidance, with the flexibility of liquid propulsion making it
possible to implement this simply without a great loss in accuracy.[53] But the price of this in-mission simplicity is a great deal
of on-ground computation, especially with multiple independently targetable reentry vehicles (MIRVs). A Soviet guidance
system will need to be preprogrammed with as many as fifty
to one hundred constants for a complex mission including
many reentry vehicles. The capacity for rapid, fully flexible
retargeting of the Soviet ICBM force, especially in wartime
conditions, must therefore be quite limited. This is a point of
some importance, because it must be expected to generate
pressure to execute a war plan before the conditions for its use
disappear, and thus be a major constraint on the Soviets fight-
ing the "limited" and "prolonged" nuclear wars envisaged in
recent American strategy.[54]

52. Delta guidance is briefly described in R. H. Battin, "Space Guidance
Evolution—A Personal Narrative," *Journal of Guidance and Control*, Vol. 5
(1982), 97–109.
53. Mathematically, delta guidance involves a Taylor series expansion around
a reference trajectory (see ibid.). The zero-order terms represent the reference path, and the task of Soviet guidance and control systems is to reduce
the first-order (linear) terms to zero. Higher-order terms are assumed negligible. Apparently a flexible liquid-fuel system makes it possible for the
Soviets to control only the linear terms and still achieve reasonable accuracy.
54. In another aspect of the need to preprogram a guidance system the
Soviets are in a worse situation than the Americans for geographical rather
than computational reasons. The issue is the "problem of the vertical"—the
need to preprogram inertial systems with knowledge of the gravity field.
Global gravity models in relatively easily computer-stored mathematical for-
mat are available, but these begin to be insufficient when high accuracies
(better than around 0.15 nautical miles or 300 meters) are demanded, as
they are for the current Minuteman force, MX, and the best Soviet missiles.
Detailed gravity survey data are therefore needed for a region several

Calibration and Error-Compensation Software

In modern American strategic ballistic missiles, the on-board digital computer does not simply carry out in-flight guidance calculations. It is also used to reduce the effect of errors in the gyroscopes and accelerometers. These instruments are calibrated *in situ,* and a mathematical model is used to compensate in flight for predictable errors in their output. So accuracy has come to depend not on the size of the "absolute" errors in gyroscopes and accelerometers but on the predictability of those errors together with the sophistication of the calibration and compensation algorithms used.

Mathematical error modeling is used considerably less in Soviet than in American systems. Even in fourth generation ICBMs it compares, in level of sophistication, to that employed in Minuteman I and II (systems of the 1960s), rather than to that of Minuteman III, or, certainly, MX, the accuracy of which is heavily dependent upon error modeling.

In large part, of course, this may be due to the restricted capacity of Soviet on-board computers. There is little spare computational capacity in some systems. But that, according to my interviewees, does not wholly account for the difference in error modeling: in some systems, computer capacity would permit the use of models more sophisticated than those being used.

Given that greater use of modeling, according to current beliefs in the United States, would lead to significantly increased missile accuracy, its "underuse" by the Soviets is a matter of some interest. Three possibilities, not mutually exclu-

hundred miles down-range of the launch point, and also behind it. For the United States this is an expensive but not intractable problem; it is also easily solved on the Soviet missile test ranges, which run east-west across the Soviet Union. But for operational trajectories from some ICBM fields, the necessary regions include parts of India, the Himalayas, and relatively inaccessible regions of Northern Siberia; see the map of ICBM fields in Berman and Baker, *Soviet Strategic Forces,* 16–17. A trade-off may be involved. The southernmost Soviet ICBM fields, where the problems of "backwards" gravity data will be greatest, have for the last decade been the least vulnerable, since they are out of range of the section of the Minuteman III force equipped with the large Mark 12A warhead rather than the original lighter Mark 12. (It is, incidentally, indicative of the difficulties faced by Western intelligence's guidance system analysts that they have to take into account, in their estimates of Soviet ICBM accuracies, the likely quality of the gravitational data possessed by the Soviet Union for operational trajectories.)

sive, suggest themselves: first, that the relevant Soviet authorities are satisfied with the accuracies their systems currently
possess. Second, there may be a difference in technical philosophy involved, a reluctance to make operational performance
too dependent on complex software and mathematical modeling. Third, it may be the result of the compartmentalization
of Soviet research and development: instrument designers
have continued myopically to seek to minimize absolute errors,
while there has not been a strong enough overall system perspective to pursue, instead, mitigation of the consequences of
these errors.

The available evidence does not enable me to offer these
explanations other than tentatively. It does, however, render
implausible the implied account in the official "Intelligence
Community Report on Soviet Acquisition of Western Technology" that the Soviet Union cannot write sufficiently good
calibration and compensation algorithms.[55] Actually, the theoretical analysis of guidance systems is, perhaps, the greatest
area of indigenous Soviet strength.

Prelaunch Alignment

An even more striking difference between American and
Soviet approaches, one much debated within the U.S. intelligence community, is prelaunch alignment of ICBM guidance
systems. Prior to launch, a missile guidance system needs to
"know where it is pointing" in the horizontal plane: a tiny error
in alignment, say three arc seconds, is incompatible with counterforce accuracy.

In early land-based missile systems, alignment was performed manually using external sources of information such
as presurveyed landmarks. In U.S. ICBMs this function subsequently was moved inside the missile, by means of gyrocompassing analogous to the way the gyro is used to find north in
a ship's gyrocompass. In Minuteman III this is accomplished
by the addition of a separate high-accuracy gyroscope, unnecessary for in-flight guidance. In the "elegant" MX guidance
system this added complexity is done away with, and the guidance system's own gyroscopes are used for gyrocompassing.

55. "Intelligence Community Report on Soviet Acquisition of Western Technology," *U.S. Export Weekly* (April 13, 1982), 69.

There is believed to be a considerable penalty to pay: gyroscopes good enough to permit high-accuracy gyrocompassing have to be two orders of magnitude more accurate (drift rates of the order of 10^{-5} degrees per hour) than required for high-accuracy in-flight guidance (about 10^{-3} degrees per hour). This, for example, is the rationale for the use of the ultra-low-drift Draper-designed third generation gyroscopes on MX. It was also a major barrier, perhaps the most significant, faced by the laser gyroscope in its unsuccessful struggle for acceptance for the Small ICBM.[56]

The Soviet Union has not taken this path, and has adopted an approach that was seen as failing when experimented with in the United States. A gyrocompass, external to the missile but within the silo, is used to determine orientation in the horizontal plane, and the information is transferred optically into the guidance system.

The Soviets seem to use a gyrocompass of Hungarian manufacture, known as MOM after the maker, Magyar Optikai Müvek (Hungarian Optical Works), which they have bought in large quantities—4,000, according to one source. The MOM gyrocompass is commercially available, so it is known to the West. It is the best commercial gyrocompass in the world; its performance level indicates that requirements other than the normal demands of survey work have shaped it. A MOM gyrocompass was purchased in Canada and subjected to detailed analysis on behalf of American intelligence. It employs a wire-suspended gyro, powered by a different kind of electric motor[57] than that used on the failed Western attempt to use the same technology.

The external gyrocompass approach offers advantages to the Soviets. It avoids the need to complicate guidance system platform design with an additional instrument for gyrocompassing, without placing on Soviet externally pressurized gas-bearing gyros the same drift rate demands as are placed on MX's Draper-designed floated gyros. Some of the disadvantages this approach would carry in the United States may not be so pressing in the Soviet Union: it requires highly skilled operators, and, given the tedium of the task and the recruitment

56. See chapter 4.
57. It is of the type known as an "induction motor" rather than the "hysteresis synchronous" motor more conventionally used in Western gyros.

difficulties of the U.S. armed services, this might prove a major problem.[58]

Nevertheless, the external gyrocompass approach may add to the vulnerability of the Soviet ICBM force. In particular, the optical transfer system could perhaps be disrupted by a nuclear explosion that was not close enough to destroy the silo or missile: that was one of the fears that led to the Minuteman guidance system being given its own internal gyrocompassing capacity. This could form a pressure for preemptive launch, or for launch on detection of an incoming attack (launch-on-warning), that has not been recognized in the Western open literature.

The implications of external gyrocompassing for the accuracy of Soviet ICBMs are controversial among guidance systems analysts. No one doubts that the MOM gyrocompass can, in principle, provide the requisite alignment accuracies. It is reckoned by Western intelligence to be an excellent instrument. One source described it to me as accurate to within two arc seconds, and the device bought in Canada gave evidence of extremely high-quality machining: the ball-bearing raceway grinding was reckoned superior to that available in the West at the time. The question is, rather, how much accuracy is lost in the process of optical transfer. If the Soviets share the belief of those Western analysts who think the loss is significant, then their retention of external gyrocompassing is of particular interest. While instrument performance limitations may block the MX path of using the guidance system's own gyros to gyrocompass, a Minuteman-style solution, a single additional high-quality gyro, would seem possible. That it has not been adopted may indicate that the accuracies achieved have been seen as "good enough," and any increase is not worth the price of a move away from a well-established solution in the direction of greater guidance system complexity.

Redundancy and "Scrunching"

A further difference between American and Soviet guidance systems is the extent of redundancy in the latter: Soviet systems use more than one device to perform the same task. While all American ICBMs have only one on-board computer, redun-

---

58. The MOM gyrocompass is undamped, so the operator has to measure the peaks of its swing.

dancy in on-board computers is universal in Soviet ICBMs, even current ones. While American missiles and space boosters have never employed more than one accelerometer for each of the three axes in space, it has been common in Soviet systems to employ more than one, with some kind of "voting" system to average their results.[59] A German guidance specialist in the United States recalls joking with a Soviet colleague that the American way was political democracy and dictatorship within guidance systems, while the Soviet way was political dictatorship but democracy within guidance systems.[60]

The guidance bureau that specializes in high-accuracy systems seems much less inclined to use redundancy: in particular, the fourth-generation hard-target-killer ICBM, the SS-18, is in Soviet terms a low-redundancy system. The other ICBM guidance bureau, by comparison, employs greater redundancy. The correlations of a "high-accuracy, low-redundancy" approach and a "medium-accuracy, greater redundancy" approach may well not be accidental. Given the expense and production difficulties of the PIGAs used in the high-accuracy approach—difficulties all too familiar to those responsible for Minuteman, MX, and Trident D5 guidance—one can understand a reluctance not to multiply the number of them that a given system requires.

Soviet guidance system design also differs from American, at least in some systems, in the optimization of instrument orientation—the Soviets place the three accelerometers, not mutually at right angles, but "scrunched" around the dominant direction of rocket thrust. The principle is known in the West; its rationale is to increase accuracy, not by adding extra accelerometers, but by having all three accelerometers measure at least a component of the acceleration in the most crucial thrust direction.[61] Its value is not universally accepted in the United States—Draper was not an enthusiast—and any advantage it possesses is reduced to the extent that one moves away from a preplanned trajectory. Its adoption in the Soviet Union, how-

59. A simple "voting" system, for example, would initiate warhead separation when the second of three thrust-axis accelerometers indicated that the required velocity had been reached.
60. Interview.
61. See G. A. Harter, "Error Analysis and Performance Optimization of Rocket Vehicle Guidance Systems," in George R. Pitman, Jr., ed., *Inertial Guidance* (New York: Wiley, 1962), 326–328.

ever, indicates that at least some specialists there believe that it increases accuracy.[62]

Guidance, Ball Bearings, and the Alert Status of Soviet ICBMs

The Soviet use of externally pressurized gas-bearing gyroscopes and accelerometers is interesting not only because it is a significant difference from current Western designs. It also has an important effect in reducing what might otherwise be a serious operational constraint on Soviet ICBMs: the speed with which the force can be brought to alert and the length of time it can be maintained there. Fluid-floated gyros would have to be brought to the correct temperature if they have cooled down—indeed early high accuracy floated gyros could not safely be cooled down—and they are known to be extremely sensitive to temperature fluctuations, while this is much less of a problem with the Soviet style of gyro.[63] The latter is also believed to have a better "turn-on-repeatability" (predictability of performance from one period of use to the next) than the fluid-floated gyro.

As we have seen, the Western solution to this has been, since the beginning of the 1960s, to maintain ICBM guidance systems in the field in continuous operation. Continuous operation does bring its problems, however. The most serious of these affected the Minuteman II force, when the early integrated circuits used began to fail at a high rate, rendering large portions of the force temporarily inoperable and necessitating an urgent and expensive replacement program. The externally pressurized gas bearing permits the Soviet Union to avoid the potential problems of continuously running guidance systems, since it means that the force can be kept "dormant" most of the time and still brought quickly to alert.

In the early years of the Soviet force, spin-axis[64] ball-bearing life appears to have been a major concern, limiting sharply the length of the time the force could be kept on alert, and poten-

62. See Ishlinskii, *Inertsial'noe Upravlenie,* for indication of detailed consideration in the Soviet Union of the effect of accelerometer orientation on accuracy.
63. Thus the gyros in Thor missiles had to be kept heated even when the missiles were being flown from the United States to the United Kingdom: Julian Hartt, *The Mighty Thor: Missile in Readiness* (New York: Duell, Sloan and Pearce, 1961), 244.
64. See figure 2.15.

tially making it highly vulnerable—in its above-ground, "soft" configuration of the early 1960s—to an American preemptive attack in a situation such as the Berlin or Cuban missile crises. It is not clear whether the Soviets have responded as have the Americans by moving from ball-bearings to spin-axis self-activating gas bearings: that shift would not show up in telemetry.[65] Certainly, though, the problems of maintaining the force on alert have diminished since then.

It is equally certain that the purchase in 1972 by the Soviet Union from the United States of 164 Bryant Centalign grinding machines had little to do with the increase in Soviet missile accuracy between the 1960s and 1970s, despite the oft-repeated story that it had.[66] The very size of the controversial purchase suggests a quite different purpose: one ball-bearing grinding machine would have supplied the inertial guidance needs of the entire Soviet Union. The Soviet Union had other sources of supply, notably a long-standing history of purchases from Swiss precision machine-tool manufacturers, and the quality of the MOM gyrocompass suggests that high-accuracy machining was also available within the Warsaw Pact. Nor has it been properly noted in the public debate on this purchase that by 1972 improving ball-bearing grinding was, according to dominant American beliefs, the wrong way to try to achieve high accuracy. The proponents of gas spin-axis bearings would claim that adopting these, and discarding ball-bearings altogether, has brought about a major improvement in the accuracy of U.S. missile gyroscopes and accelerometers.[67]

65. Again, these latter should not be confused with the output-axis externally-pressurized gas bearings used in Soviet missiles. Radin's *Miniature Bearing Technology in the USSR*, 24, notes an emphasis on (spin-axis) ball bearings in Soviet missile guidance bureaus, at least until the end of the 1970s, well after the Americans moved to spin-axis self-activating gas bearings.

66. See, e.g., Charles Levinson, *Vodka-Cola* (Horsham, West Sussex: Biblios, 1980), 223. A version of the story, admittedly with significant qualifications, is even to be found in the official "Intelligence Community Report." For a discussion of the ball-bearing episode, see Julian Cooper, "Western Technology and the Soviet Defense Industry," in Bruce Parrott, ed., *Trade, Technology and Soviet-American Relations* (Bloomington, Indiana: Indiana University Press, 1985), 183–184.

67. So the most relevant area of potential technology transfer might not have been ball-bearings but, for example, ceramics that could be used in self-activating gas bearings and technology for high-accuracy machining of ceramics.

Reentry Vehicle Design

A final accuracy-relevant feature of Soviet ICBMs is their reentry vehicles. Unlike Soviet guidance system design, changing Soviet reentry vehicle design has been relatively widely noted in the West[68] and so needs only brief mention here. In all currently deployed ICBMs, and also submarine-launched ballistic missile systems, Soviet and American, reentry vehicles are unguided. If reentry is not as predicted (for example, if the reentry vehicle's shielding burns off unevenly and causes unforeseen aerodynamic forces or if a major unpredicted air current in the upper atmosphere is encountered), then errors will be introduced.

Historically, this was a second-order concern to the designers of reentry vehicles. The original goal was simply to prevent the contents of the reentry vehicle being destroyed or damaged by the heat generated during reentry. For this, a "blunt" reentry body (one with a low "beta," or ballistic coefficient) came to be believed to be better than a streamlined one, since the heat generated is dissipated better in slow reentry. High beta, on the other hand, means faster reentry. This imposes tougher requirements for heat protection, but increases accuracy.[69]

Like the Americans, the Soviets have moved from low-beta to high-beta reentry vehicles, though the Soviet effort to do so was later than the American and met with early difficulties. For some time the Soviets had to return to low beta, before a successful return to high beta. This was an important factor in the increased accuracy of Soviet fourth generation ICBMs.

Submarine Navigation and Submarine-launched Ballistic Missile Guidance

As in the United States, the development of a Soviet land-based missile force was followed by the development of a submarine-launched force (table 6.2). The term "submarine-launched," though, contains an ambiguity when applied to the earliest Soviet missile of this type, the SS-N-4, first deployed in 1958 (or perhaps 1959), because the submarine carrying it had

68. See, for example, R. J. Smith, "An Upheaval in US Strategic Thought," *Science*, Vol. 216 (April 2, 1982), 30–34.
69. Fast reentry also brings benefits in terms of evading endo-atmospheric anti-ballistic missiles.

Table 6.2
Soviet Submarine-launched Ballistic Missiles

Year first deployed	Missile	Range (nautical miles)	Guidance
1958	SS-N-4 (surface-launched)	350	inertial
1963	SS-N-5	700	inertial
1968	SS-N-6	1,300	inertial
1973	SS-N-8	4,200	stellar-inertial
1978	SS-N-18	3,500	stellar-inertial
1983	SS-N-20	4,500	?
1987	SS-N-23	5,000	?

to come to the surface to fire it. Only with the SS-N-5, first deployed in 1963, did submerged launch become possible, and only with the 1968 SS-N-6 did the Soviet Union possess a system fully comparable to America's Polaris.[70]

All three of these missiles, the SS-N-4, 5, and 6 were equipped with an inertial guidance system (the difficulties of radio guidance from a submarine are evident). But like the Americans, the Soviets had to face the problem of providing the missile's guidance system with knowledge of its orientation, position, and velocity at the time of launch. In the United States, as we saw in chapter 3, this problem was solved by the development of highly accurate submarine inertial navigators (Ships Inertial Navigation Systems, or SINS), which are periodically updated using the land-based radio navigation system Loran-C, the satellite-based Transit system and sonar images of seabed features.

The Soviets appear to have had great difficulty with SINS technology. One source suggests Soviet ballistic-missile submarines may not have had SINS as late as 1966 or even 1976. This is implausible. But even in the 1980s the Soviets seem to have lagged significantly behind the United States in SINS technology. This lag parallels one in the other main non-missile, nonspace application of inertial technology: aircraft navigation.[71]

70. Wright, *Soviet Missiles*, especially p. 33.
71. K. J. Moore, Mark Flanigan and Robert D. Helsel, "Developments in Submarine Systems, 1955–1976," in Michael MccGwire and John McDonnell, eds., *Soviet Naval Influence: Domestic and Foreign Dimensions* (New York: Prae-

Why should the Soviets do relatively well in one version (missile guidance) but badly in other versions (submarine and aircraft navigation) of what in Western eyes is essentially the same technology? One possible explanation is that the Soviets have awarded these latter areas lower priority. But another explanation is compartmentalization: that aircraft and submarine navigation are not aspects of the same technology as missile guidance in Soviet practice. American SINS technology has benefited enormously from direct inputs from other areas of inertial technology. The first U.S. operational SINS was essentially a converted cruise missile guidance system, the directly developed SINS having proved unsatisfactory. The major subsequent revolution in SINS technology—the move to gyroscopes where the spinning rotor is supported by an electrostatic field—involved the adaptation of a system that had been designed as an aircraft navigator, with the variant directly developed as a SINS again proving a partial failure.[72] If compartmentalization of the Soviet research, development, and production system inhibits such "horizontal" transfers of technology, this may be the cause of Soviet difficulties with SINS technology.

Stellar-inertial Guidance

In the early 1960s the Americans began to experiment with a technology that promised to make submarine-launched missiles as accurate as silo-based ICBMs. In this technology, stellar-inertial guidance, the attempt is made to correct uncertainties in initial launchpoint and orientation by taking a fix, in flight, on a star or stars.

As noted in chapter 5, stellar-inertial guidance did not win immediate acceptance in the United States. No ballistic missile

ger, 1977), 157, 170; "Intelligence Community Report," 69. On our current understanding of the possibilities for underwater navigation, without a black-box navigator ballistic missile submarines would suffer severe limitations on operational flexibility. Either they would have to come to or near the surface for some minutes before firing in order to take a stellar or radio navigation "fix," or they would be restricted to areas where seabed features had been mapped accurately. If it were true that Soviet submarines lacked inertial navigators, it seems unlikely that these limitations would not have been noted in the literature, but they have not been.

72. See chapter 5.

using it was deployed until Trident I (C4) in 1979. Stellar-inertial guidance is thus unusual among weapons technology in that it was used by the Soviet Union before it was by the United States. The first Soviet stellar-inertial ballistic missile was the SS-N-8, which became operational in 1973.[73]

The other main novel characteristic of the SS-N-8 was a quantum leap in range. With nearly three times the range of its predecessor (4,200 nautical miles, or 7,800 kilometers, compared to the 1,300 to 1,600 nautical miles of the SS-N-6), the SS-N-8 made it possible for Soviet ballistic missile submarines to strike the United States from the relatively safe waters of the Barents Sea and Sea of Okhotsk, rather than open ocean areas closer to the United States, where they would be more vulnerable to Western antisubmarine warfare.[74]

Stellar-inertial guidance seems to have been used by the Soviets to achieve this increase in range rather than to increase the SS-N-8's accuracy, which may be little better than that of the SS-N-6.[75] To achieve the same accuracy at tripled range in the absence of stellar-inertial guidance would have meant a roughly threefold increase in the performance of both the missile guidance system and the submarine's SINS. The SS-N-8's stellar-inertial system was a simple one, directed at correcting a single source of error (orientation in the horizontal plane), but in terms of the much greater range it permitted, it must be counted a successful technology.

The successor to the still operational SS-N-8 was the SS-N-18. This first Soviet MIRVed submarine-launched missile entered service in 1978. It too was equipped with stellar-inertial guidance, but of a much more sophisticated type than that of the SS-N-8. The SS-N-18's system has the capacity for multiple star-sightings.[76]

73. The first published reference to Soviet stellar-inertial guidance that I have found is William Beecher, "SIG: What the Arms Agreement Doesn't Cover," *Sea Power* (December 1972), 8–11.
74. Wright, *Soviet Missiles*, 50–51, 272.
75. Anon., "Soviets Test New MIRV Warhead ICBMs," *Aviation Week and Space Technology* (February 25, 1974), 20, quotes unnamed U.S. officials who believed that "stellar-inertial guidance . . . has done little to improve the accuracy of the missile." See also Wright, Soviet Missiles, 50.
76. Richard T. Ackley, "The Wartime Role of Soviet SSBNs," *US Naval Institute Proceedings* (June 1978), 41, describes the SS-N-18's (an unfortunate but obvious misprint has this as "SS-N-8") guidance system as having the

There is a very interesting difference in approach between the Soviets and Americans. As noted in chapter 5, the Americans have designed both the Trident C4 and also the new D5 stellar-inertial guidance around the unistar principle, taking only one fix on one star. The Soviets, on the other hand, appear to design for both sightings of two stars and multiple sightings of the same star. This means increased mechanical complexity, and, if the American theory is correct, cannot bring with it enhanced accuracy; indeed a proponent of unistar would argue that the Soviet design could reduce missile accuracy, because greater mechanical complexity reduces the stability of the optical system and hence the accuracy of the star-sight.

Again, ignorance cannot be the cause of the Soviet choice here: the unistar principle was described in an unclassified publication as early as 1971, while the SS-N-18's tests did not begin until 1975, and there has been no sign of the Soviets subsequently shifting to the American design.[77]

At least two possible rationales for the divergence suggest themselves. One is that the relevant Soviet design bureau does not accept the American theory and is pursuing maximum possible accuracy by avoiding the "trap" into which the Americans have fallen. It is, for example, acknowledged by the proponents of unistar that an optimally located star will generally not be available, and the resultant enforced use of a nonoptimum star will lead to a reduction (albeit a small one, they would argue) in accuracy.

Another possible explanation is distrust of complex software and sophisticated mathematical algorithms. The American system requires the maintenance, within the submarine, of a large computerized star map, since it is important to have a near-optimum star always available. It also relies on a priori knowledge of system error statistics, as noted in chapter 5. The Soviet system, on the other hand, demands less in terms of star map "bookkeeping" (since any two bright enough stars sufficiently separated will suffice), and attempts a deterministic, rather

"capability for two celestial observations." Interviewees confirmed the use of multiple star-sightings in Soviet stellar-inertial guidance.
77. David G. Hoag, "Ballistic-Missile Guidance," in B.T. Feld et al., eds., *Impact of New Technologies on the Arms Race* (Cambridge, Mass.: MIT Press, 1971), 91–94.

than a statistical, correction. By taking two sightings at different points in time, errors in initial information can be separated from in-flight gyro drift (since the effects of the former are constant, while the latter vary with time).

Despite the early success of the SS-N-8 star-tracker system, the quantum leap in sophistication thereafter attempted led to problems, but these appear to have been solved. Nevertheless, the Soviet Union's submarine-launched ballistic missile force does not (at least yet) appear to have developed in the same direction as the Americans'—toward the prioritization of counterforce capability. Despite their use of stellar-inertial guidance, the Soviets do not seem likely to deploy a "hard-target killer" submarine-launched missile.[78] This may be because of difficulties in constructing such a system, but perhaps that has not been seen as the appropriate role for this portion of the Soviet strategic force.

Conclusion

In its development, Soviet strategic ballistic missile guidance resembles the U.S. pattern much more strongly than either the French or Chinese patterns. As in the United States, but not in France and China, increasing missile accuracy has been a very high Soviet priority. The general development of inertial instruments, the quality of the MOM gyrocompass, the introduction of high beta reentry vehicles, and the increased sophistication of Soviet stellar-inertial guidance all point in the same direction.

The goal of enhanced accuracy was achieved differently in the Soviet Union than it was in the United States. There are some similarities—roughly the same pattern of use of inertial and stellar-inertial technology, for example—but many differences. Soviet inertial systems use different gyroscopes and accelerometers, different guidance mathematics, more redundancy, different levels of error compensation, and different

78. See Wright, *Soviet Missiles,* especially 76–77, on developments more recent than the SS-N-18. Department of Defense, *Soviet Military Power, 1988* (pp. 48–50), hedging its bets by the use of the conditional, states that "Improved accuracy of the Soviets' latest SLBM [submarine-launched ballistic missile], as well as possible efforts to increase SLBM reentry vehicle size and warhead yield, would confirm Moscow's plans to develop a hard-target kill capability for its SLBM force."

means of prelaunch alignment. Soviet stellar-inertial guidance systems differ from American in the fundamental matter of the number of star-sightings.

Going beyond noting differences to explaining them is tricky. Here the lack of direct interviews with Soviet missile guidance designers has its consequence. Without access to their technical beliefs, which cannot be assumed to be the same as American, explanation of a given technical difference becomes deeply problematic. The Soviets may do things differently from the Americans because they have to, for example through lack of access to a particular technology. Or the reason may be that Soviet goals are different. But we must always bear in mind a third possibility, often neglected in Western writing with its confident assumption that technical differences represent a Soviet "lag" behind the West. The Soviets may do things differently because they believe their way is correct, and U.S. technology is wrong. We have, for example, seen some tentative evidence, notably in the use of multiple star-sightings in stellar-inertial guidance, of a Soviet preference for hardware solutions rather than complex software.

The lack of direct interviews also makes explaining the Soviet goal of increased accuracy a perilous business. We can, I think, dismiss the idea that it was simply a natural and inevitable direction of technical change. The general reasons given in chapter 4 count against this explanation. The cases of France and China also show that in different circumstances enhanced accuracy did not appear a natural goal. And the difficult circumstances of Soviet guidance system development also count against any idea of enhanced accuracy easily or automatically achieved. There was no dynamic, wider inertial industry that missile guidance could simply draw upon, nor any burgeoning indigenous computer industry creating "sweet" technologies that present themselves effortlessly to guidance system designers. Enhanced accuracy seems to have been achieved only with great cost and considerable difficulty.

If not a natural trajectory of technology, the effort to increase missile accuracy could still be the result of the "capturing" of Soviet national policy by ambitious guidance engineers. This explanation cannot be dismissed a priori, but no evidence in its favor exists, and it would not be keeping with the general pattern of Soviet weapons development, where entrepreneurship by technologists seems less important than initiative from the political and military leadership.[79]

So the most plausible explanation of the high priority awarded missile accuracy is simply that the Soviet military and political elite have desired the counterforce capability that accuracy brings. Such evidence as exists points to this. Thus as long ago as the early 1960s, when the effort to increase accuracy was only in its infancy, Marshall Sokolovsky's authoritative *Nuclear Strategy* made it quite clear that this elite saw counterforce targeting as perfectly proper.[80] Nuclear preemption, a strike based on reliable intelligence that the Soviet Union was about to be attacked, has been seen in much Soviet strategic writing as a proper strategy, and effective preemption clearly requires counterforce accuracy.[81]

Some of this had already changed under Brezhnev, and it is all now changing fast under Gorbachev. The idea of nuclear war being winnable has been disavowed, a doctrine of "sufficiency" more akin to finite deterrence has gained weight, and, most important, there seems to be the willingness to see these changes translated into major cuts in Soviet nuclear forces—even the weapons most suitable for preemption such as the SS-

79. Matthew Evangelista, *Innovation and the Arms Race: How the United States and the Soviet Union develop New Military Technologies* (Ithaca, N.Y.: Cornell University Press, 1988).

80. V. D. Sokolovsky, ed., *Nuclear Strategy: Soviet Doctrine and Concepts* (New York: Praeger, 1963), for example, p. 280: "The targets in a modern war will be the enemy's nuclear weapons, his economy, his system of government and military control, and also his army groups and his navy in the theaters of military operation."

81. There is considerable literature on the nature of Soviet nuclear strategy. While this contains many disagreements, I find no well-supported argument that runs counter to the view expressed here. See, for example, Raymond L. Garthoff, "Mutual Deterrence and Strategic Arms Limitation in Soviet Policy," *International Security*, Vol. 3, No. 1 (Summer 1978), 112–147; Donald G. Brennan, "Commentary," *International Security*, Vol. 3, No. 3 (Winter 1978), 193–198; Benjamin S. Lambeth, "The Political Potential of Soviet Equivalence," *International Security*, Vol. 4, No. 2 (Fall 1979), 22–39; John Baylis and Gerald Segal, eds., *Soviet Strategy* (London: Croom Helm, 1981); John Erickson, "The Soviet View of Deterrence: A General Survey," *Survival*, Vol. 24, No. 6 (November/December 1982), 242–49; Holloway, *Soviet Union and the Arms Race*, chapter 3; Michael MccGwire, *Military Objectives in Soviet Foreign Policy* (Washington, D.C.: Brookings, 1987); Stephen M. Meyer, *Soviet Theatre Nuclear Forces*, Adelphi Papers Nos. 187 and 188 (London: International Institute for Strategic Studies, Winter 1983/84).

20 intermediate-range ballistic missile and SS-18 heavy ICBM.[82]

But even when the idea of nuclear preemption carried greater weight than now, it did not amount to a Soviet commitment to "first strike," at least as far as one can infer from the shape of the Soviet arsenal. Two crucial episodes seem to have posed to the Soviets a choice between increasing the accuracy of their arsenal and increasing its security. The episodes were the choice between radio and inertial guidance and the decision to make a substantial proportion of the ICBM force mobile. In each case the Soviets took the path that, according to Western technical beliefs at least, gave increased security at the cost of accuracy, and therefore emphasized retaliatory over first-strike capability.

Differences within the Soviet ICBM force are also interesting in this respect. The existence in one generation of ICBMs of both high- and medium-accuracy systems, together with differences in warhead yields, suggests a more differentiated approach to nuclear targeting than any simple reliance on a "first strike" would imply. The simplest interpretation of this pattern is the allocation of different types of weapons system to different types of target: high-accuracy, large-yield ICBMs for hard targets such as missile silos and command and control bunkers; and medium-accuracy, often lower-yield, ICBMs for softer military, economic, and population targets.

So if strategic requirements have shaped the Soviet arsenal, these requirements are diverse. The arsenal includes the high-accuracy missiles needed for a preemptive strike against Western forces. But the presence in greater numbers of medium-accuracy systems would indicate that the Soviet authorities do not believe that a war would terminate after counterforce strikes alone. And the measures taken to protect Soviet forces would indicate no confidence that the Soviet Union would be able to preempt. This picture, interestingly enough, is very similar to what an outsider, deprived of detailed information on the processes that have created it, would deduce from inspection of the current U. S. arsenal: that it departs from

82. See Meyer, *Theatre Nuclear Forces*, for a useful discussion of Brezhnev's pledge not to be the first to use nuclear weapons, and the likely connection of that pledge to a strategy of conventional preemptive attack on NATO nuclear forces.

the requirements of finite retaliatory deterrence but not in a way wholly consistent with full priority to first-strike capability.

At the level of the assessment of Soviet intentions, even before the Gorbachev era, the analogy is perhaps reassuring. But it ought also to stand as a methodological warning. What we have learned from investigation of the processes that have led to the current U.S. arsenal is that there are cases in which no consistent strategic philosophy underlay its shaping.[83] So making inference about "American strategic goals" from this arsenal would be quite misleading. We must thus be cautious in concluding that the Soviets' differentiated arsenal is simply the result of rational shaping of different portions of that arsenal to have the optimum characteristics for attacks on different portions on the spectrum of Western targets. It could be, for example, that the Strategic Rocket Forces contain factions embracing different strategic philosophies, and that a differentiated arsenal is an outcome of their conflict, rather than the result of logical deduction from a single set of targeting requirements. We simply lack the evidence necessary to be sure.

What of the influence on guidance system design of the Soviet structure of research, development and production? From organizational sociology, we can predict that a stable, highly compartmentalized structure would tend to produce an incremental form of technical change, involving continuous refinement of the different components of a system, but little change in the system's overall configuration.[84] Though this form of change is not universal in Soviet military technology,[85] it certainly seems prevalent in the guidance sphere, with, most noticeably, a history of some forty years refinement of the same basic type of inertial sensor.

This pattern, however, is to be found almost as strongly in the United States. The Draper fluid-floated instrument has a history of continuous evolution from the 1940s onward that is similar to the German-Soviet gas-supported instrument. Despite their many differences, in their overall "evolutionary"

83. Trident I (C4) is the best example, being essentially a compromise between the proponents of assured destruction and those of counterforce.
84. See, for example, Tom Burns and G. M. Stalker, *The Management of Innovation* (London: Tavistock, 1961).
85. Holloway, "The Soviet Style."

form of technical change the similarities between the Soviet and American histories of strategic ballistic missile guidance technology strike one more forcibly than the differences.

Significant differences between the United States and the Soviet Union in the form of technical change are thus to be found primarily in the applications of inertial technology to areas other than missile guidance, such as submarine and air-craft navigation. In these there appears to be a much greater Soviet disadvantage, and it could well be that compartmental-ization is part of the problem. In contrast, applications in the West have been characterized by major changes in basic tech-nology and by technology transfer from one application to the other.

Finally, what have we learned about the significance of tech-nology acquired from the West in enhancing Soviet missile accuracy? I think we must dismiss the more extravagant West-ern claims about Soviet dependence on technology "stolen" from the West, and we must therefore also be skeptical that even the tightest set of technology export controls would be effective in preventing increases in Soviet missile accuracy. Only one episode of technology transfer was crucial to the development of Soviet missile guidance—access in the 1940s to German technology and technologists. Even then, a power-ful Soviet indigenous tradition of research existed that made possible assimilation of the knowledge gained. Since then, acquisition of Western technology in missile guidance has not been crucial. The key apparent instance of it—the famous "ball-bearing" case—is spurious.

Much of the debate on Soviet acquisition of Western tech-nology is not rooted in the realities of technology transfer, especially in an area such as high-accuracy missile inertial guid-ance, where production processes are far from routinized and tacit knowledge, embodied in people rather than documents or blueprints, is irreplaceable.[86] The "Intelligence Community Report on Soviet Acquisition of Western Technology" asserts that "The Soviets will . . . give top priority to acquiring infor-mation on the latest generation of U.S. inertial components upon which the MX ICBM and the Trident SLBM guidance

86. The significance of tacit, embodied knowledge in science and technology is a robust finding of the sociology of science and technology. See chapter 2, note 40.

systems are based."[87] This may well be so; but even were the Soviets to acquire actual MX or Trident instruments, the effect on Soviet programs would not necessarily be large. There is no reason to expect such an acquisition to disturb the well-established Soviet preference for sensor technology of a different type. And even if it did, it is far from certain that possession of information on these components, even actual samples of them, would allow the Soviets successfully to replicate the delicate "labor-sensitive" processes of production that generate them.

Indirect evidence of this is provided by the same report when it asserts that " the Soviets have yet to demonstrate a capability to deploy reliable, accurate airborne inertial navigation systems for long-range navigation and weapons delivery. Thus, while long used in the West, these systems are still prime candidates for acquisition."[88] But such Western systems have been used in aircraft in service with the world's airlines since the early 1970s, so it is hard to believe that Soviet acquisition of a state-of-the-art Western inertial navigator was impossible. If technology transfer in the inertial area was the straightforward matter the report seems to imply, then the Soviet difficulty referred to surely would have vanished by now.

To say that technology acquired from the West has not been of major significance to the development of Soviet ballistic missile accuracy since 1950 is not to deny that the Soviets have had to deal with constraints absent in the United States. The most significant have been in microelectronics and digital computing, areas where Soviet shortcomings are now publicly acknowledged in the Soviet Union.

Yet even an apparently fundamental constraint such as this can be "designed around," for example by adopting guidance mathematics of a type such that most of the calculations are done in advance, rather than in flight. It is also true that different technical styles, such as the apparent Soviet distrust of complex software solutions to guidance problems, can make what one set of designers would see as a great constraint seem much less onerous to another. It is dangerous to assume that the technical world looks identical to Soviet and American guidance system designers.

87. "Intelligence Community Report," 69.
88. Ibid.

The Construction of Technical Facts

It is time to make explicit an issue that has been implicit in the book thus far, but becomes inescapable when we ponder the extent to which it is safe to assume that Soviet designers with the same goals as their American counterparts would make the same technical decisions. How deep does the flexibility of the technical go? If we dig deep enough, can we not find a solid foundation of technical fact, matters that rationally cannot be disputed? Is there not, ultimately, a sphere of the technical that is genuinely insulated from politics and the clash of organizational interests?

In this chapter I look for an answer to this question in three stages. In the first stage I examine perhaps the most crucial fact of all about nuclear missiles. Do we know that they would work: that is, would their warheads explode? Nowadays, there is concern over the more restricted issue of missile reliability, but the most radical form of doubt—that perhaps no missile would work—seems to have vanished. In the United States of the early 1960s, however, thoroughgoing doubt about whether missiles would work could be found at the heart of the military establishment. We will examine both the grounds for the doubt and the constituency to which skepticism appealed.

Examining whether we know missiles will work raises this chapter's central theme: the testing of technologies. For here, surely, are to be found the solid factual foundations underlying technological politics.

Testing, however, turns out not to be the simple matter it appears at first sight. This is particularly the case in the issue that will form the second stage of my discussion: our knowledge of missile accuracy. Missile accuracies very often are taken to be unproblematic facts—that is how I have generally treated them in previous chapters. But just how is the accuracy of a missile known? The obvious test is to fire the missile at a particular target and see how far away it lands. But this is not taken by insiders as on its own revealing the accuracy of a missile. Nor can it reveal to the Americans how accurate a Soviet missile is, because the former cannot know the target.

In the early 1980s, public controversy erupted in the United States as to whether missile accuracies were facts at all. I examine this controversy, trying to tease apart the very different strands of skepticism and again investigating the different constituencies to which skepticism appealed. I also discuss the beliefs of those closest to the production of missile accuracy figures, and argue that their position is distinctive. They do not share the radical skepticism of the public critics, nor do they believe missile accuracy figures to be straightforward facts.

There are strong legal and political constraints on the possible forms of missile testing, and thankfully, there has been no operational use of nuclear missiles against which to compare the record of testing. In the third stage of my discussion I shift inside the black box of guidance to the inertial sensors: the gyroscopes and accelerometers. There are no similar contraints on the testing of these and no barrier to their use. Yet we shall see that even here testing is a complex matter, and the facts it generates are facts only within a wider web of assumptions and procedures. So even the apparently simple question "which is the most accurate gyroscope?" cannot be answered in a way that compels consensus, if a skeptic is determined and resourceful enough.

Let me emphasize that none of what follows should be read as criticism of inertial guidance. The historical or sociological literature that exists on technological testing suggests to me that broadly similar conclusions could be reached for other fields of technology, including fields much less dramatic and particular than nuclear missiles.[1] And, as noted in chapter 1,

1. See Walter G. Vincenti, "The Air-Propeller Tests of W. F. Durand and E. P. Lesley: A Case Study in Technological Methodology," *Technology and Culture*, Vol. 20 (1979), 712–751; Edward W. Constant, *The Origins of the Turbojet Revolution* (Baltimore: Johns Hopkins University Press, 1980), chapter 1; Edward W. Constant, "Scientific Theory and Technological Testability: Science, Dynamometers, and Water Turbines in the Nineteenth Century," *Technology and Culture*, Vol. 24 (1983), 183–198.

recent analyses have shaken received images of scientific knowledge as the simple result of human rationality's encounter with
reality. So, although undoubtedly there are specific circumstances that affect the construction of technical knowledge of
the characteristics of nuclear missiles, the features of technological knowledge discussed here may actually be quite general.

Will Nuclear Missiles Work?

The best single statement of why it might be wrong to be
certain that nuclear ballistic missiles would work came in 1961
from the Armed Services Committee of the U.S. House of
Representatives:

Who knows whether an intercontinental ballistic missile with a
nuclear warhead will actually work? Each of the constituent elements
has been tested, it is true. Each of them, however, has not been tested
under circumstances which would be attendant upon the firing of
such a missile in anger. By this the committee means an intercontinental ballistic missile will carry its nuclear warhead to great heights,
subjecting it to intense cold. It will then arch down and upon reentering the earth's atmosphere subject the nuclear warhead to
intense heat. Who knows what will happen to the many delicate
mechanisms involved in the nuclear warhead as it is subjected to
these two extremes of temperature?[2]

By 1961 the United States had conducted many nuclear
weapons tests; it had also conducted many flight tests of ballistic
missiles. But all U.S. missile tests had used missiles without
warheads, while the bombs in ordinary nuclear weapons testing
had either been detonated in fixed positions or dropped from
towers or aircraft.[3] The challenge mounted by the Armed Services Committee was therefore this: testing the components of
a nuclear missile system separately was not adequate against
the risk that some feature of the way they were combined
would prevent the nuclear missile from working.

The challenge was not treated as absurd. Indeed, doubts
were perhaps strongest at senior levels of the U.S. armed ser-

2. Quoted in K. Johnsen, "Senate Vote emphasizes Proven Weapons," *Aviation
Week and Space Technology* (May 22, 1961), 22.
3. Some nuclear warheads were lofted by missiles into the upper atmosphere
or space for explosion there, but critics could argue that this was different
from being carried over the full trajectory of a missile, including reentry.

vices. The Joint Chiefs of Staff pressed for missile tests to be conducted with warheads on board, and sought presidential permission to fire an Atlas ICBM with a live warhead from Vandenberg Air Force Base in California.[4] The trajectory would take the missile "near some populated areas," but, though it was argued that "the need to proof-fire the warhead outweighs the risk of the Atlas exploding during launch or in flight and damaging surrounding communities," the Kennedy Administration was reluctant to authorize such a test.[5]

A politically more palatable alternative was also being canvassed, however, where the risks were seen as less, and were to Pacific Islanders or members of the armed services, not American civilians. The administration was prepared to authorize this alternative. On May 6, 1962, the nuclear submarine USS *Ethan Allen* fired a live Polaris missile on a 1,200-mile trajectory towards the nuclear testing ground at Christmas Island.[6] Operation Frigate Bird, as it was called, was a success. *Aviation Week* reported that the shot "hit right in the pickle barrel."[7] The warhead exploded with a force estimated at half a megaton.

This looks like as clearcut a case as one could get: an argument put forward, subjected to a test, and decisively refuted. But there were several ways skeptics could minimize the effect of Operation Frigate Bird on their case. One would have been to argue that its success was a fluke, which we now believe may in one sense actually have been the case.[8] Another was to point out that it was a test of a relatively short-range submarine-

4. Debate over the Atlas test is recorded in the notes of the National Security Council meeting of April 18, 1962. I am grateful to Graham Spinardi for a copy of these notes, which are in the Lyndon Baines Johnson Library.
5. Anonymous, "Washington Roundup," *Aviation Week and Space Technology* (May 14, 1962), 25.
6. Lockheed Missiles and Space Company, Inc., *The Fleet Ballistic Missile System: Polaris, Poseidon, Trident* (Sunnyvale, Calif.: Lockheed Missiles and Space, n.d.), 20; Chuck Hansen, *U.S. Nuclear Weapons: The Secret History* (Arlington, Texas: Aerofax, 1988), 85.
7. Anonymous, "Live Polaris Launch," *Aviation Week and Space Technology* (May 14, 1962), 35.
8. The W47 warhead used on Polaris A1 and A2 had a checkered history involving corrosion of the fissile materials and problems with the mechanical arming system. By 1966 it was being estimated by the Livermore nuclear weapons laboratory that between half and three quarters of W47 warheads would fail to detonate. Hansen, *U.S. Nuclear Weapons*, 205.

launched missile, not an ICBM with its quite different trajectory. A third was to suggest that the modifications to the missile needed to minimize the risk of the test invalidated the latter.

In actuality, Operation Frigate Bird was not treated as settling the matter. Skeptics did mobilize these arguments to claim that doubts about missiles working should not be regarded as eliminated. "There are some unknowns and uncertainties you should know about," Air Force Chief of Staff Curtis LeMay told the Defense Appropriations Subcommittee of the House of Representatives in 1964. "One is that we have only had one test, it was not under fully operational conditions, we fired one Polaris out in the Pacific with a warhead on it. It was not truly operational. It was modified to some extent for the test."[9]

The matter even became an issue in the 1964 campaign for the presidency between Republican challenger Barry Goldwater and Democratic incumbent Lyndon Johnson. As noted in chapter 3, the Kennedy Administration had greatly expanded American ballistic missile forces. But, argued Goldwater, what was the use of this if we did not know they would work? Implicitly drawing the distinction between Frigate Bird's Polaris and an ICBM, Goldwater noted in his platform *Where I Stand,* "I have raised, and will continue to raise until all the facts are in, fundamental questions about the reliability of our intercontinental ballistic missiles. It is not a question of theoretical accuracy. The fact is that not one of our advanced ICBMs has ever been subjected to a full test (of all component systems, including warheads) under simulated battle conditions."[10]

By 1964, however, a live ICBM test like the one that had aroused the qualms of the Kennedy Administration was impossible. In the face of substantial domestic opposition, the Democratic administration in 1963 had committed the United States to the Partial Test-Ban Treaty, prohibiting nuclear tests in the atmosphere, underwater, or in outer space. So, Goldwater continued, "[t]he Administration's decision to enter into a test ban treaty precluding all atmospheric nuclear explosions means

9. G. C. Wilson, "GOP to Capitalize on LeMay's Charges," *Aviation Week and Space Technology* (April 20, 1964), 26.
10. "Goldwater Defense Philosophy," *Aviation Week and Space Technology* (August 31, 1964), 11.

that we cannot properly test even our present missile systems
. . . we are building a Maginot line of missiles."[11] LeMay and
Goldwater had an alternative to propose to reliance on missiles:
the manned bomber. By the early 1960s, the missile revolution
described in chapter 3 had so successfully challenged the hegemony of the bomber that it was far from certain whether there
would be any new U.S. strategic bomber to succeed the 1950s
B-52. The bloodless theoreticians of the Kennedy Administration (as Air Force generals such as LeMay saw them) had
concluded that ballistic missiles were superior weapons to
bombers, and were not inclined to authorize the large expenditure needed to deploy a supersonic bomber to replace the
subsonic B-52.

Doubt that missiles would work was thus an argument for a
new bomber. The statement by the House Armed Services
Committee was a justification for voting $337 million to con-
tinue manned bomber production the Kennedy Administration wished to close down. LeMay was the leader of the fight
to protect the future of the manned bomber. Even Goldwater
was not simply seeking a way to criticize Democratic defense
policy. As well as senator for Arizona, he was an Air Force
Reserve major general, who had, according to *Aviation Week*,
"long identified himself with the bomber faction."[12]

So the constituency of the challenge was clear: it was what
we might call the "bomber lobby." Since the mid-1960s, however, doubts about missiles working have almost disappeared,
at least in their strong form such as the suggestion that perhaps
no ICBM will work. Senator Goldwater still notes on occasion
that "we have never fired a missile with a warhead."[13] Sometimes the possible consequences of this are remarked upon.
"Does a test conducted dead still at the bottom of a hole provide
reliable assurance that the same weapon will work after travelling several thousand miles an hour in a reentry vehicle

11. Ibid.

12. Anonymous, "Fixed planned for Minuteman Deficiencies," *Aviation Week
and Space Technology* (February 2, 1964), 26–27.

13. Senate Armed Services Committee, *Department of Defense Authorization
Hearings, Fiscal Year 1979, Part 9, Research and Development* (Washington, D.C.:
U.S. Government Printing Office, 1978), 6470. Operation Frigate Bird had
been forgotten to such an extent that the Air Force general being questioned
by Senator Goldwater did not contest Goldwater's assertion.

through extremes of temperature?"[14] But now the challenge seems to have no force behind it.

I can only speculate on why. It can hardly be put down to new evidence. There have been no further American experiments like Operation Frigate Bird, so in that sense matters remain as they were in 1964.[15] The challenge's constituency, the bomber lobby, did not go away. It finally achieved its goals in the 1980s with the deployment of the B-1 bomber and advanced development of the B-2 "Stealth" bomber. Though that success might be seen as having removed the need to dispute the missile's claims, the challenge had declined well before then.

An immediate cause of the issue's decline from public view may have been the way the nuclear issue turned into electoral disaster for Goldwater, who was branded as irresponsibly bomb-happy in the 1964 campaign. Related to this, but perhaps of greater long-term significance, was the way the ban on nuclear testing in the atmosphere proved extremely popular with the public in the United States and elsewhere, ending years of fear about the consequences of fallout from atmospheric tests. It would have been extraordinarily dangerous politically for a U.S. administration since then to have abrogated the test ban.

So it could be that one, paradoxical, reason for the decline of the challenge's plausibility is precisely this: the impossibility now of testing its truth by firing a missile with a live warhead. It has been noted in science that the plausibility of a hypothesis can be raised simply by the existence of moves to test it, and

14. A. L. Meyers, "Nuclear Testing," *Bulletin of the Atomic Scientists*, Vol. 43, No. 8 (August-September 1986), 66–67.

15. The People's Republic of China conducted a live missile test, apparently successfully, in 1966, and another in 1976. The Soviet Union had also conducted such tests prior to the 1963 Partial Test-Ban Treaty. See Richard W. Fieldhouse, "Chinese Nuclear Weapons: An Overview," in Stockholm International Peace Research Institute, *World Armaments and Disarmament: SIPRI Yearbook 1986* (Oxford: Oxford University Press, 1986), 104, and F. Hussain, *The Future of Arms Control: Part IV. The Impact of Weapons Test Restrictions*, Adelphi Papers, No. 165 (London: International Institute for Strategic Studies, 1981), 53fn. 35. There is no evidence that these overseas tests were important, before or after 1963, in influencing beliefs about U.S. missiles, however, and it is unclear how much detail was known in the United States about the procedures used in them, or indeed how widely it was known they had taken place at all.

perhaps the reverse also is true.[16] Perhaps, too, the transformation of the issue, already beginning to be evident in 1964, into the esoteric (and classified) calculus of missile reliability blunted the challenge. It was on this terrain that the Democratic administration, especially Secretary of Defense Robert McNamara, chose to reply to LeMay and Goldwater.[17]

Constructing Accuracy

The issue raised by this first, largely vanished challenge was whether the separate testing of missiles and their weapons was sufficient to prove that the latter would work in use, when the two would be combined. Whether testing was sufficiently like use to allow inferences to flow was also the central issue in the more recent controversy over whether missile accuracies were facts about missiles or artifacts of the processes of testing them.[18]

Since the dispute has concerned ICBMs in particular, let me begin by describing how these are tested. American ICBMs are flight tested by being fired west over the Pacific from Vandenberg Air Force Base on the coast of California.[19] The target is nearly always a point in the large enclosed lagoon, some ten miles wide and sixty miles long, at Kwajalein Atoll in the Marshall Islands.[20] There is a host of radars and other sensors around Kwajalein Lagoon, so much more information can be gained from tests with their target there, though ICBMs are

16. Bill Harvey, "Plausibility and the Evaluation of Knowledge: A Case-Study of Experimental Quantum Mechanics," *Social Studies of Science*, Vol. 11 (1981), 95–130. I owe the suggestion of a parallel to Harry Collins.

17. See, e.g., "McNamara Discusses Retaliatory Forces," *Aviation Week and Space Technology* (March 2, 1964), 74–82. McNamara also broadened the question of reliability to "dependability," enabling him to bring to bear against the bomber matters such as the "survivability of the soft bombers and their ability to penetrate enemy defenses" (ibid., 78).

18. Note the pervasive role in these matters of "similarity judgments": see Barry Barnes, *T. S. Kuhn and Social Science* (London: Macmillan, 1982).

19. In the past ICBMs also were tested by being fired east over the Atlantic from Cape Canaveral in Florida.

20. For a description of the Kwajalein Missile Range, first established in 1959, and an account of the controversy surrounding its impact on the lives and rights of Marshall Islanders, see Giff Johnson, *Collision Course at Kwajalein: Marshall Islanders in the Shadow of the Bomb* (Honolulu, Hawaii: Pacific Concerns Resource Center, 1984).

still sometimes fired toward other targets, such as points in the Pacific at greater distances from Vandenberg.

Early tests while a missile is being developed are normally above-ground "pad launches." But later tests are from a silo at Vandenberg. To test missiles that have already been deployed, one is selected at random. If it can successfully be brought to alert status in its operational silo, its reentry vehicles are removed and transported to a facility in Texas, where the live nuclear warheads are taken out and replaced by telemetry equipment. The missile is taken to the test silo at Vandenberg. The reentry vehicles are replaced, a mechanism for exploding the missile if it goes off course is added, and the guidance system is aligned and calibrated. The missile is brought to alert and fired by a crew from the operational ICBM fields. Ground-based radars, telescopes, and instruments within the reentry vehicles track the test, and a "splash net" of hydrophones is used to record where the reentry vehicle hits the surface of the lagoon at Kwajalein.[21]

Statistical analysis of these impacts provides a circular error probable (CEP), officially defined as the radius of the circle around the target within which 50 percent of warheads will fall in repeated firing. Since flight testing will involve trajectories with different geometries and to some extent different ranges, in fact a set of circular error probable figures will be produced. And since the number of tests is finite, figures for the circular error probable will have a statistical uncertainty: they will have a "confidence interval" associated with them.[22]

What the Air Force needs to know, of course, is not the accuracy of an ICBM when fired at Kwajalein, but its accuracy when fired at a particular target in the Soviet Union. Since the latter patently cannot be discovered directly, it has to be extrapolated from the results on the test range. This is no straight-

21. The description in this paragraph is drawn from Matthew Bunn and Kosta Tsipis, *Ballistic Missile Guidance and Technical Uncertainties of Countersilo Attacks*, MIT Program in Science and Technology for International Security report No. 9 (Cambridge, Mass.: MIT Program in Science and Technology for International Security, 1983), appendix C.

22. See R. A. Moore, "The Evaluation of Missile Accuracy," in R. F. Kiddle et al., eds., *An Introduction to Ballistic Missiles, Volume IV: Guidance Technologies*, revised edition (Los Angeles: Air Force Ballistic Missile Division and Space Technology Laboratories, Inc., 1960), 111–227.

forward empirical matter, because errors with different causes are understood to extrapolate differently. Some, for example, will vary more or less linearly with range to the target. Others will not. So what is needed is knowledge not just of overall error but of its causes.

The mathematical model embodying such knowledge is known as the missile's error budget, so called because in it different proportions of overall error are apportioned to different causes.[23] The precise process of its construction has not been described in open literature, but it is clear that the data used include not just the results of test firings, but also of information from laboratory, rocket-sled, and ultracentrifuge tests of components, estimates of the errors in mapping and in models of the gravitational field, and so on.[24]

It is also worth noting that a mathematical model of error processes is used not just to construct an operational accuracy figure. It is also used in the actual process of guidance, since with modern missiles (especially American), the on-board computer seeks to correct for errors that are believed to be predictable. In American stellar-inertial guidance, for example, a model of error processes is needed for the statistically optimum guidance correction to be made following the single star-sight.

Because of this centrality of the theoretical understanding of errors, much more is expected from the test range than just a final, total miss distance. Thus test-range instrumentation is important. In developing and seeking to verify the theoretical model, it is most helpful to have independent sources of data (such as from ground-based sensors) on the position and velocity of the missile at all times. These can then be compared with

23. Strictly, the apportionment is of the square of overall error, equivalent of the statistician's "variance," since that is mathematically much more tractable. For simple unclassified examples of error budgets see D.G. Hoag, "Ballistic-missile Guidance," in B. T. Feld et al., eds., *Impact of New Technologies on the Arms Race* (Cambridge, Mass.: MIT Press, 1971), 19–108 and Bunn and Tsipis, *Technical Uncertainties.*

24. See Moore, "Missile Accuracy"; A. N. Drucker, "Performance Analysis of Rocket Vehicle Guidance Systems," in G. R. Pitman, ed., *Inertial Guidance* (New York: Wiley, 1962), 329–391; and General A. Slay, testimony to House of Representatives Armed Services Committee, *Hearings on Military Posture, Fiscal Year 1979* (Washington, D.C.: U.S. Government Printing Office, 1978), part 3, book 1, 299–358.

what the guidance system reports, through telemetry, as the condition of the missile.

The Soviet Union possesses an interesting, little-discussed advantage over the United States in these matters: the ability to conduct ballistic missile test flights over land. The inability of the United States to do likewise is, of course, a matter of politics and legality, not geography. But, at least until the advent of accurate satellite-based tracking systems, first employed in the U.S. Navy's Improved Accuracy Program of the 1970s, land-based tracking stations in fixed, known positions were seen as a considerable advantage.

Soviet investment in test-range instrumentation, such as land-based tracking stations, has certainly been heavy, with some of the technology allegedly acquired illicitly from the West.[25] This would seem to indicate that Soviet procedures for the construction of missile accuracy figures are similar in nature to American. A final miss distance is not seen as an adequate measure of accuracy, so considerable effort has to be devoted to understanding the causes of those errors.

A theoretical understanding of errors is at least equally central to the construction of American knowledge of Soviet missile accuracies. True, the United States can often tell where Soviet missile test flights have landed. "Key Hole" intelligence satellites can (cloud cover permitting) film the small craters made by Soviet reentry vehicles as they crash down on the Kamchatka Peninsula, the end of the main Soviet missile test range, which runs from west to east across the Soviet Union.[26] Intelligence vessels can record the splashdown in the Pacific of longer-range tests. Large radars at Shemya Island in the Aleutians, at Kwajalein, and also on a naval vessel, the USS *Observation Island,* can track Soviet reentry vehicles, at least up until the point where the earth's curvature interferes.

The problem is to know what the Soviets were aiming at. "They don't paint a bullseye on Kamchatka," said more than

25. This is one of the allegations in "Intelligence Community Report on Soviet Acquisition of Western Technology," *U.S. Export Weekly* (April 13, 1982), 58–70.
26. Not all Soviet tests end here—longer range ones continue and splashdown in the Pacific. Most ICBM test launches are from the main Soviet space and ballistic missile center at Tyuratam near the Aral Sea, but there is a second test center at Plesetsk, used mainly for testing intermediate-range missiles (Bunn and Tsipis, *Technical Uncertainties,* 131–132).

one interviewee. This does not actually stop one from drawing inferences about accuracy from patterns of impact points, but the basis of them has not been revealed, and insiders seem to lack wholehearted confidence in them.[27]

So the analysis of impact points is less significant by itself than as one of many factors in the construction by American analysts of an error budget for the Soviet missile in question. The American error budget for a Soviet missile will be simpler than that for a U.S. missile, and clearly there will be greater difficulty in finding values for the component parts of the budget. But the construction of it is still the central task of U.S. intelligence's guidance system analysts.

As we saw in chapter 6, intercepted telemetry from Soviet

27. For example, see Undersecretary of Defense for Research and Engineering William J. Perry's testimony to the house of Representatives Armed Services Committee, *Hearings on Military Posture, Fiscal Year 1980* (Washington, D.C.: U.S. Government Printing Office, 1979), 100:

Mr. Dickinson How do we know what they [the Soviets] are shooting at?

Dr. Perry Well that is the—

Mr. Dickinson I remember Robin Hood, so that he wouldn't tip off his hand as to how accurate he was, he would shoot at a different point in the bullseye.

Dr. Perry That is a fundamental question you have to answer in this analysis. There is a satisfactory answer to it, but it is very complicated to try to explain.

Mr. Dickinson I couldn't understand it anyway. But you do have confidence you know where they are shooting, rather than them building in a bias to offset it?

Mr. Perry Yes; I do. It has to do with the way they fly their [deleted]. You see the [deleted].

Mr. Ichord Any further questions on the [deleted].

Mr. Dickinson I have more questions on that.

Mr. Hughes Yes, sir.
[Deleted]

Dr. Perry [Deleted.]

The furthest my interviewees from the intelligence community could be drawn on this was to assert that methods of impact point analysis take into account the possible deliberate disguising of accuracy, and to point out that if impact points in repeated tests of the same missile form a cluster, then an upper bound on the missile's circular error probable could be deduced from the size of the cluster. The missile may of course be more accurate than the size of the cluster indicates, since targets may be being changed from test to test, but it is unlikely to be less accurate.

missile testing is probably the most important single input into this, though it is vital to be able to correlate this with the results of radar tracking of trajectories. The key original site for both telemetry interception and the tracking of the early part of test trajectories was in the north of Iran, reasonably close to Tyuratam. The loss of these facilities as a result of the Iranian revolution was a major blow to the United States. Monitoring from the U.S. base at Diyarbakir in Turkey has grown in significance since then, and, interestingly, a covert U.S. monitoring station is now being operated in the People's Republic of China.[28] American intelligence satellites have also been playing an increasing role in monitoring Soviet missile tests. It is even possible sometimes to film Soviet reentry vehicles in flight: that is the role of a large airborne telescope carried by an RC-135X aircraft, flown by the Sixth Strategic Reconnaissance Wing based in the Aleutians.[29]

Challenging Accuracy

The specific problem of knowing the accuracy of the other side's missiles was only one theme in the controversy that developed over whether missile accuracies were facts. Though the issue was not a new one, it received its greatest prominence in the early 1980s. A variety of critics publicly cast accuracy into doubt, and their skepticism received widespread press coverage. They questioned whether the Soviets themselves knew the accuracy of their ICBMs and, above all, whether the proud accuracy specifications of American ICBMs, especially the new MX, had any real validity outside the range from Vandenberg to Kwajalein.

Their grounds for skepticism can, somewhat arbitrarily, be divided into two.[30] The first set of grounds is interestingly

28. Bunn and Tsipis, *Technical Uncertainties*, 138–143. William M. Arkin and Richard W. Fieldhouse, *Nuclear Battlefields: Global Links in the Arms Race* (Cambridge, Mass.: Ballinger, 1985), 73. Although it is now dated, a useful review of U.S. means of monitoring Soviet missile tests is F. J. Moncrief, "SALT Verification: How we monitor the Soviet arsenal," *Microwaves* (September 1979), 41–51.
29. Craig Covault, "Alaskan Tanker, Reconnaissance Mission Capabilities Expanded," *Aviation Week and Space Technology* (June 6, 1988), 70–72.
30. For brevity, I am lumping together the arguments of various authors who by no means all agree, using the following statements of the critics' case:

the inverse of a common form of skepticism in science. A frequent reaction to experiments with heterodox results is the suggestion that the experiments have been performed incompetently.[31] In the dispute about missile accuracy, the skeptics argue that the "experiments" involved—the test-range firings, both American and Soviet—are performed, not incompetently,

J. Edward Anderson, Letter to Editor, *Scientific American*, Vol. 222 (May 1970), 6; J. Edward Anderson, "First Strike: Myth or Reality," *Bulletin of the Atomic Scientists*, Vol. 37 (November 1981), 6–11; J. Edward Anderson, "Strategic Missiles Debated: Missile Vulnerability—What you Can't Know!" *Strategic Review*, Vol. 10 (Spring 1982), 38–42; Andrew Cockburn and Alexander Cockburn, "The Myth of Missile Accuracy," The *New York Review of Books* (November 20, 1980), 40–42; James Fallows, *National Defense* (New York: Random House, 1981), 139–170; Fred Kaplan, *The Wizards of Armageddon* (New York: Simon and Schuster, 1981), 374–376; E. Marshall, "A Question of Accuracy," *Science*, Vol. 213 (September 11, 1981), 1230–1231; Arthur G. B. Metcalf, "Missile Accuracy—The Need to Know," *Strategic Review*, Vol. 9 (Summer 1981), 5–8; Thomas Powers, *Thinking about the Next War* (New York: New American Library, 1983), 88–97; James R. Schlesinger, testimony to the Subcommittee on Arms Control, International Law and Organization of the Senate Committee on Foreign Relations, *U.S. and Soviet Strategic Doctrine and Military Policies* (Washington, D.C.: U.S. Government Printing Office, 1974). I found interviews with the following helpful: J. Edward Anderson, Minneapolis, March 9, 1985; Richard Garwin, New York, October 23, 1984; Arthur Metcalf, Waltham, Mass., November 6, 1984; James Schlesinger, Washington, D.C., September 22, 1986; Hyman Shulman, Santa Monica, Calif., January 13, 1987; Kosta Tsipis, Cambridge, Mass., November 2, 1984. I am also grateful to Professor Anderson and Dr. Metcalf for providing me with copies of unpublished papers and correspondence dealing with the episode. Published rebuttals by defenders of the facticity of accuracy are fewer, but see Charles Stark Draper, "Imaginary Problems of ICBM," *New York Times*, September 20, 1981, section 4, 20; General Robert T. Marsh, USAF, "Strategic Missiles Debated: Missile Accuracy—We Do Know!" *Strategic Review*, Vol. 10 (Spring 1982), 35–37; and R. T. Marsh, "A Rebuttal by General Marsh," *Strategic Review*, Vol. 10 (Spring 1982), 42–43. Various rebuttals were also put to me interviews, especially by Major General Aloysius Casey (U.S. Air Force), Mike Gorman, and Ed Rae, San Bernadino, Calif., February 25, 1985. Arguments from both sides can be found in the testimony to the Townes Panel on MX Basing, reported in P. S. Mann, "Panel Reexamines ICBM Vulnerability," *Aviation Week and Space Technology* (July 13, 1981), 141–145.

31. Where what the findings "ought" to be is controversial, an "experimenters' regress" can set in. For example, believers in a controversial phenomenon can dismiss as incompetent those experiments that fail to detect it, while those who do not believe it exists can dismiss those experiments that do detect it. See Harry Collins, *Changing Order: Replication and Induction in Scientific Practice* (London: SAGE, 1985).

but too competently. In the hundreds of flights of different missiles over a given test range, and even in the dozens of firings of a particular missile, learning is possible, while in wartime there would be no opportunity to learn and no second chance. It has also been suggested that a missile about to be tested will receive special attention not received by the bulk of operational missiles. One retired general went as far as to claim that "about the only thing that's the same [between the tested missile and the Minuteman in the silo] is the tail number; and that's been polished."[32]

The second set of grounds for skepticism concerns what, again drawing an analogy from studies of science, we might call the "auxiliary" assumptions involved in the extrapolation from test range to operational accuracy.[33] Most often raised was an auxiliary assumption concerning gravity. Because of the "problem of the vertical" in inertial guidance, extrapolating from test range to operational trajectories involves the assumption that our knowledge of the earth's gravitation field over those trajectories (e.g., in Arctic regions) is not seriously at fault. But a range of other physical phenomena were also cited as possible sources of unpredicted effects on operational accuracies.

Skeptical arguments of both kinds found common expression in an attempt to redefine missile accuracy. As noted above, the standard measure has been the circular error probable, understood as the radius of the circle round the target within which 50 percent of warheads will fall in repeated firing. The critics argue instead that missile accuracy has two components.

32. Quoted in Arthur T. Hadley, "Our Ever-Ready Strategic Forces: Don't Look Closely If You Want to Believe," *Washington Star*, July 1, 1979. This assertion is denied by those close to ICBM testing: letter to author from Major General John W. Hepfer (USAF, ret.), December 27, 1988. Again notice the role of similarity judgments (note 18 above) to this sort of dispute. No one claims that the tested missile is identical to the one in the silo. The issue is whether the differences are of any significance in the determination of accuracy.

33. The significance of auxiliary assumptions in science was pointed out by Pierre Duhem, *The Aim and Structure of Physical Theory* (Princeton, N.J.: Princeton University Press, 1954; first published in 1906). Auxiliary assumptions, for example about apparatus being used, are involved in making empirical predictions from theories. If a prediction fails, the problem may lie in one of the auxiliary assumptions, not in the theory. So there is a potential degree of openness in the conclusions to be drawn from a failed prediction.

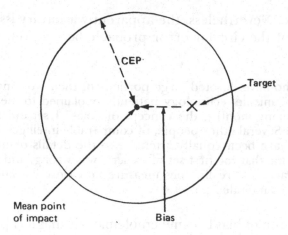

Figure 7.1
"Bias" and the Critics' Definition of Circular Error Probable (CEP)

The circular error probable, according to them, defines a circle around the mean point of impact. But that mean point of impact will also be offset from the target. So systematic "bias" would be present as well as random error, as illustrated in figure 7.1.

"Bias" was highlighted as two critics "deconstructed" accuracy for the readers of the *New York Review of Books* in November 1980:

The propositions and conclusions that follow . . . deal with a problem that has led a semi-secret existence for a decade. The shorthand phrase often used to describe this problem is "the bias factor." Bias is a term used to describe the distance between the center of a "scatter" of missiles and the center of the intended target. This distance, the most important component in the computation of missile accuracy, accounts for the fact that the predictions of missile accuracy cited above ["that a missile fired 6,000 miles can land within 600 feet of a target no more than fifty yards in diameter"] are impossible to achieve with any certainty, hence the premises behind "vulnerability," the MX, and Presidential Directive 59 are expensively and dangerously misleading.[34]

Defenders of the facticity of accuracy accept the possibility of bias, but say that biases are known, understood, small, and, most important, are included in their definition of the circular

34. Cockburn and Cockburn, "Myth," 40.

error probable. Nevertheless, the apparently arbitrary issue of the meaning of the circular error probable became central to the debate.

Several men who have devoted large portions of their working lives to the details of missile technology patiently explained to me, over the course of many months, the concept of "bias" I set out in the next few pages. Several other people, of comparable intelligence and integrity, who have been equally attentive to the details of nuclear policy, assured me that the first set of experts was wrong, and that a missile's "accuracy" . . . really does measure how close its warheads would come to its intended target.[35]

Why the notion of bias became emblematic is that it captured both the "too competent" and the "auxiliary assumptions" critique. The critics argued that in repeated test-range firing bias was artificially set to zero by "laborious adjustments"[36] of guidance systems and software, leaving only random dispersion. "What is probably happening in the case of both countries is that repeated tests over the same range have internalized a very substantial bias which remains unknown to the owner of the missiles."[37]

But because there were no previous firings to base these adjustments on for the novel trajectories of wartime, bias would reappear operationally. And errors in the assumptions on which wartime trajectory calculations were based, such as "uncorrected gravitational field anomalies, errors in the model of the terminal atmosphere, magnetic drift due to unknown electronic charge on the reentry vehicle—and anything else that has not been thought about" would also appear in the form of systematic bias.[38]

Furthermore, if these considerations applied both to American knowledge of the accuracy of American missiles, and Soviet knowledge of the accuracy of Soviet missiles, they applied even more strongly to American knowledge of the accuracy of Soviet missiles. "[I]f the Russians themselves cannot be sure what the operational bias of their missiles would be if

35. Fallows, *National Defense,* 150.
36. Cockburn and Cockburn, "Myth," 42.
37. Tsipis, "Precision and Accuracy," 4.
38. Anderson, "First Strike," 10.

they fired them against MM [Minuteman] silos, how does the U.S. Department of Defense purport to know?"[39]

Defenders of the factual status of missile accuracies denied that what went on was ad hoc test-range "tweaking" of guidance systems so as to cancel out any systematic tendency to overshoot, undershoot or fall to the left of right of the target. They asserted that bias-reducing adjustments were never made without the causes of the bias (and thus the form it would take in wartime firings) being understood. In the words of Charles Stark Draper: "We have at times noted biases in impact patterns for some missiles. In each case we have analyzed the problem to the degree that, where necessary, design or calibration corrections were made commensurate with the performance required. No magic was involved, because the nature of the problem was understood and applicable universally."[40]

Defenders also argued that the auxiliary assumptions involved in moving from test-range to operational firings were secure. The gravity field over the North Pole, at least at the heights of ICBM trajectories, was well known from observation of satellite orbits; the effects of unpredicted winds and other climatic variations was negligible with modern, fast reentry vehicles, the build-up of electrical charge on a reentry vehicle would have to be implausibly large for a significant effect to be exerted by the earth's magnetic field, and other potential effects would be vanishingly small in their consequences.

In sum, wrote General Robert T. Marsh, Commander of the Air Force Systems Command, "this pseudo-technical world [of the critics] says one cannot predict ballistic trajectory accuracy a priori, i.e., prior to actually firing some missiles on the expected trajectory. Such reasoning defies all the demonstrated rigor of physical laws from Newton to Einstein. It is akin to saying a gun may not be fired on a new azimuth without a prior shot in that direction. We certainly cannot allow such logic to affect important strategic decisions."[41]

Or as Charles Stark Draper concluded: "There is indeed grave risk in using ballistic missiles, but that risk is not uncertainty of accuracy."[42]

39. Tsipis, "Precision and Accuracy," 4.
40. Draper, "Imaginary Problems."
41. Marsh, "We Do Know!" 37.
42. Draper, "Imaginary Problems."

The Uses of Uncertainty

MX was the particular issue of the late 1970s and early 1980s that was central to the dispute. If the Soviets could not know how accurate their missiles would be in use, then it could be argued that there was no real threat to the existing Minuteman force. So the "window of vulnerability" argument for MX was spurious. And if the Americans could not know that MX's accuracy was not a test-range artifact, then the missile's claimed counterforce capability was make-believe.

Though its bearing on the fate of MX gave the dispute its public focus, and its public prominence, to different groups of critics uncertainty of accuracy nevertheless had quite different implications. Several threads were present in the criticism that need to be disentangled.

One thread to be found was, essentially, a revival of the bomber lobby's campaign in the 1960s against overreliance on the ballistic missile. The central figure in this was Arthur Metcalf, a Second World War Army Air Forces lieutenant colonel, Boston University professor of physics, and wealthy president of the Electronics Corporation of America.[43] The implication Metcalf drew, and the audience for whom he was drawing it, was clear right at the start of his 1981 editorial, "Missile Accuracy—The Need to Know," in the widely read *Strategic Review* that he publishes:

To the extent that the Air Force argues in the name of "invulnerability" to commit a large portion of its budget to MX missiles buried in holes in the ground in a basing mode, the wisdom of which is seriously questioned, and fails to signal loudly and clearly its requirement for a manned offensive bomber at the forefront of aeronautical technology, to that measure does it confuse the public understanding of what airpower is all about and undermine the very rationale for an independent Air Force at all!

The great air leaders of the past warned against what could well be a blind jeopardizing of the hard-won achievement of a separate service for the projection of airpower. And there are with us great air leaders who still warn against throwing away the experience gained at the cost of much blood and treasure to hazard the national safety

43. There is a description of Metcalf in James Parton, *"Air Force Spoken Here": General Ira Eaker and the Command of the Air* (Bethesda, Maryland: Adler and Adler, 1986), 478–484.

on untried "shell-game" war-deterrence schemes, tied to the ground, which put no enemy military force at risk.[44]

Metcalf was not an opponent of MX, though this kind of language sometimes made it seem as if he was. Rather, he argued that the new missile should simply be deployed in existing Titan II and Minuteman silos, which were at no real risk, and the money thus saved spent on strategic bombers, which unlike MX provided real accuracy and counterforce capability.[45]

Metcalf had strong links to leading Air Force circles, and retired generals, especially Ira Eaker, pioneer of Second World War daylight bombing and former Commander of the Eighth Air Force, were frequent guests on Metcalf's yacht, *Veritas*.[46] One such link was to General Bruce Holloway, former Commander of the Strategic Air Command and an editorial board member of Metcalf's *Strategic Review*. Though involved in the formation of the operational requirement for MX, Holloway saw force in Metcalf's line of argument. He arranged for one of the leading critics, J. Edward Anderson, professor of mechanical engineering at the University of Minnesota, to brief Verne Orr, Secretary of the Air Force, and Air Force staff officers, on the criticism.[47]

Senator Goldwater, too, extended his critique of whether missiles would work to whether they would be accurate, telling a 1985 interviewer that he had "never been convinced of the accuracy" of either American or Soviet ICBMs: "I have seen nothing except computer language that will testify to the accuracy."[48] With an impeccably hawkish reputation, and by 1984 in a powerful position as Chair of the Senate Armed Services Committee, Goldwater's belief that the Pentagon "can get along . . . without the MX missile," was a significant factor in MX's troubles, as arguably, were "deals" he was prepared to enter into to defend manned bomber programs.[49]

44. Metcalf, "Need to Know," 5.
45. Ibid.
46. See Parton, *Air Force Spoken Here*, 478–484, for Metcalf's links to Eaker.
47. Marshall, "A Question of Accuracy."
48. Anonymous, "Interview with Goldwater," *Aviation Week and Space Technology* (February 25, 1985), 112.
49. Goldwater as quoted in Harold Jackson, "Goldwater Attack may spell MX end," *The Guardian* (London), (December 7, 1984), 1. In 1976 Goldwater

But this hawkish, probomber thread in the challenge was less apparent than a "liberal" thread. Here the implication of uncertainty of accuracy was straightforwardly that MX should be canceled. There was no desire to foster the bomber instead. Professor Anderson, for example, made clear his overall position when, in what is to my knowledge the first public attack on the facticity of accuracy, he wrote: "I agree fully that it is of the utmost urgency that the arms race be stopped and have made my own contribution by leaving the field of missile-guidance engineering for engineering work in non-military fields."[50] Certainly, it was the liberal interpretation of uncertainty that was most prominent in press reportage of the controversy that stretched from the *New York Times* to the Elk River *Star News*, and most of the technical critics and commentators who took up the issue could, in broad terms, be classed as liberal.[51]

But there were at least two more threads to the challenge. The first predates considerably the public controversy of the early 1980s. It was centered at the Rand Corporation, and expressed in particular by a Rand analyst with considerable guidance system expertise, Hyman Shulman.[52] Bias was the issue that most exercised Shulman, who had much experience of analyzing data from Minuteman test firings, though he was also concerned at the small sample sizes on which many conclusions about accuracy were drawn.[53]

was persuaded to agree to prohibiting the Air Force from making an initial deployment of MX in silos, in return for Senator McIntyre (New Hampshire), who along with Senator Brooke was the main Senate opponent of counter-force, dropping opposition to the B-1 bomber. See John Edwards, *Super-weapon: The Making of MX* (New York: Norton, 1982), 119–121.

50. Anderson, letter. Anderson had worked on inertial navigation for Honeywell, for example on the electrostatic gyro navigator system proposed for Polaris (Anderson interview).

51. Tom Wicker, "Rethinking the MX," *New York Times*, August 25, 1981; Jeanne Hanson, "ICBM Inaccuracy makes MX System Unnecessary," *Star News* (Elk River), June 25, 1981. I am grateful to Professor Anderson for showing me his extensive file of reports on the controversy.

52. See chapter 5 above. Shulman's influence is described, though his name misspelled, by Kaplan, *Wizards*, 374–376. He was not the only Rand critic: see the testimony of William Hoehn of Rand to the Townes Panel on MX basing (Mann, "ICBM Vulnerability").

53. Shulman interview.

Shulman's solution was radical. Bias could be greatly reduced by correcting an inertial system with radio guidance, especially immediately prior to warhead separation. "Not part of the inertial guidance mafia," Shulman had since the late 1950s advocated continued use of radio guidance for strategic missiles.[54] He felt jamming to be little real risk because of the enormous size of jammer needed, and vulnerability could be removed by building vast redundancy into a network of ground-based transmitters.

A similar argument in favor of radio guidance was also put forward by another critic, physicist Richard Garwin. In his case it was a contribution to the debate about where to base MX missiles. He favored basing them in small non-nuclear submarines, perhaps carrying two missiles each, which would patrol waters close to the United States. Radio guidance would be needed to preserve the accuracy of MX in such a basing mode. Garwin proposed use of the new, very accurate, satellite-based Global Positioning System, supplemented by a radio beacon system that would be switched on if the Global Positioning System was destroyed.[55]

But neither Shulman nor Garwin met with success in their arguments in favor of radio guidance. Use of the Global Positioning System for both MX and Trident guidance briefly was considered by the Air Force and Navy, but was dropped, and the radio beacon scheme was never taken up.

Shulman's critique of the certainty of the accuracy achievable with an all-inertial system however found an influential convert in Rand Director of Strategic Studies, James Schlesinger. When Schlesinger became Secretary of Defense he expressed these doubts in a 1974 statement that was to be quoted many times in the early 1980s controversy:

I believe there is some misunderstanding about the degree of reliability and accuracy of missiles . . . As you know, we have acquired from the western test range [Vandenberg-Kwajalein] a fairly precise accuracy, but in the real world we would have to fly from operational bases to targets in the Soviet Union. The parameters of the flight

54. Ibid.
55. Sidney D. Drell and Richard Garwin, "Basing the MX Missile: A Better Idea," *Technology Review* (May/June 1981), 20–29; letter to author from Richard Garwin, December 21, 1988.

from the western test range are not really very helpful in determining those accuracies to the Soviet Union.

We can never know what degrees of accuracy would be achieved in the real world.[56]

Despite the radicalism of this last statement, Schlesinger was not an all-out skeptic. As we have seen, as Secretary of Defense he initiated programs to increase the accuracy, and remove doubts about the accuracy, of Minuteman and Trident. He also believed that to implement his policy of small, selective nuclear strikes it would be possible to have an elite force of missiles that "really had the guidance systems tweaked up so that for 10 or 20 or 50 missiles you weren't worried about degradation [of accuracy]"[57]

But Schlesinger also concluded that uncertainty of accuracy meant that the move in the 1960s to lower-yield weapons on both ICBMs and Fleet Ballistic Missiles was mistaken.[58] The formal calculus of nuclear strikes (see appendix B) had been used to argue that "small" warheads could be employed for counterforce strikes, because accuracy was more crucial than yield in the mathematics. But precisely because accuracy was more crucial than yield, operational degradations in accuracy were that much more serious. As Schlesinger told the House of Representative Armed Services Committee, "nobody knows what the accuracies will be when one goes from operational silos to operational targets. Accuracies are inherently subject to degradation. By contrast throw-weight megatonnage is not subject to degradation. So one builds into our force structure a much higher degree of uncertainty with regard to accomplishments."[59] The comparison, of course, was with the Soviet

56. Schelsinger, testimony, 15.
57. Schlesinger interview.
58. He felt that a particular error of the McNamara years had been to terminate the deployment of the high-yield Titan II well short of the originally planned force. See James R. Schlesinger, "The Changing Environment for Systems Analysis," in Stephen Enke, ed., *Defense Management* (Englewood Cliffs, N.J.: Prentice Hall, 1967), 89–112.
59. Quoted in Joseph Albert Cernik, *Strategy, Technology, and Diplomacy: United States Strategic Doctrine, with Emphasis on the Tenure of James R. Schlesinger as Secretary of Defense, 1973–1975* (Ph.D. dissertation, New York University, 1982), 272.

Union, which had, as we saw in chapter 6, retained much larger missiles with greater "throw-weight" and larger-yield warheads.

It is difficult to assess just how important this line of reasoning has been in the gradual return to high-yield warheads in U.S. ballistic missiles since 1974. Schlesinger believes "that's the reason we moved to bigger warheads," starting with the new high-yield W78 warhead/Mk 12A reentry vehicle combination for Minuteman III begun under him; but others would contest this.[60]

Certainly, however, we have a remarkable instance here of the plasticity of implications.[61] The same premise, uncertainty of accuracy, was drawn on to argue for manned bombers, for the cancellation of MX and a more "dovish" defense policy, for radio guidance and for larger missile warheads!

The same plasticity can be found in the implications drawn from claimed uncertainty in American knowledge of the accuracy of Soviet missiles. As we have seen, one conclusion was that neither the Soviets nor the Department of Defense knew how accurate these missiles were, and that therefore the threat they posed to Minuteman was unreal. But uncertainty also could be used to argue that the threat was greater, not less, than assumed. That was the position of the 1970s "B Team" of hawkish critics of the Central Intelligence Agency's "Soviet estimates," and of John Brett, the guidance expert in its missile-accuracy group. "B Team members argued from the theoretical point that the accuracy of Soviet missiles was essentially unknowable and that there was no evidence to support CIA's position that Soviet ICBMs were less accurate than US ones. Indeed the B Team held it was possible that Russian missiles could be even more accurate than our own."[62]

Though these were the main implications drawn from uncertainty in accuracy, they do not entirely exhaust the possibilities. Thus anyone familiar with the controversy then raging could

60. Schlesinger interview.
61. See Barry Barnes, "On the Implications of a Body of Knowledge," *Knowledge: Creation, Diffusion, Utilization*, Vol. 4 (1982), 95–110.
62. John Prados, *The Soviet Estimate: U.S. Intelligence Analysis and Soviet Strategic Forces* (Princeton, N.J.: Princeton University Press, 1986), 251–252; interview with John Brett, Wayne, N.J., January 15, 1987. Brett was no general critic of the facticity of accuracy: indeed, as Under-Secretary of Defense for Strategic Systems he was held to be a counterweight to Shulman's influence on Schlesinger (interview data).

hardly avoid reading the following 1981 statement from a captain in the U.S. Navy Special Projects Office as asserting the greater certainty of submarine-launched ballistic missile accuracy over ICBM accuracy:

SLBM modernization engineering development can confidently proceed to provide ICBM accuracies with the mobile and survivable SLBM platform. Future tests of such an SLBM system would continue to benefit from the fact that the submarine's broad based ocean area launches enjoy the ability to test over diverse launch latitudes, trajectory ranges and flight azimuths. This provides what is the best proof for national confidence that the submarine launched ballistic missile operational test performance truly represents performance under wartime conditions.[63]

But, despite its potential for being used in this way, uncertainty was never fashioned into an explicit argument for the submarine-launched balllistic missile as against the ICBM.

The Controversy Passes

For all its heat, this controversy, like the first, passed quickly. The verb is appropriate, since neither side can be said to have won it. At most, the defenders' strategy of seeking to place critics in a position where they "had either to quit or start a new controversy about a still older and more generally accepted fact" caused some of the claims initially made about gravitational uncertainties and, especially, magnetic effects to be backed away from.[64] It was not that it was impossible in principle to dispute well-entrenched parts of the network of knowledge, like gravitational and electromagnetic theory. The controversy over the claimed detection of gravitational radiation, and more recent reports of gravity field anomalies that are hard to account for through standard theory, show it could be done.[65]

63. Captain Robert L. Topping (U.S. Navy), "Submarine Launched Ballistic Missile Improved Accuracy," paper AIAA-81–0935 read to American Institute of Aeronautics and Astronautics Annual Meeting and Technical Display, May 12–14, 1981, Long Beach, Calif., 7.
64. The quotation is from Bruno Latour, *Science in Action: How to Follow Scientists and Engineers Through Society* (Milton Keynes, Bucks.: Open University Press, 1987), 77.
65. For the former, see Harry M. Collins, "The Seven Sexes: A Study in the

The critics were, however, in no position credibly to dispute accepted theory, but that did not undermine the entirety of their position. In interviews conducted between 1984 and 1987, both sides were still asserting strongly the correctness of their point of view. But by then not nearly as many people were listening.

Media attention span is doubtless part of the explanation: the readers of the *New York Review of Books* could hardly be expected to digest more than one article on gravitational anomalies and the correct statistical definition of missile accuracy.[66] But what was also important was the Reagan Administration's 1983 decision to base MX in Minuteman silos. Suddenly, silo vulnerability switched from being an argument for MX to being an argument against it. The liberal audience for the argument that because of uncertainty of accuracy silos were not really vulnerable—the dominant theme in 1980–1981—therefore vanished.

So from 1983 it was the Air Force, not the critics of MX, that had to argue that silos were safe after all. Air Force missile proponents did not, however, become an audience for the belief that the missile's accuracy was unknowable. A more modest and less far-reaching reinterpretation of the conclusion of silo vulnerability took place. As noted in chapter 4, "auxiliary assumptions" about the geology of the rocks under Warren Air Force Base, where MX is being deployed, were reassessed, and the conclusion was reached that silos there were much less vulnerable than had at first been calculated.

Uncertainty and the Production of Knowledge

The controversy passed, but the work of production of knowledge of accuracy goes on. Most of those who took part in the controversy were at at least one step removed from that work. Crucial though his historical role had been, by 1981 even

Sociology of a Phenomenon, or the Replication of Experiments in Physics," *Sociology*, Vol. 9 (1975), 205–224, reprinted in Barry Barnes and David Edge, *Science in Context: Readings in the Sociology of Science* (Milton Keynes: Open University Press, 1982), 94–116; also Collins, *Changing Order*, chapter 4. For the latter, see Tim Radford, "Physicists bring Newton's Theory down to Earth," *The Guardian* [London], (August 3, 1988), 7.
66. Cockburn and Cockburn, "Myth."

Charles Stark Draper was not an active participant in the production of this knowledge. Those I interviewed who were directly involved in that production seemed to have a point of view interestingly different from that of either side of the controversy. They were certainly not critics; none, for example, believed that accuracy was in principle unknowable. But neither were they prepared straightforwardly to defend accuracy figures as facts.

The reason appears to be their intimate understanding of the human vagaries involved in the production of accuracy. A good instance is former Air Force Major General John Hepfer, who was deeply involved with the Minuteman program and regarded by many as the "father" of MX. Hepfer dismisses the critics' "technical" arguments. Thus any claims about uncertainty in knowledge of the gravitational or magnetic fields are "weak because technically there's no phenomena in . . . the earth's gravity or anything else that should have a big effect on it [accuracy firing North as distinct from West.] . . . [W]e usually shoot him [J. Edward Anderson] down with his technical arguments."[67]

"But," General Hepfer continues, "I wouldn't argue too much that it [operational accuracy] might not be quite as good as the theoretical capability." The reason he gives is "human error." A striking example he quotes concerns the alignment of the Minuteman guidance system, a process crucial to its accuracy: "aligning a Minuteman missile to accomplish its intended mission is like threading a needle 400 feet away."[68] The mirrors used to do this had to be adjusted manually with extreme precision. "They had a tube that went down into the silo, and then they had a mirror, and they would have guys go out in sub-zero weather, minus 30 degrees . . . sighting on the stars and transferring . . . an azimuth alignment. . . . You're talking of arc seconds to align to . . . I had some of the young fellows work for me and they would say, 'well, we got kind of cold, and your only desire was to get out of the cold, you didn't

67. Interview with Major General John Hepfer (U.S. Air Force, rtd.), Washington, D.C., July 29, 1985.
68. J. M. Wuerth, "The Evolution of Minuteman Guidance and Control," *Navigation: Journal of the Institute of Navigation*, Vol. 23, No. 1, (Spring 1976), 71.

really care if was 10 arc seconds or 30 arc seconds,' and that really would negate the accuracy of the system."[69]

These conditions were obviously quite different from those at Vandenberg Air Force Base, with its pleasant Californian coastal climate, and thus could well have led to operational performance not being as good as that on the test range. This is a past issue: with Minuteman III, alignment became sufficiently automated that optical alignment crews were no longer needed.[70] But Hepfer was also well acquainted with the possibility for human error in matters such as entering the figures from surveys of both U.S. silos and Soviet targets. "[T]here is some degradation due to human error in surveying all our sites, all the Soviet sites, and putting all that data into the system. We found numerous times, that no matter how careful we are, somebody will make a mistake. And of course we'd catch it on our range [Vandenberg-Kwajalein] when something goes wrong, but when you have a thousand missiles out there. . ."[71]

According to Hepfer, human concerns also entered into what was done with the figures for accuracy once they had been produced. As noted above, because of the finite (and, recently, quite small) number of missile test flights, what statisticians would call a confidence interval, rather than a single circular error probable, is produced. Thus there is flexibility as to how to construct the single number:

We [the Air Force Ballistic Missile Office] used to tell them [the Strategic Air Command and Air Force Headquarters] that you can use a better number [i.e. lower circular error probable] than you are using and they wouldn't change it. But that trend has changed over in the past ten years. They'll use the best number you can give them now . . . they were very conservative and now they're probably going the other way, the reason being political. They want to get the best accuracy they can possibly get to make sure they stay better than submarines. There are many political forces that say, "You know, submarine's invulnerable under the water." The Air Force response is, "Yes, but it's not very accurate, because it's on a moving base. . . . We have to kill hard targets and the Navy can't do that" . . . so [the

69. Hepfer interview.
70. Wuerth, "Minuteman Guidance," 73.
71. Hepfer interview. Other interviewees also told me of this, citing cases where digits had been transposed in data entry.

Strategic Air Command and Air Force Headquarters] go from the pessimistic to the very optimistic side of the equation now to show they're still more accurate.[72]

Apart from these different forms of "human factor," the other main issue that concerns these insiders is the mathematical models of error processes, the "error budgets." As explained above, an error budget is necessary to construct an operational accuracy. Errors of different types—guidance and control, gravity and geodesy, reentry dispersion and the like, and the suberrors within these, especially the first of them— are extrapolated differently as one moves to different trajectories, ranges, and reentry angles.

To separate out different error sources is a difficult procedure, necessitating very good tracking of test firings to be compared with the telemetry from the missile's guidance system. Thus even though the Navy had geographical advantages over the Air Force—tracking stations could be built in the Bahamas for the Eastern (Atlantic) test-range that was used by the Navy, while the Air Force Western (Pacific) one was more exclusively over water—this did not prevent dispute breaking out over Fleet Ballistic Missile error budgets. The availability of very large amounts of money to improve range instrumentation helped resolve the dispute, but even with this,

the analyst interprets the results of these comparisons and draws conclusions about model validity, potential model improvements, and the causes of system inaccuracy. For this step, there are no well defined methodologies. The analyst must rely heavily on experience, insight, judgment, and good communications with experienced colleagues in the Fleet Ballistic Missile Weapon System community.[73]

On the Western test range, the problems seem on occasion to have been worse (though not, to my knowledge, to have resonated with an internal organizational split as they did in the Navy case, as noted in chapter 5). In a remarkable piece of Congressional testimony in 1976, John B. Walsh, Deputy Director of Defense Research and Engineering, described the recent history of the Minuteman error budget:

72. Ibid.
73. Topping, "Improved Accuracy," 5.

In 1971 the gravity and geodesy term decreased significantly and the accuracy was reduced. Notice at the same time our understanding of the guidance and control got better, which showed we had a greater guidance and control error—that in 1970 we were just wrong.

In 1973 we took out some of the mistakes we had made in 1970 and the guidance and control error went down, and the total error went down.

But now look: as these other large terms go down we began to wonder how to account for the errors we were observing, and concluded the reentry dispersion was probably greater than we previously thought.

In 1973 the reentry error was carried as a larger number. . . . Meanwhile the gravity and geodesy term goes down, which is supported by all the research the whole world is doing in the field of gravity. . . . But right now we carry no improvement in the reentry dispersion because it is unknown. Maybe we will understand it better later. It is not clear that we can make it go away.[74]

More recently, instrumentation on the Western test range has also been improved. Even with this, though, you "just can't deduce individual errors" unproblematically. "But you can do it well enough to bound guidance system error and know there are no unusual errors in there and that you are meeting specifications."[75]

Similar issues arise in regard to American knowledge of the accuracy of Soviet missiles, epitomized by one of the intelligence community's guidance system analysts I interviewed. He was scathing about the public critics, describing their position as "the Luddites' argument," and would defend the factual status of some of the products of his esoteric trade, in interview presenting qualitative conclusions about Soviet guidance technology based on the analysis of telemetry with considerable certainty.

At the same time, though, he did not regard the process of reaching a circular error probable for a Soviet missile as producing unequivocal fact. Different methods were used for analyzing impact points—"cheater conservative" or "monitor conservative"—involving different assumptions as to the degree of deliberate disguising of accuracy taking place. Con-

74. House Armed Services Committee, *Hearings on Military Posture, Fiscal Year 1977: Part Five, Research and Development* (Washington, D.C.: U.S. Government Printing Office, 1976), 199.
75. Interview.

structing an error model to account for observed trajectories was a difficult process, "even with our own missiles we can have big arguments over what impact points mean."

Some aspects of the error budget were much better known than others: the classical problem was that a lot was known about Soviet accelerometers (though even there some people had "weird ideas"), but less about Soviet gyroscopes. Constructing the error budget for a Soviet missile was not just an analytical task, and involved for example talking to the geophysics and geodesy community about assumptions necessary in order to construct that component of it.

The result was a range of "reasonable disagreement" in estimates of the circular error probable of a Soviet missile that had, according to this analyst, stayed reasonably constant in relative terms through time: if an average estimate was x, then the bound of disagreement would be about "a quarter or a half x."[76] There were, he said, also "big ambiguities" in estimates of the yield of Soviet warheads: even if one knew the nuclear mass of a warhead (which was not exactly known), the yield/mass ratio for Soviet weapons designs was known only approximately. Silo hardness also was subject to "variation." So with all these uncertainties in the figures going into the mathematics used to calculate it (see appendix B), the probability of silo destruction was, he said, a "very mushy" number.

The process of estimating, too, was seen by him as subject to political pressures. Peer group processes among guidance system analysts employed by different agencies were healthy: "you can always get an argument. There's a lot of interchange, interaction, across the board." But at the "higher levels," interagency disagreements in estimates were sometimes "plain inside the Beltway balls-out political."[77]

The Certainty Trough

There seems, therefore, to be uncertainty of two quite different kinds about the facticity of missile accuracies. To stereotype,

76. Though we did not discuss specific numbers in this interview, this is roughly consistent with what is known of different agencies' estimates of Soviet missile accuracies (see appendix A).
77. Interview. The "Beltway" is the ring of freeways around Washington, D.C.

the first is the uncertainty of the alienated and those committed to an alternative weapon system: the manned bomber. The second, perhaps more surprising, form of uncertainty, is that of those closest to the heart of the production of knowledge of accuracy. Rejecting the public critics' arguments, the latter group nevertheless find in their intimacy with this process of production reasons for doubt of a more private and more limited, but nevertheless real, kind. We have here, I would suggest, an analog for technology of what has been found in the sociology of science. "Certainty about natural phenomena . . . tends to vary *inversely* with proximity to the scientific work . . . proximity to experimental work . . . makes visible the skilful, inexplicable and therefore potentially fallible aspects of experimentation, it lends salience to the web of assumptions that underlie what counts as an experimental outcome . . . distance from the cutting edge of science is the source of what certainty we have."[78]

Between those very close to the knowledge-producing technical heart of programs, and those alienated from them or committed to opposing programs, lie the program loyalists and those who simply "believe what the brochures tell them."[79] These lie in what one might call the "certainty trough" of figure 7.2.[80]

It is possible, indeed, that this schematic and impressionistic diagram *might* describe (I am merely speculating) the distribution of certainty about any established technology. I emphasize "established," however, because in the early days of a technology that faces widespread skepticism there are often to be found charismatic technical figures, close to its heart, who evince great certainty, such as Charles Stark Draper or Edward Teller.

78. Harry Collins, "The Core Set and the Public Experiment," typescript.
79. Schlesinger interview. Dr. Schlesinger suggests that to "believe what the brochures tell" you is a distinctively American trait, not found in the Soviet Union, whose leaders "have a deep-seated feeling that nothing is going to work . . . here's a society in which you can't make light bulbs or you can't construct buildings . . . so they've got a deep-seated caution built in . . . . So I think it [was] particularly useful to make that speech [his 1974 testimony denying the factual status of accuracy figures] to the people in the United States" (ibid.).
80. I have to thank Harry Collins for a useful discussion of the appropriate name for the certainty trough.

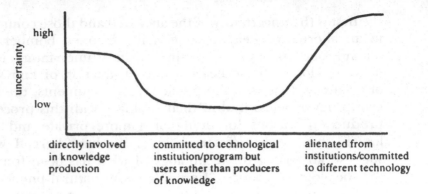

directly involved in knowledge production	committed to technological institution/program but users rather than producers of knowledge	alienated from institutions/committed to different technology

Figure 7.2
The "Certainty Trough"

Are There Facts Inside the Black Box?

How much of what we have found so far is due to the particularities of the nuclear ballistic missile—the political, legal, and economic constraints on testing, and the absence of a record of use? To answer this question, let us look inside the black box of guidance, and consider the status of knowledge of the technical characteristics of the components to be found there.

Here there are no political or legal restrictions on testing, and though test equipment is expensive, a gyro or accelerometer can be tested many times, or for years on end, without being destroyed, and thousands of each can be tested. Gyros and accelerometers can also be used as well as tested. Participants seemed to treat a missile test flight as equivalent to use from the point of view of guidance components, and these components can also be used, if desired, in vehicles other than missiles.[81] So do all the vagaries of fact construction we have found for entire missiles disappear for the small components inside their guidance black boxes?

81. One question, of course, is the effect of different azimuths (north-firing as distinct from west-firing) on components. I asked interviewees connected to the Central Inertial Guidance Test Facility about this effect in aircraft flight-tests of inertial navigation systems. They said that azimuth effects were found, "some systems look good going North-South, then bad East-West," for example, but these tended to be more a software than a hardware problem. Interview with Dr. Martin Jaenke and Peter Zagone, Holloman Air Force Base, New Mexico, September 19, 1986.

Certainly, testing was regarded by all those concerned with inertial sensor development as an utterly central matter. The development of these sensors went hand-in-hand with the means of determining their characteristics. Without these means, the gyroscopes and accelerometers would have become black boxes to those seeking to improve them. "The testing of these instruments in terms of techniques, equipment, and data processing has had to keep in step with the accuracy of the instruments to be tested, and in large measure has had to lead the way."[82]

There was a certain skepticism in the armed services about test results from the producers of inertial sensors: one interviewee told me of his maxim that "all test results from gyro manufacturers are test results from liars."[83] This has led to the creation of specialized facilities for inertial component and system testing, to check, as another put it, how sensors and systems got on "away from mother."[84] The most important of these is the Central Inertial Guidance Test Facility, established in 1959 at Holloman Air Force Base, New Mexico.[85] Although there have been efforts to make all the U.S. armed services do their testing there, in practice it tends to be regarded as an Air Force facility, and the Navy has maintained its own facility at the Naval Air Development Center, Warminster, Pennsylvania.[86]

The existence of these facilities means simple "rigging" of tests by manufacturers would be detected, and this does not, indeed, appear to be an issue since their establishment. But that does not mean that here we have found factual bedrock.

The essence of the issue is that testing inevitably involves "the construction of a background against which to measure

82. William G. Denhard, "The Evolution of Testing of Precision Gyros and Accelerometers," AGARD Lecture Series No. 30 on "Inertial Component Testing—Philosophy and Methods," Paris (June 24–28, 1968), 3.
83. Interview with Thomas A. J. King, Arlington, Virginia, April 2, 1985.
84. Interview with William G. Denhard, Cambridge, Mass., February 18, 1985.
85. Anonymous, *6585th Test Group: Facilities and Capabilities* (Holloman Air Force Base, New Mexico: Air Force Systems Command, Armament Division, 6585 Test Group, n.d.), 11–34.
86. Interview with John McHale, Silver Spring, Maryland, September 22, 1986.

success."[87] For testing missile accuracy as a whole that was unproblematic: a "splash net" of hydrophones locates the ocean impact points of reentry vehicles, and though its results could potentially be challenged, they do not appear to have been.[88]

For gyroscope and accelerometer testing, however, "construction of a background" was much more difficult. For inertial guidance and navigation these instruments have to be extraordinarily sensitive. Even a basic second generation inertial gyroscope should "drift" by only a hundredth of a degree per hour, while a 1960s missile accelerometer had to be sensitive to accelerations on the order of a hundred thousandth of earth's gravity.[89] Constructing a stable background against which to measure this sort of sensitivity was a demanding task. Artificially to provide a rotational input, and then to check the gyroscope's response to it, would require enormously accurate knowledge of that input, so the rotation rate of the earth—a secure piece in the network of knowledge—was generally taken as the input. Similarly, the gravity field could be used as a test input to both gyroscopes and accelerometers.[90]

This did not remove the need for high precision in test tables and other equipment. Increased accuracy specifications in sensors necessitated increased accuracy in the test equipment. "The basic dilemma here is not new: The instruments to be tested are the best which precision production technology can turn out and to conduct a meaningful test the tester is confronted with the necessity to provide a test method, a standard, which is even better."[91]

87. John Law, "Technology and Heterogeneous Engineering: The Case of Portuguese Expansion," in Wiebe E. Bijker, Thomas P. Hughes, and Trevor Pinch, eds., *The Social Construction of Technological Systems: New Directions in the Sociology and History of Technology* (Cambridge, Mass.: MIT Press, 1987), 126.

88. The splash net system could be said to be in error. Thus in 1963 the "limit of accuracy" of the Missile Impact Location System was "about 0.15 nautical miles": George Alexander, "Atlas Accuracy Improves as Test Program is Completed," *Aviation Week and Space Technology* (February 25, 1963), 54. Even "where a missile lands" is not unproblematic!

89. C. S. Draper, "Wisdom versus Completeness in Testing," paper read to Air Force Operations Analysis Technical Conference, Eglin Air Force Base, Florida, April 13, 1965, 8 (copy in Charles Stark Draper papers, Library of Congress, Washington, D.C., box 17).

90. Denhard, "Evolution of Testing."

91. Martin G. Jaenke, "Introductory Remarks: Test Technology Trends," in

The "stability" of this background, in the literal sense, is problematic. Here we do indeed find bedrock, but it is not good enough. Gyro test tables are placed on pillars sunk into solid rock, but even so "microseisms and human cultural activities" interfere.[92] At Ferranti's gyro production and test facility in Edinburgh, for example, the systematic movement of workers that takes place at the end of each working day has a noticeable effect. In very high-accuracy testing, "[e]xperience has indicated that the most successful way of accomplishing this [improving the environment in which the gyro is to be tested] is to isolate the gyro-test-station from the human environment."[93] Though attempts have been made to combat the problem technologically, by developing a test table with an "active control system" to measure and counter movements of the ground, by 1986 it was reckoned that the limits even of the Central Inertial Guidance Test Facility's New Mexico desert site had been reached, and the search was on all over the United States for the new site needed "if they do indeed want to increase the accuracy of the guidance systems of the future."[94]

As with the missile's error budget, much more is involved in gyroscope and accelerometer testing than simply the production of a single accuracy figure such as a drift rate. Again, a mathematical model of error processes is constructed. In the case of a gyro, for example, this involved even in the 1950s "components insensitive to the earth's gravity field, those proportional to it, and those proportional to the square of it," deduced from test data by the mathematical technique known as Fourier analysis.[95]

As accuracy specifications increased, so too did the complexity of the mathematical model. Even the behavior of the supposedly stable background was beginning to be incorporated

Testing Philosophy and Methods of Guidance and Control Systems [sic], AGARD Lecture Series no. 60 (Neuilly sur Seine: NATO Advisory Group on Aerospace Research and Development, October 1973), 1/2.
92. Ibid.
93. Joseph G. Walsh, *Summary of Digital Gyro Test Data Acquisition Techniques* (Cambridge, Mass.: MIT Instrumentation Laboratory, July 1970, E-2515), 2.
94. Interview with Jaenke and Zagone.
95. William G. Denhard, *Test and Evaluation of Inertial Components* (Cambridge, Mass.: MIT Instrumentation Laboratory, March 1960, R-200), 11.

into the model. By 1973: "the error models for gyroscopes used by various investigators now include in addition to the 10 'classical,' gravity dependent terms, additional terms describing the error contributions of temperature effects, float motion, magnetic effects, power supply variations and of parasitic angular motions of the test stand." Not surprisingly, this caused problems akin to those in constructing a missile error budget in separating out the various error sources: "The development underway is concerned with the techniques of measuring the driving functions of these terms and of separating them analytically."[96]

One obvious limitation of using the earth's rotation and its gravity field as input was that testing was only for a particular range of conditions. By altering the orientation of a device being tested it could be subjected to different components of gravitational acceleration [g], but it could not be subjected on an ordinary test table to more than one g. In ballistic missile use, however, the guidance system was subject to accelerations of as much as 30 g. The mathematical error model allowed extrapolation from behavior at one g and under to behavior at higher accelerations, but this was not considered on its own entirely adequate.[97]

Two large and expensive types of test equipment have been constructed with which inertial sensors can be subjected to high accelerations under controlled conditions. The first was the precision centrifuge, of which there are three in the United States, in which devices could be subect to very high but exactly known accelerations—up to 100 g at the 260-inch radius centrifuge at the Central Inertial Guidance Test Facility. The other is the high speed test track. The track at the Central Inertial Guidance Test Facility is a ten-mile "railway" laid out with great precision. Along it hurtle rocket sleds, their velocity measured to a thousandth of a foot per second by a photoelectric cell

96. Jaenke, "Test Technology Trends," 1.2. The ten "'classical" terms are listed in Peter J. Palmer, "Gyro Error Model Development," *Electromechanical Design* (February 1970), 32–39, which contains useful history of gyro error modeling.

97. "The g^2 sensitivity [of the MX gyroscope and accelerometer] is so small that it is not excited in a one-g test": K. R. Wernle, "Missile X/Third Generation Inertial Instrument Test Program," *Navigation: Journal of the Institute of Navigation*, Vol. 26, No. 2 (Summer 1979), 108.

system, with the position of "interrupters" in this system placed along the track measured to within a thousandth of an inch.[98]

Each instrument had disadvantages. "A centrifuge is very nice, except you always come back to where you started from," which is patently not the case in missile flight, while even a ten-mile track allows only a short period for errors to build up and be measured at rocket-sled speeds. Some dispute has indeed taken place as to which provided the better "background against which to measure success."[99]

But even moving entirely to "the real thing"—missile flight—was not wholly unproblematic. True, telemetry of what the guidance system "believed" to be the missile's position and velocity could be compared with what those "actually were," but only if what they "actually were" was known with sufficient accuracy. Especially with an entirely over-ocean range, this was problematic. While there was a clear element of subterfuge in the development of what became the MX guidance system as the Missile Position Measurement System, it was not, at the time, a ridiculous idea that the best "background" against which to measure the performance of a missile guidance system was another, presumably better, missile guidance system.

The development of the SATRACK system for tracking missile tests using Global Positioning System navigation satellites has meant that testers now have what is agreed to be a much better background at least potentially available to them.[100] But again the problem surfaces of knowing exactly what a given deviation between "belief" and "actuality" is caused by: the consequences, say, of initial misalignment of the stable platform can be hard to disentangle from in-flight gyro drift. With enough time to make measurements, the question could be resolved to the satisfaction of all concerned, since the effect of gyro drift will increase with time. But enough time is not

98. Interview with Colonel Leonard R. Sugerman (United States Air Force, rtd.), Las Cruces, New Mexico, September 18, 1986; Anonymous, "6585th Test Group," 17, 37–38; interview with Jaenke and Zagone.
99. Interview with Norman Ingold, Holloman Air Force Base, New Mexico, September 19, 1986; see John M. Buchanan, *Inertial Guidance Equipment Testing on the MIT Precision Centrifuge with some comparisons to Sled Testing* (Cambridge, Mass.: MIT Instrumentation Laboratory, June 1963, E-1223), and the dissenting comments by supporters of sled testing quoted on 5–6.
100. One interviewee told me that Air Force use of this "background" was less than Navy use of it.

necessarily available. "There is inadequate time from launch to burnout for SATRACK or most of the other added instrumentation to really do a good job indentifying an average gyro bias, much less how it changes in flight, so we go on guessing and giving statistics."[101]

A further issue in missile flight testing is that, at least inside the "defense establishment," a flight test is a public event. Naturally, those performing it want the best possible result, the highest possible accuracy. They are, therefore, not necessarily going to permit trajectories to be varied in the ways that those seeking to build error models would find most helpful.

Contrary to missile flight trajectories, for which all error sources must be minimized to arrive at the best possible C[ircular] E[rror] P[robable], the guidance sled run trajectories are deliberately tailored to promote the growth of specific error terms to allow comprehensive evaluation and correction of error sources. This is particularly valuable for separating nonlinear terms in error model equations. Sled tests in the past have proven extremely useful because for practically every system tested they have led to the discovery and correction of design deficiencies and unexpected error quantities.[102]

Conclusion

What we find, therefore, is that there is no simple continuum whereby the closer we approach to "use" the less problematic becomes the knowledge generated. Knowledge is indeed a network wherein different kinds of test are performed against differently constructed backgrounds, with no one test—not even "use"—and no one background being accepted by all as the ultimate arbitrer. It takes a gruesome imagination to think of it, but I could envisage guidance analysts in the aftermath of a nuclear war—if any were left alive—deciding that missile performance in it had been "unrepresentative" and "poorly recorded," and that the knowledge of missile accuracies gained in peacetime had been superior.

Some of the broader consequences of the nature of "technical fact" must be left to the next chapter, but here I will draw out one that bears in particular on the preceding chapters. In

101. Letter to author from guidance technologist, March 9, 1987.
102. Anonymous, "6585th Test Group," 39.

those, we have several times come across controversy between, for example, the proponents of different kinds of inertial instrument, or between the proponents and opponents of stellar-inertial guidance. Some of the time at least, these controversies concerned not whether accuracy was desirable, but how best to achieve it. When it was the latter was at issue, I can imagine the reader asking "Why didn't they just try it and see? Wouldn't that have cleared the matter up?"

There is of course an analogous question to be asked of science: if scientists believe different theories, will experiment not show which is correct? A whole literature in the history, philosophy, and sociology of science has revealed why this is not always so, and why, in principle, it might never be so.[103]

What we have seen shows why it is not so in technology either. A range of tests is available, and the results of one are not necessarily in harmony with those of another. A poor test result might always be the result of a problem with the "background against which to measure success." The folklore of inertial guidance testing is full of tales where it was eventually decided that the device being tested was correct and the test equipment, or other assumptions in the test procedure, were wrong.[104] Use, not testing, can be claimed to be the proper ultimate arbiter, and differences between the two cited to deny that inference can flow from testing to use.[105] On the other hand, testing that resembles use, or even use itself, can be said to be too haphazard, too poorly controlled, or too poorly "instrumented" to be appropriate for the generation of hard fact.[106]

103. The single most influential argument to this effect was, of course, Thomas S. Kuhn, *The Structure of Scientific Revolutions* (Chicago: University of Chicago Press, 1962).
104. Thus at the high-speed track at the Central Inertial Guidance Test Facility "data points that didn't fit" could sometimes be traced to interrupters that had been surveyed incorrectly (Jaenke and Zagone interview).
105. As we have seen this is the strategy of the critics of the facticity of missile accuracy.
106. Jack M. Arabian, *Error Analysis for Precise Laboratory Testing,* Technical Report MDC-TR-67–23 (Holloman Air Force Base, New Mexico: Air Force Missile Development Center, 1967), 3, writes, "There is a compatibility compromise which must be met between Precision Testing Equipment (PTE), which operates in a benign but precise laboratory environment, and Operational System Equipment (OSE), which duplicates the rugged environment

Though it may be possible in the abstract to make any or all of these arguments, it is important to note that it may not always be socially credible or feasible to do so. The "accepted practices" of testing are strongly entrenched and largely, if not entirely, consensual, and patterns of belief in the technical community are not infinitely flexible.[107]

Nevertheless, testing should not be expected to close issues of controversy to the satisfaction of all concerned. One interview discussion I had is of particular interest here. It was with one of the key figures in the development of the dry tuned-rotor gyroscope, which was, until the advent of the laser gyroscope, the main competitor of Draper's floated gyro. This guidance technologist quoted the opinion that the process of incremental improvement of the floated gyro was "like polishing a turd." When I asked him about the extremely low drift rates achieved by Draper's third generation gyroscope, he said they were "bullshit." You could achieve "a ten-thousandth of a degree per hour" in a laboratory, "but put it in a system and start shaking it" and the results would be quite different.[108]

MX's "baby in the womb" guidance design means that the third generation gyroscope in a sense carries its vibration-protected, finely temperature-controlled laboratory with it. So these remarks, even if true, would by no means damn the device or prove that a tuned-rotor gyroscope would be superior. But they do show the potential for dispute as to what the most accurate gyroscope or accelerometer is, even with the capacity to test as freely as one wishes.[109]

The relative accuracy of different gyroscope and accelerometer designs is also made a more complex issue by the way the meaning of the "accuracy" of a gyroscope or accelerometer has changed over the years. As error modeling has developed, and as on-board digital computers have become more powerful, the size of "brute" error has become less important than

specs but is not necessarily as precise," and goes on to develop arguments against the latter.
107. This is the phenomenon that Constant, *Turbojet Revolution* and "Technological Testability," refers to as "traditions of technological testability."
108. Interview.
109. One potential competitor of the Draper PIGA is Kearfott's vibrating beam accelerometer. Proponents of that also argue that it would have advantages in more demanding conditions, such as when rapid start-up from dormancy was required (Brett interview).

how well the errors of a device can be predicted mathematically. The "inaccuracy" of a missile guidance inertial sensor has become, in effect, the residual after the guidance system software has corrected for all predictable error processes. What has thus to be evaluated is not merely the device itself, but the whole network of knowledge surrounding it.[110] If this is true for the components of systems, it is even more true for the systems themselves. In the case of unistar stellar-inertial guidance, for example, we saw how a statistical model of error processes was vital to the system's functioning.

Nowhere in this complex process of modeling and testing do unchallengeable, elementary, "atomic" facts exist. This does not mean that accuracy is a mere fiction, an "invention" in the pejorative sense, for this absence of "atomic" fact is characteristic of all scientific knowledge.[111] It does mean, however, that the more deeply one looks inside the black box, the more one realizes that "the technical" is no clear-cut and simple world of facts insulated from politics.

110. Several interviewees attributed the accuracy of the MX guidance system less to the hardware as such than to what one described as the "crackerjack" calibration and error compensation algorithms used.
111. A clear exposition of why this is so can be found in Barnes, *Kuhn.*

8

Patterns in the Web

It is time to draw together the threads. What can be learned from the half century of the development of guidance technology reviewed here? For the sake of simplicity, I shall divide the lessons to be learned into five areas: technology, politics, the paradoxical ordinariness of the technical and political world of nuclear weaponry, the relationship between technology and politics in that world, and facts. These are lessons that I hope have a wider bearing on how we should understand other areas of technology and other questions of weaponry and war. They also bear, I believe, upon how we should act. An analysis of a matter as close to the future of humankind as nuclear missile guidance should not, after all, rest content with an understanding detached from the issue of what should be done about it all.

I have no wish to try to lay down particular courses of action, or to cajole the reader into being antinuclear or pronuclear. Rather, my goal is to demystify: to show how this book's findings undermine the big determinisms, technological and political, that paralyze action. What the previous chapters have revealed, I believe, is how matters that look set and beyond influence, including matters of technology and of fact, are nothing of the kind.

As this concluding chapter proceeds, I will give some particular examples where I believe I can show the mistakenness of the assumption that nothing can be done to change things. But the whole tenor of the analysis is to suggest that many more such instances must exist. If this book encourages readers to look for them, then it will have served its purpose well.

About Technology

The most pervasive and most paralyzing determinism of all is technological determinism. In our bleakest moments, the nuclear world has seemed to be a technological juggernaut out of control, following its own course independent of human needs and wishes.[1] Few people might want to endorse that view without reservations, and when this is diagnosed as the overall condition of the nuclear world it is typically in order to condemn it.[2] But particular manifestations of technological determinism are everywhere to be found in discussions of the arms race. Think, for example, how often the introduction of a new weapon is described as "modernization," as if it were the natural and unproblematic outcome of technological progress.

The rhetoric of "modernization" is typically a resource for those who wish to justify weapons development. But opponents of nuclear weaponry often explain changes in similar fashion, though they obviously deplore rather than welcome them. Growing missile accuracy as the natural direction or "natural trajectory" of technological change is an assumption that crosses political divides. Even those who do not want missile accuracies to increase assume that they will. "There seem no

1. This type of thinking clearly is not unique to questions of the arms race: see Langdon Winner, *Autonomous Technology: Technics-out-of-Control as a Theme in Political Thought* (Cambridge, Mass.: MIT Press, 1977).
2. See Ralph Lapp, *Arms Beyond Doubt: The Tyranny of Weapons Technology*, (New York: Cowles, 1970); Hans Bethe, "The Technological Imperative," *Bulletin of the Atomic Scientists*, Vol. 41, No. 7 (August 1985), 34–36; Herbert York, "Multiple-Warhead Missiles," in Bruce M. Russett and Bruce G. Blair, eds., *Progress in Arms Control? Readings from Scientific American* (San Francisco: W. H. Freeman, 1979), 122–131; Lord Zuckerman, "Science Advisers and Scientific Advisers," *Proceedings of the American Philosophical Society*, Vol. 124 (1980), 241–255; Dietrich Schroeer, *Science, Technology and the Nuclear Arms Race* (New York: Wiley, 1984) and Schroeer, "Quantifying Technological Imperatives in the Arms Race," in David Carlton and Carlo Schaerf, eds., *Reassessing Arms Control* (London: Macmillan, 1985), 60–71; Deborah Shapley, "Technology Creep and the Arms Race: ICBM Problem a Sleeper," *Science*, Vol. 201 (September 22, 1978), 1102–1105; Marek Thee, *Military Technology, Military Strategy and the Arms Race* (London: Croom Helm, 1986); Edward Thompson, "Notes on Exterminism, the Last Stage of Civilization," *New Left Review*, No. 121 (May/June 1980), 3–31; John Turner and Stockholm International Peace Research Institute, *Arms in the '80s: New Developments in the Global Arms Race* (London: Taylor and Francis, 1985).

insurmountable obstacles to continuing improvement in C[ircular] E[rror] P[robable] at present rates for another twenty-seven years. . . . Policy makers clearly will have to cope with almost unbelievably accurate ICBMs in the future."[3]

The single most important lesson of this book is the fallacy of this technological determinism. What we have found is that technological change is social through-and-through. Take away the institutional structures that support technological change of a particular sort, and it ceases to seem "natural"—indeed it ceases altogether.

Different phases in the history of missile guidance show the social nature of technological change in different ways. Examining the emergence of black-box navigation shows the shortcomings of two of our most common notions of how new technologies are created, notions that underpin, in quite different ways, the idea of technological change as an autonomous, asocial process. First, rather than being the result of a sudden flash of individual inspiration, black-box navigation came into being as the result of the work of many people over several decades. That work, too, was heterogeneous. Changing people's perceptions and gathering resources (at both of which Charles Stark Draper excelled) was at least as important a part of the process as writing equations and drawing blueprints. Second, rather than being the application of science, inertial navigation came into being in the face of an apparently secure deduction from established scientific knowledge—the theory of relativity—that it was impossible. To prove it possible, that perception had to be changed, and a process of production of sufficiently accurate gyroscopes had to be created. The latter, again, was as much a "social" as a "technical" matter.

The "social engineering" involved in creating inertial navigation was, however, fairly specific in its scope. Draper and the other proponents of inertial navigation did not have to prove a black-box system desirable, since the value of an autonomous means of navigation was clear to anyone concerned with the

3. Schroeer, "Technological Imperatives," 64. The precision of the "twenty-seven years" has to do with Schroeer's predictions for the development of computer power, which he sees as the most likely cause of increasing missile accuracy. He does propose perfectly sensible arms control measures to inhibit the growth in accuracy. The point, however, is that in his analysis these are artificial barriers to the natural course of technological change.

projection of military power over long distances; their task, rather, was to prove it possible. By comparison, creating the vehicle that was to form the most significant application of inertial technology, the long-range ballistic missile, involved much more extensive "social engineering," at least in the United States. The reason was that there already existed, in large and growing numbers, what many felt to be a perfectly adequate means of delivering nuclear weapons: the bomber.

In what I called the "missile revolution," organizational and technological change thus went hand-in-hand. With considerable skill, missile proponents created the social conditions in which ICBMs could be built even "without support from either SAC [the Strategic Air Command] or most of the 'establishment.'"[4] It was also crucial to the success of the Polaris submarine-launched ballistic missile program that a similar sphere was created within which they were relatively autonomous. Eventually, too, national strategy had to be reworked as well as military organization. To bolster Polaris and preserve their control of it, Navy leaders challenged dominant views of nuclear war with the then heretical idea of finite deterrence.

The appeal of the idea of an autonomous logic of technological change, however, ultimately lies less in its account of the early phases in the development of a technology, where its flaws are apparent, than in its account of its maturity. For here we do typically find the continuous, predictable, apparently inexorable technological change that has so impressed those who have examined the growth of missile accuracy. Here we find the phenomenon that gives the notion of a "natural trajectory" of technology its plausibility.

The phenomenon cannot be denied, but its explanation as a natural trajectory can be. The empirical argument against the notion as applied to missile guidance is clear-cut. An alternative form of technological change exists, which is no less progressive (on some conventional criteria, such as its use of novel inertial sensors, it is more so), but where progress has a quite different meaning. Its institutional base is civil and military air navigation, where extreme accuracy is little prized, but reliability, producibility, and economy are.

What appears to be a natural trajectory ought instead to be seen, I suggest, as an institutionalized pattern of predominantly

4. The phrase is taken from a letter to the author from one of those involved.

incremental technological change involving, centrally, a self-fulfilling prophecy. The incremental nature of increasing missile accuracy is clear. Improvement has mainly come about by identification of the barriers to accuracy in existing systems and painstaking search for ways of altering those systems (rather than discarding them) so as to remove these barriers. Extrapolating that process of incremental change into the future, the proponents of inertial guidance have prophesied what they will be able to achieve if given the resources to do so: "Such performance [ICBM accuracies of 30 meters or better] is possible. It will become reality if the government and military incentive cause resources to be committed to the challenge."[5]

To talk of the "institutionalization" of a pattern of technological change means several things.[6] First, it indicates the existence of a relatively stable organizational framework within which the technological change takes place. The centerpiece of that framework in the United States has been Draper's MIT Instrumentation Laboratory, later the Charles Stark Draper Laboratory, Inc., with its forty years and more of commitment to increasing the accuracy of inertial guidance and navigation by predominantly incremental means. Similar continuity, if not quite the same longevity, has also been found in the Autonetics Division of North American Aviation (later Rockwell International), and in the group of German guidance specialists working for the U.S. Army. The military organizations with which the technologists have had to deal—for strategic missiles, chiefly the Air Force Ballistic Missile Office and Navy Special Projects Office—have a continuous history now stretching back over thirty years. Soviet missile guidance likewise has devel-

5. David Hoag of the Draper Laboratory, as quoted in chapter 4. I do not suggest that Hoag's purpose in saying this was to accumulate resources. The setting was a 1970 symposium on the impact of new technologies on the arms race, one of the "Pugwash" meetings originally suggested by Bertrand Russell, Albert Einstein, and other scientists concerned with the threat of nuclear war. In that context, Hoag's prediction counted more as a warning than as a promise. In other contexts, however, similar predictions were promises.
6. I am here drawing on the sociological meaning of "institution," which is wider than formal organization, and includes any stable, trans-individual pattern of social behavior. See, for example, Peter L. Berger and Thomas Luckmann, *The Social Construction of Reality* (Harmondsworth, Middlesex: Penguin, 1971), 65–109.

oped within a stable organizational framework, at the heart of which have been three missile guidance bureaus, with lengthy histories and relatively well-defined "niches."

Second, institutionalization implies the channeling of resources to support this organizational framework and its activities. Though the Draper Laboratory sought other means of support for its preferred form of technological change, for example from civil air navigation, ultimately only one "interest" could be found to sustain it—the desire for counterforce accuracy in ballistic missiles.

As far as one can tell, if that interest had not existed, the technological trajectory towards greater guidance accuracy would have ceased, given the absence of other sources of support of commensurate size. The "natural" form of change in inertial technology would appear to be that sustained by the different demands of civil and military air navigation. Growing reliability, producibility, and economy, rather than greater accuracy, might then seem to us to correspond to an inherent developmental logic of that technology, as it moved from "wet" to "dry" gyroscopes, from mechanical to optical sensors, and from mechanical complexity plus simple computation to mechanical simplicity plus complex computation.

Did Draper and the other guidance technologists create the "interest" in counterforce accuracy in ballistic missiles, precisely so as to institutionalize the form of technological change they wanted? There are many examples of this from the history of other technologies, and the existence of a "guidance mafia," often former pupils of Draper, in strategic locations in the armed services, especially the Air Force, lends weight to the suggestion.[7]

The answer to the question is both "yes" and "no." The affirmative answer arises from the third meaning of the institutionalization of a form of technological change: the credibility of the prophecy that is at its core. Organizations are created and sustained, and resources flow, to the extent that it is

7. See, for example, Thomas P. Hughes, "Technological Momentum in History: Hydrogenation in Germany, 1898–1933," *Past and Present*, No. 44, 106–132, though the same ambiguity is present in that case as here. Systematic reflections on the issue can be found in Bruno Latour, *Science in Action: How to Follow Scientists and Engineers through Society* (Milton Keynes, Bucks.: Open University Press, 1987), chapter 3.

believed that the predicted change in technology will become, or at least has a chance of becoming, a reality. As noted in chapter 4, there is not always a need for there to be a desire that it happen: the need is for a belief that it can and will. Thus an argument always existed to persuade those in the United States who felt it would be for the best if missiles remained too inaccurate for anything other than countercity retaliation: that Soviet missiles would increase in accuracy, whatever the United States did, so the country should devote resources to prevent its arsenal lagging in counterforce capability.[8]

Nothing guaranteed that the prediction of extremely accurate ballistic missiles would be believed. The ICBM was originally seen as a countercity weapon, with counterforce argued to be the province of the bomber—in good part because of the missile's perceived inaccuracy. When members of the guidance mafia in the early 1960s predicted very accurate ICBMs, some felt, as noted in chapter 4, that "they must be smoking opium."

So the sense in which guidance technologists did create the interest in counterforce accuracy in missiles is that their activities were crucial to making the promise of the latter credible—and therefore a potentially real goal rather than an irrelevant fantasy. The confidence inspired by Draper, who could convince even a skeptic that he "knows more by far about the science and technology of inertial-guidance systems than anyone else in the Western world,"[9] was a central factor, with his persuasive skills bolstered by an impressive record of delivering on his promises.

The limits to Draper's powers of persuasion are, however, also indicative of the sense in which guidance technologists such as he did not create the interest in counterforce accuracy.

8. See Colin S. Gray, *The Future of Land-Based Missile Forces,* Adelphi Paper No. 140 (London: International Institute for Strategic Studies, 1977), 5–6. My interviews made it clear that this argument was deployed in matters such as the struggle for the acceptance of stellar-inertial guidance, described in chapter 5.

9. Herbert York, *Race to Oblivion: A Participant's View of the Arms Race* (New York: Simon and Schuster, 1970), 86. York was involved in the ballistic missile decisions of the 1950s, and met with Draper's predictions of extreme accuracy. The passage quoted continues: "He [Draper] is a great optimist, and, although he is usually right, he sometimes seems to know even more than is so."

The Air Force had its own reasons, organizational and strategic, for desiring counterforce capability, and therefore the promise of its possibility fell on fertile ground. The Navy originally lacked an organizational and strategic rationale for counterforce missiles, at least in the view of the key leaders of the Special Projects Office, with their considerable autonomy. Draper tried to persuade them otherwise, telling them: "fleet ballistic missiles offer many well-known advantages, but will surely be handicapped in competition for national support unless they can be fired with accuracy levels comparable to those of land-based missiles."[10] But he failed, and for almost two more decades the key Navy decision makers were unconvinced that, for them, there was anything either natural or appropriate in the pursuit of extreme accuracy.

Draper's own reflections on the general role that he played are of interest here, even though they do not refer specifically to counterforce accuracy:

CSD I had always been interested in people that I had to deal with and their mental attitudes and why they had these mental attitudes and choosing something that the people in charge had to have.

DMcK Had to have?

CSD Had to have, yeah.

DMcK Even if they didn't know they had to have it?

CSD Well, they found out pretty damn quick, well, that's right.[11]

Several aspects of this are of interest. First is the breadth of Draper's conception of his role as a technologist, which firmly included "mental attitudes." It cannot wholly be coincidental that the first degree of the supreme "heterogeneous engineer" of inertial guidance was in psychology.[12] A second interesting

10. C. S. Draper, "Submarine Inertial Navigation—A Review and Some Predictions," originally classified paper presented to Polaris Steering Task Group, October 22, 1959 (copy in Library of Charles Stark Draper Laboratory, Inc., CSD-107).

11. Interview with Charles Stark Draper, Cambridge, Mass., October 12, 1984.

12. Draper received a B.S. in psychology from Stanford University in 1922, followed by an S.B. in electrochemical engineering in 1926, an M.S. (without specification) in 1928, and a Ph.D. in physics (with a minor in mathematics) in 1938, these last three degrees all from MIT (Charles Stark Draper, "Ori-

aspect is to whom Draper oriented himself: "the people in charge." But most interesting of all is Draper's orientation, not to their expressed wishes, but to what we might call their interests, or, to be more precise, Draper's view of their interests: "choosing something that the people in charge *had to have.*" He wanted his ultra-improved gyros, and the black-box navigation they made possible, to become indispensable.

It is worth noting the similarity of Draper's self-perception to the analysis of historian of technology David Noble: "technical people strive continuously to anticipate and meet the criteria of those in power simply so that they may be able to practice their calling."[13] There is no evidence that Draper, or his peers and rivals at Autonetics and elsewhere, particularly wished to make U.S. ballistic missiles more accurate. They would have been just as happy making civil air inertial navigators more accurate. But because "the people in charge" had an interest in the former, and not the latter, that was what Draper had to do to "practice his calling" of pursuing ultimate accuracy in inertial instruments. Many of his peers, in corporations tied to the air navigation market, had to define or redefine their calling so as not to include commitment to ultimate accuracy.

Noble's formulation is a mite too passive. Technical people sometimes also seek to shape, as well as to anticipate and to meet, the criteria of those in power.[14] They may even seek to alter power and who holds it.[15] But Noble is right to emphasize that technical people are not always the all-powerful manipulators that authors such as Lord Zuckerman sometimes seem to suggest.[16]

gins of Inertial Navigation," *Journal of Guidance and Control,* Vol. 4, 1981, 449). The fullest account of Draper's life is his own, in interview with Barton Hacker, January 19, February 2, March and April 5, 1976 (MIT Archives and Special Collections, Oral History Collection, Charles Stark Draper, MC 134, box 1).

13. David F. Noble, *Forces of Production: A Social History of Industrial Automation* (New York: Knopf, 1984), 43.

14. See Latour, *Science in Action,* chapter 3.

15. See Edwin T. Layton, Jr., *The Revolt of the Engineers: Social Responsibility and the American Engineering Profession* (Cleveland, Ohio: Case Western University Press, 1971).

16. Zuckerman, "Science Advisers." Zuckerman seems to have in mind particularly the scientists in the nuclear weapons laboratories, such as Los Alamos

In short, to think of a matter of technological change as a "natural trajectory" is to miss everything that is interesting about it and which makes it possible: the interplay of interests, the flow of resources, and the credibility of predictions. The reason why a pattern may nevertheless appear natural was noted in chapter 4. It lies in the self-fulfilling nature of the prophecy at the core of a trajectory. If it comes to be believed that there is only one way to'advance a technology, then that one way has at least a chance of becoming a reality. The others do not, and soon their disadvantage may become irreversible. An example in miniature, one where the Air Force guidance mafia dictated to Draper, rather than him influencing them, was the replacement of ball-bearings by self-activating gas bearings in high-precision floated gyros. The former still had their proponents, but starved of development funds they lacked any real chance to prove their case. It is hard now not to think that the gas bearing is simply inherently better.

Thus it is fortunate for my argument that there does exist the bifurcation of technological change in inertial guidance and navigation between the pursuit of high accuracy, using traditional, extremely "labor-sensitive" technology, and the pursuit of reliability, producibility, and economy using novel sensors.[17] As noted above, in the absence of either path, the remaining one might indeed appear "natural," the product simply of following the inherent possibilities of the technology. But I believe that the above analysis of the nature of technological trajectories holds even where there is no actual, existing alternative to which one can point.

That may indeed shortly be the situation of inertial guidance and navigation. Though there are those who would still stand by the prediction, it is now widely doubted that further substantial increases in missile accuracy by the incremental refinement of existing technology are possible. As noted at the end of chapter 4, there is disagreement as to whether the barrier that has been reached is a fundamental constraint of the phys-

and Livermore National Laboratories, who do, admittedly, seem to have had remarkable success as "heterogeneous engineers."

17. This is an oversimplification. At least two other forms of technological change exist, associated with submarine navigation and tactical missiles respectively. As noted in chapter 5, the first has become the "home" of the electrostatically suspended gyroscope. The second may become the "home" of the fiber-optic gyro.

ical world or the social constraint of rising costs at a time when even nuclear weapons budgets are under pressure. But whatever the grounds for the skepticism, there is certainly deep doubt that further investment in the refinement of traditional floated gyros and accelerometers will have a meaningful payoff. By the mid-1980s, that form of technological change seemed, accordingly, to have lost much of its support and impetus.

This is one of the cases where an understanding of the processes of technological change provides a lever for affecting those processes. A correlate of the belief that missile accuracy could be increased greatly by the incremental improvement of existing technology was a belief that it could not be brought within the scope of arms control agreement: "there is no way to get hold of it, it is a laboratory development, and there is no way to stop progress in that field," as one of those quoted in chapter 1 put it. The demise of the belief in the possibility of continuous evolutionary increases in missile accuracy could give rise to the demise of the belief that control over accuracy is impossible. As noted at the end of chapter 4, there is a good chance that increasing missile accuracy by means other than the traditional incremental ones could be prevented by a properly constructed arms control agreement.[18]

Unpicking the "technological trajectory" of increasing missile accuracy, and comparing it to the other major form of change in inertial technology, together provide one window on the nature of technological change in guidance. Another window is provided by cross-national comparison. The cases of France and the People's Republic of China show us that, under different circumstances from that of the United States, the pursuit of extreme accuracy in missile guidance did not seem at all natural.

A rather different comparison is with the Soviet Union. Increasing missile accuracy appears to have been as high priority a goal there as it has been in the United States. As we saw in chapter 6, however, the means used to achieve that goal have been different, both in the design of components and in that of overall systems. These differences cannot all be inter-

18. This argument is spelled out in more detail in D. MacKenzie, "Missile Accuracy—An Arms Control Opportunity," *Bulletin of the Atomic Scientists*, Vol. 42, No. 6 (June/July 1986), 11–16.

preted as a Soviet "lag" behind the West, with that term's implicit assumption that the Soviet Union is following the same technical path, only more slowly. More than thirty years after Draper-style fluid-floated gyroscopes and accelerometers became widely known, the Soviet Union shows no sign of having adopted them in its missile guidance. Soviet designers have followed the same path of incremental improvement, but it has been incremental improvement of different devices.

What happens when we move from the overall patterns of technological change to its details? Once a goal (such as greater accuracy) is established and a means (such as incremental improvement) is institutionalized, is there anything left for social analysis? I believe that there is, since there are always, at least potentially, issues of what should be improved, and how, indeed of what incremental improvement consists in.[19] Should one continue to improve ball-bearings, or discard them in favor of gas bearings? Should one continue to use electromagnets in the torque generators, or shift to permanent magnets?[20]

To change metaphors, the black box of technology is actually a Russian doll. Inside one black box (a guidance system, say) lie others: accelerometers, gyroscopes, and so on. Inside gyroscopes are bearings, torque generators and much else; inside torque generators are magnets; inside magnets are. . . .[21] There are many nested black boxes to open before one reaches one that no participant has opened. Only the analyst's energy and the reader's patience provide a limit.

In an institutionalized form of technological development and with a stable division of labor, we will normally change social scale as we proceed in this fashion deeper into the black box. At any point dispute is possible, and questions of bearing or magnet design can raise as strong passions among those involved as questions of national nuclear strategy. But to whom the disputes matter typically changes. Deeper inside the black

19. To repeat a necessary warning: what looks like incremental change to one person may seem a radical departure to another.

20. The torque generator in a floated gyro can be seen in figures 2.14 and 2.15. I owe my knowledge of the significance of this issue, which I have not discussed in the text, to an interview with Ralph Ragan, Cambridge, Mass., October 4, 1984.

21. The topology of the Russian doll metaphor should not be taken too literally here. The magnets form part of the torque generator, but the device has no actual inside or outside.

box, different actors are involved—that indeed is simply what the division of labor means. Some issues involve the Strategic Air Command, Congress, and the Department of Defense; others the Ballistic Missile Office and a guidance contractor; others are contained within the guidance contractor; and others within a single department of that contractor.

This might lead one to the conclusion that there are instrinsically "big" issues, on which the analysts of technology should focus, and matters of "detail" they can safely ignore. But that would be a dangerous conclusion. For there is nothing absolute about these differences in social scale. Those involved can turn big issues into matters of detail and matters of detail into big issues.

The latter is of particular interest here. If stymied at one level of the division of labor, actors can always seek to get what they want by recruiting allies who would not, normally, have anything to do with the issue in question. The central example is the development of stellar-inertial guidance in the United States. We saw how its proponents, feeling themselves blocked at the level of guidance contractors and the Navy's Special Projects Office, took the matter to wider circles of defense intellectuals and higher levels in the Department of Defense.

Stellar-inertial guidance could have been presented as a small matter of detailed improvement to the guidance systems of submarine-launched missiles. Instead, it was presented as a big issue: a means of giving these missiles counterforce capability and enabling them to rival in accuracy the Air Force's ICBMs. So instead of remaining a matter of controversy only within the different departments of the Special Projects Office, and between guidance contractors such as Kearfott and the Draper Laboratory, it became an issue for the top levels of the Navy and a matter of debate on the floor of Congress.

That this was a contingent development, rather than in some sense intrinsic to the issue, is shown by how differently the proponents of stellar-inertial guidance approached the issue after their efforts backfired on them and the stellar-inertial system for Poseidon was canceled. For the next missile system, Trident I, stellar-inertial guidance was presented much more as a matter of detail, connected to the relatively uncontentious issues of missile range and times between necessary updates to the submarine's inertial navigator. The black box having been opened so publicly, it was hard to shut it again completely, and stellar-inertial guidance remained at least mildly controversial in the wider political arena for years. But the more cautious tack taken to justify its use in Trident I was successful.

I shall return shortly to this question of the relationship between technology and politics. Before, turning to this, however, I need to complement my discussion of what we have learned about the first term in the dichotomy, "technology," with what we have learned about the second, "politics."

About Politics

Technological determinism is one way of thinking about weapons development that inclines us to passivity. Another is what might be called political determinism. According to this, what happens is a result of decision making by a state. Sometimes a particular person (the President, perhaps) or a collectivity (the "political-military elite," perhaps) is seen as representing the state. But in all cases, the form of explanation is the same. The state is conceived of as akin to an individual, rational human decision maker. It has a goal, and chooses means such as nuclear strategies and weapons systems so as to fufill that goal.

The goal may be believed to be benign, such as protecting one's population from possible attack. That, of course, is how weapons decisions are typically legitimated: as the rational means for a state to maintain or improve its security. This version is the most pervasive form of political determinism, and one that underlies much academic discussion of arms races, though there it has also been common to note the possibility of detrimental consequences when two or more states are doing this competitively.[22] Academic discussions, too, have often hypothesized that the goals of states and their leaders may be less benign, albeit still rationally pursued. The position known as "realism," for example, has as its premises: "that states are the key actors in world politics; that they seek power, either as an end or as a means to other ends; and that they rationally seek to advance their interests."[23] And, of course,

22. This is the well-known "prisoners' dilemma" of game theory, around which a large literature has developed. For one clear exposition of the basic point, see Barry Barnes, *About Science* (Oxford: Blackwell, 1985), 134–137.
23. Robert O. Keohane, "Alliances, Threats, and the Uses of Neorealism," *International Security,* Vol. 13, No. 1 (Summer 1988), 173fn.

it is very common indeed to impute benign motives to one's own state and a power-seeking hegemonic drive to other states.

This way of thinking can lead to passivity even when account is taken of the deleterious consequences of benign motives or the possibility of less benign ones. True, states, their goals, and the international system of states within which they operate are generally seen as human creations, not reflections of nature. So the passivity is less deep than in the stronger versions of technological determinism. But the division of the world into separate, competing states is patently not an easy thing to change. If it is believed that nothing can be altered fundamentally without world government, for example, then maybe nothing very much will be altered at all.

Fortunately, far too radical an oversimplification is built into the political determinism that attributes all major patterns of development in military strategy and weaponry to the security-oriented or power-oriented decisions of states. The key flaw is the anthropomorphism of regarding a state as akin to an individual decision maker.

The previous chapters have examined several such major patterns of development: whether to have a counterforce or an assured destruction nuclear strategy, whether to build missiles or bombers, whether to pursue extreme accuracy in missile development, and so on. In explaining each, I have always had to disaggregate "the state," identifying the often conflicting preferences of its different parts such as different armed services and even subgroups within these services. So the state should not be though of as unitary. I have also typically had to disaggregate "decision," identifying instead various different levels of policy process, each leading perhaps to a result, but not necessarily to any overall coherence.

Let me make the discussion more concrete by considering in particular the first of these issues, whether national nuclear strategy should simply be "assured destruction" retaliation or one of the many variants of counterforce. This, of course, is the strategic issue to which missile accuracy most closely links, for the reason outlined in chapter 1, and an issue that is agreed by those who debate it, whatever their particular positions, to be central to national security and perhaps national power.

Contest over the issue within the American state is evident. Different armed services have had different positions, the Navy inclining to assured destruction, and the Air Force, since the early 1960s, to counterforce. Even within particular armed services there has not always been agreement, as witnessed by the repeated conflicts between the Navy Special Projects Office, with its primary commitment to assured destruction, and the planners in the Office of the Chief of Naval Operations who favored counterforce. During the late 1960s, Congress became engaged with the issue, and for a while a powerful and well-organized anticounterforce lobby there was a significant barrier to those who wished to see the accuracy of U.S. missile increased. The civilian defense establishment, Secretaries of Defense, Directors of Defense Research and Engineering, and even Presidents have been involved.

It might be imagined that the issue could simply be decided by a President, who is, after all, at the pinnacle of U.S. state power. It is interesting to note, however, that the first President to come to office with clearly expressed views on the matter, Jimmy Carter, was forced by circumstances to change his position. He wished to see nuclear weapons eliminated, or at least reduced to the level of an assured destruction finite deterrent, and was skeptical of MX in particular. But during the four years of his presidency he was pushed toward endorsement of MX and toward an explicitly procounterforce statement of national nuclear strategy, Presidential Directive 59. Other political actors, in particular a resurgent right, the growing sense of a Soviet threat, and the impact of external events in Iran, Afghanistan, and elsewhere, all forced him to set aside his original convictions.[24]

Secretaries of Defense have the advantage that they do not have to engage with the full range of issues that a President has to, and at least two Secretaries, Robert McNamara and James Schlesinger, have focused personally on issues of nuclear strategy. Circumstances such as his relationship to the Air Force helped push McNamara in the opposite direction from Carter, from counterforce to assured destruction. Schlesinger managed to maintain a consistent direction, playing an important role in promoting a commitment to greater accuracy and counterforce capability, especially in U.S. submarine-launched ballistic missiles, but even he could not simply command that

24. For one account of Carter's enforced transformation, see John Edwards, *Superweapon: The Making of MX* (New York: Norton, 1982).

commitment. The most he could do, in the term chosen independently by himself and by the man some claimed to be the main obstacle he faced, was to "push" for it.[25]

So the actors involved in dispute between assured destruction and counterforce are many, and none is all-powerful. In part as a consequence, the issue never seems to be decided. Under Reagan, the balance of forces was more overwhelmingly in favor of counterforce than at any other point in time. He had been elected on a Republican Party platform that explictly stated that assured destruction "is simply not credible and therefore is ineffectual."[26] Hawks were in the ascendancy in Washington and nuclear "warfighting" rhetoric was in the air. Even the Navy Special Projects Office had, however reluctantly, embraced counterforce. Yet Reagan could secure deployment of only fifty MX missiles, a number insufficient to threaten more than a portion of the Soviet Union's nuclear forces.

The other main reason why the issue between counterforce and assured destruction never seems finally resolved is that it is not one issue, but (at least) three. At one level, it is a question of what is called "stated posture," the official rationale of the U.S. nuclear arsenal. At another, it is a question of operational planning for nuclear war. At a third level, it is a question of the criteria according to which new nuclear weapons technologies ought to be designed and judged.

One might imagine that all three levels of policy would be consistent. "In theory, the strategy should define the target system and determine the design of the weapon systems."[27] In actuality, nothing guarantees this. Quite different actors and considerations are involved at each level. "Stated posture" is a matter primarily for those at the pinnacle of formal power— the President, Secretary of Defense, Chiefs of Staff, and the like—and their eyes must be both on domestic politics, especially Congress, and the outside world. Operational planning, on the other hand, is much more a military insiders' business, and, because of its secrecy, does not have to keep that wider

25. See chapter 5.
26. 1980 Republican Party platform, as quoted in Robert Scheer, *With Enough Shovels: Reagan, Bush and Nuclear War* (London: Secker and Warburg, 1983), 127.
27. Richard Lee Walker, *Strategic Target Planning: Bridging the Gap between Theory and Practice,* National Security Affairs Monograph Series, 83–89 (Washington, D.C.: National Defense University, 1983), 12.

audience in mind. We saw, for example, how 1960s operational planning remained in essence unchanged, despite the shift in stated posture from counterforce to assured destruction. Weapons system design, finally, involves a different constituency yet again. It often involves different branches of the armed services: the Strategic Air Command, central to operational planning, was quite deliberately excluded from the process of the design of early U.S. ICBMs. Corporations, too, can have an economic interest in weapon system questions while they do not have much an interest, at least in any direct sense, in stated posture or war planning.

It is particularly important to bear matters such as this in mind when we consider questions of nuclear strategy, war planning, and weapons design in the Soviet Union. Because our evidence is fragmentary, there is a constant temptation to make inferences of the form "since their weapons design is this, their operational strategy must be that." But, as noted at the end of chapter 6, those inferences would often mislead if applied to the United States, where, for example, weapon systems such as Trident I can be found whose design is the result of compromise between opposed priorities rather than any clear strategic goal. We should not assume a priori that matters are any more consistent in the Soviet Union.[28]

Those who know the literature on state policy, especially that on foreign policy and defense, will recognize in what I am saying elements of the position known as "bureaucratic politics."[29] Often, it is true, the proponents of this position have

28. The American state is close to being unique in its multiple and overlapping jurisdictions, division of powers, and relatively permeable boundary between the state and surrounding civil society. It could therefore be that the potential in the United States for bureaucratic politics, in the narrow sense, is greater than in more hierarchically organized states such as the Soviet Union, the United Kingdom, and France. The problem is that the correlate of these features of the American state is its relative openness to empirical study. The vast bulk of studies of bureaucratic politics is not merely by Americans but of America. Hence it may be that bureaucratic politics is no more prevalent in the United States than elsewhere, just better known.
29. Perhaps the most famous single formulation is in Graham T. Allison, *Essence of Decision: Explaining the Cuban Missile Crisis* (Boston: Little, Brown, 1971). Among the position's intellectual roots is the thinking of Herbert Simon on matters such as "bounded rationality." See, for example, Simon, "A Behavioral Model of Rational Choice," *Quarterly Journal of Economics,* Vol. 69 (1955), 99–118.

made the mistake of overemphasizing the influence on state policy of processes within and between formal organizations.[30] But the position does contain the crucial insight. "America," or, putting it more generally, "the state," intends nothing, decides nothing, does nothing. Policy is not "decision," with that term's connotation of an anthropomorphic decision maker, but "outcome."

I do not mean to imply that the only determinants of state policy are domestic. Events in the external world matter. In particular, what happens in the Soviet Union does affect what happens in the United States and vice versa. The nature of those external events is, however, often unclear, and the construction of an agreed picture of them involves a process of negotiation. U.S. intelligence estimates of strategic developments in the Soviet Union are a well-documented case in point.[31] Domestic concerns in other words, structure how the outside world is perceived. In chapter 3, for example, we saw how, until the "missile revolution" in the United States, it was assumed that the Soviets were giving the same low priority to ballistic missiles as the United States, and how the American proponents of the ballistic missile helped change that perception of Soviet development.

Furthermore, even a consensual view of the nature of the outside world by no means compels any single course of action. In showing how branches of the armed services typically advocate courses of action that favor their interests—for example, that will increase their budgets—the bureaucratic politics literature might be seen as imputing cynical and self-serving motives to those involved. This book may also be seen as having done so on occasion. But no such imputation is necessary. All the training, traditions, experience and esprit de corps of military organizations will tend to lead their members to the conclusion that they have a crucial part to play in meeting the "external threat." An Air Force officer will be led to see the

---

30. Typical critiques include Robert J. Art, "Bureaucratic Politics and American Foreign Policy: A Critique," *Policy Sciences,* Vol. 4 (1973), 467–490; Desmond Ball, "The Blind Men and the Elephant: A Critique of the Bureaucratic Politics Theory," *Australian Outlook,* Vol. 28 (1974), 71–92.
31. Lawrence Freedman, *U.S. Intelligence and the Soviet Strategic Threat* (London: Macmillan, 1977); John Prados, *The Soviet Estimate: U.S. Intelligence Analysis and Soviet Strategic Forces* (Princeton, N.J.: Princeton University Press, 1986).

role that aircraft, or perhaps ICBMs, can play; a Navy officer the role of aircraft carriers or submarines, and so on.

The question of the role of the external world in shaping contested outcomes blends into the wider question of the influence on these outcomes of what we might call "structure": the place of the United States or Soviet Union in the world system of states; their different characteristics as societies; and so on. A shortcoming of the bureaucratic politics approach, which it shares with American "pluralist" political science more generally, is that, in focusing on the observable influence of particular actors on contested outcomes, it misses the structural processes that define agendas, actors, and their interests.[32]

Here is not the place to address this difficult and contentious issue in its full generality. I shall restrict myself to pointing out one structural influence on my chosen example of a contested outcome: the balance between assured destruction and counterforce. Since the Second World War, the United States has become deeply involved in political developments, not just in the Americas, but also in Europe and the rest of the world. It has become not just a state but a superpower.

This involvement in global events, exemplified by the continued presence of American troops in Europe and the wars fought in Asia, broadens the range of considerations that enter into the making of defense policy. If it did not exist, that policy could have a very simple objective: since a conventional invasion of the United States is an eventuality that few have considered likely, it could concentrate on deterring a nuclear attack on the United States. But U.S. global involvement does exist.

Proponents of counterforce draw an inference from this. Assured destruction, they argue, might suffice as a strategy if deterrence of a nuclear attack on American territory were all that was at stake. But an arsenal shaped according to the precepts of assured destruction cannot deter more than that. With American cities vulnerable to Soviet retaliation, no American President could credibly threaten to destroy Soviet cities in response to events in Europe or the Third World. So more than an assured destruction arsenal is needed. As the 1980 Republican Party platform put it, summarizing the arguments

32. See, with reference to "pluralism" more generally, rather than to bureaucratic politics in particular, Steven Lukes, *Power: A Radical View* (London: Macmillan, 1974)

of many hawkish analysts: "An administration that can defend its interest only by threatening the mass extermination of civilians . . . dooms itself to strategic, and eventually geo-political paralysis."[33]

Two main versions of the inference exist. One has been that the United States could hope to gain political leverage from the capacity for counterforce attacks on the Soviet Union. Hawkish civilian strategist Colin Gray spelled out this version in 1982. "If U.S. and U.S.-allied forces are overmatched, or even just matched, around the rimlands of Eurasia, how is the U.S. to enforce a tolerable crisis or wartime outcome in the event of a theater struggle? In extremis, the United States has to be able, not incredibly, to threaten central nuclear employment against the Soviet homeland." More than just the capability for hard-target counterforce was needed to make a credible threat a reality. But, asked Gray, "if the United States could not dominate a process of nuclear escalation to coerce the U.S.S.R., how could Soviet gains in a theater be reversed?"[34]

The other version was, as noted in chapter 4, much the more widely propounded, in the 1970s in particular. It was in a sense simply the mirror image of the first: the fear of the consequences of a Soviet advantage in counterforce capability. No one seriously worried that the Soviet Union could wholly disarm the United States, denying the latter any retaliatory capability, but there was much speculation as to how the Soviets might turn a counterforce advantage into political gains or, worse, into a triumph in a "limited" nuclear war.

The inference that a superpower's global involvement requires a counterforce arsenal can of course be contested.[35] One counter is the argument that to limit the nuclear arsenal

33. Quoted in Scheer, *With Enough Shovels*, 128.
34. Colin S. Gray, "The Idea of Strategic Superiority," *Air Force Magazine* (March 1982), 63.
35. My criticism of the authors who seek to explain counterforce primarily by superpower global involvement is that they do not fully take into account the implications of its contestability. See Alan Roberts, "Preparing to Fight a Nuclear War," *Arena* (Melbourne), No. 57 (1981), 45–93, reprinted in part in D. MacKenzie and J. Wajcman, eds., *The Social Shaping of Technology* (Milton Keynes, Bucks.: Open University Press, 1985), 279–294; and Earl C. Ravenal, "Counterforce and Alliance: The Ultimate Connection," *International Security*, Vol. 6, No. 4 (Spring 1982), 26–43.

to the relatively small size required for assured destruction frees money for the conventional forces that constitute the practical way to project national power.[36] (It is worth noting that the heyday of assured destruction as stated posture coincided almost exactly with the Vietnam War.) Another counter is the claim that any use of nuclear weapons, counterforce as much as countercity, carries with it such a high risk of escalation to all-out war that no sane leader would ever contemplate such a use, and thus the threat of it would lack all credibility. Therefore, it has been argued, even the most massive superiority in counterforce capability is meaningless, so long as it falls short of the capability to destroy effectively all of an opponent's nuclear forces, a capability impossible realistically to hope for in the age of large, diversified arsenals and nuclear submarines. The United States could not profit from such superiority and need not fear the Soviet Union gaining it.[37]

So the connection between superpower status and a counterforce arsenal is contestable. Nevertheless, were it not for the global involvements of the United States, the argument that an assured destruction arsenal might suffice for deterrence against nuclear attack on one's territory, but is of little use for anything else, would lose its point. This feature of the United States's position in the world system of states thus has an effect on the balance between assured destruction and counterforce approaches to nuclear strategy. It is not the only feature affecting that balance (to assert that would be to return to a version of the political determinism I have been arguing against), but it is one. In particular, it seems to have had an effect in the swing in stated posture away from assured destruction towards counterforce during the 1970s and 1980s.

An Ordinary World

Though the question of the relationship of U.S. nuclear strategy to the country's global position cannot and should not be avoided, there is a risk of misleading the reader by raising it.

36. This has of course been one reason for the attractiveness of assured destruction to the Navy.
37. For a useful discussion of the changing, disputed meaning of "superiority" in matters nuclear, see Lawrence Freedman, "Strategic Superiority," *Coexistence*, Vol. 21 (1984), 7–21.

Because it concerns precisely the sort of matters that the politics of the nuclear world is usually thought to be about—global power, deterrence, national survival, and the like—it obscures how little direct role these "grand" matters often play. Overall, what is perhaps most surprising about the politics of the nuclear world, given the awesomeness of the matters at stake, is its ordinariness and familiarity.

What we have found is that nuclear politics is influenced by much the same sort of processes and range of factors as any other kind of politics.[38] The typical day-to-day concern is not Armageddon, or even the Russians, but getting the job done, preserving one's autonomy and good reputation, negotiating the next contract, minimizing pressures on the budget, keeping the others off one's turf, and so on. Because the nuclear world lacks any single central determining axis, technological or political, these are not mere perturbations. Their cumulative interactions have major effects. We saw, for example, how a combination of the Polaris program and Air Force hegemony pushed the Navy toward an assured destruction strategy, how the Air Force in response embraced counterforce, and how much this shaped the current U.S. arsenal.

Apart from the general prejudice that the true driving forces must be grander things, there is a particular difficulty for those who are not intimately familiar with it in keeping the ordinariness of nuclear politics in mind. Like the politics of any office, to an outsider it seems intricate, devious, and boring, difficult to understand in comparison with the memorable but misleading simplicities of technological and macropolitical determinism. So it is perhaps worth substituting a single striking example for all the details. It may not even be altogether accurate, as it is not well documented, but if true, it vividly shows how the same sort of mundane considerations can shape nuclear war plans as can shape, for example, a local authority budget. If nuclear war had broken out early in 1961, Moscow was to have been the target for no fewer than 170 American

38. For example, the range of explanatory factors drawn on in the study of foreign economic policies is similar to that drawn on in "political" explanations of arms dynamics, with similar disputes as to their relative weight (though technological determinism is naturally not found in the former). See, for example, Peter J. Katzenstein, "International Relations and Domestic Structures: Foreign Economic Policies of Advanced Industrial States," *International Organization*, Vol. 30 (1976), 1–45.

nuclear weapons. This was not because that many were needed to destroy it: even a tiny fraction of that number would have been more than sufficient. Nor was it primarily because of worries that technical failure, a Soviet preemptive strike, or Soviet defenses might lead to attrition of the attacking force. The chief reason was the reluctance of the various branches of the armed services to give up the Soviet capital for a less prestigious target.[39]

The ordinariness of nuclear politics needs emphasis, not simply because it is counterintuitive, but because it is indicative of the potential for successful intervention in the nuclear world. Because that world is not determined by the behemoth of technology or by the logic of the system of states, small and unexpected events can, in the right circumstances, derail the most awesome programs, just as they can in all the other ordinary worlds of politics. "Missile X is a nuclear weapon of unprecedented destructive power. Once deployed, it could destroy all Soviet land missiles in one thirty-minute barrage. If based on land, it will require a construction effort greater than the Panama Canal or the Great Wall of China."[40] Yet a crucial part of the derailing of MX was local, environmental protest in two sparsely-populated American states. As we saw in chapter 4, the ultimate counterforce ICBM program has been unable to recover from this mishap.

This ordinariness, rather than the missile as phallic symbol, also seems to me the key to the masculinity of the world of nuclear weaponry. Masculine it certainly is. Only one woman's name appears in the list of well over a hundred interviewees at the end of this book, and she was Charles Stark Draper's secretary.[41] That extraordinary disproportion of gender is, I think, a fair reflection of the virtually complete absence of women from roles of any power in the processes I have

39. Gregg Herken, *Counsels of War* (New York: Knopf, 1985), 144; Herken's source is a calculation made at the time on the basis of a briefing on the Single Integrated Operational Plan, by Kennedy Administration strategist, and later "Pentagon papers" leaker, Daniel Ellsberg. The explanation of the concentration on Moscow is Ellsberg's.
40. I am quoting from the dust-jacket of John Edwards, *Superweapon: The Making of MX* (New York: Norton, 1982).
41. One woman technologist in Britain declined to be interviewed, saying her area of expertise was not relevant, and a clash of schedules prevented me from interviewing a former Department of Defense official who is a woman.

described: they assembled the gyroscopes but had no say in either their design or what was subsequently done with them.[42]

There is a strongly personal flavor to much of the day-to-day politics of nuclear weaponry. Even if it is later formally ratified between their organizations, much of what takes place is the result of informal, face-to-face contact between men who have interacted and known each other, often wearing different organizational "hats," for years or decades. Even though women are now starting to enter technical jobs in this area, it will be some time before they can operate effectively in the world of the technico-political "mafias" and the wheeler-dealing of weapons acquisition.[43]

If male psychology matters, it may be less through the extravagances of psychoanalytic symbolism than in the capacity (or handicap) of divorcing work from life, means from ends, and the details from the overall picture. If men do this more than women,[44] then they may find it easier to treat their role in the nuclear business as "just a job." Asked about attitudes in his office to questions of nuclear strategy and disarmament, an officer in one of the two main organizations reponsible for U.S. strategic ballistic missile development replied: "We don't talk about those things much. They're not germane to your job. You talk about those kind of things more with your neighbors than you do with the people you work with."[45] He was by no means untypical; indeed he was a particularly thoughtful man.

This sense of ordinariness is as much true of the "technical" as of the "military-political" aspects of the nuclear world. By far the dominant view of their work that nuclear missile guidance engineers have is that it is an interesting, challenging, and rewarding technical task. Only a minority of interviewees spontaneously mentioned either moral qualms (these were very

42. The production processes of inertial instruments have high representations of women workers.
43. I have no figures for guidance technologists and base this remark on informal impressions alone. It is interesting to note, however, that while in 1965 only 3.1 percent of National Aeronautics and Space Administration scientists and engineers were women, by 1987 10.2 percent were (JoAnn H. Morgan and James M. Ragusa, "Women: A Key Work Force in Preparation for Space Flight," typescript, June 26, 1987).
44. I know of no definitive evidence to this effect, so can make the point only tentatively.
45. Interview.

rare) or pride in contributing to defense against the "Soviet threat." One interviewee at the Draper Laboratory even welcomed the Laboratory's regular Monday morning peace movement picket as reminding him and their colleagues what their job was actually about.

A pervasive feature of the ordinary non-nuclear worlds of both politics and technology is, of course, a concern with money. At the time of the "missile revolution" and immediately thereafter, strategic missile design was substantially insulated from financial concerns. The designers of Polaris were guaranteed: "if more money is needed, we will get it."[46] Even then, however, the conception of Minuteman was strongly shaped by considerations of economy, and since that "golden age" (as it now seems to many who participated in it) financial constraints and all the associated paperwork have multiplied. Budgets may still be generous by other standards, but they matter more and more.

For a long time a chief distinction between the two "trajectories" of inertial technology was that costs mattered enormously in aircraft navigation and little in strategic missile guidance. But that is changing. In the 1960s, the Air Force forced Autonetics to reduce the costs of Minuteman guidance systems by threatening to introduce a second, competitive, supplier. In the late 1980s, they started going even further down the path of competitive production of MX's beryllium baby. True, both these episodes concern the production process of systems whose design has already been fixed. If, however, the 1990s or beyond see a new U.S. strategic ballistic missile guidance system, which is far from certain, it seems very likely that economic pressures will shape its design much more forcibly than they have those of its predecessors.

These are of course all matters where economic considerations are accepted as wholly legitimate, even if, to the designer, they are sometimes irksome. What of the sphere of illegitimate economic considerations—programs initiated and continued,

46. Memorandum from [Chief of Naval Operations] Arleigh Burke for Rear Admiral Clark and Rear Admiral Raborn, Subject "ICBM—IRBM," December 2, 1955, reproduced in Lockheed Missiles and Space Company, Inc., *The Fleet Ballistic Missile System: Polaris, Poseidon, Trident* (Sunnyvale, Calif.: Lockheed Missiles and Space, n.d.), 6–7.

or designs selected, because of corporate and personal financial gain?

It is difficult to assess the importance of this as a shaping factor. Because it is illegitimate, indeed illegal, it has to be surreptitious, so an interviewer is not going to be told about it if it does go on. That I have no evidence of bribery and corruption does not prove that none exists, and that I know of few episodes of the covert wielding of corporate power does not show that it has not happened.

My instincts, however, remain with the ordinary. I suspect that few of the developments I describe are actually to be explained by corporate executives passing envelopes full of large-denomination notes, or suitcases full of small-denomination ones, to senators, defense officials, or generals. The role of private economic gain has, I suspect, been much less dramatic.

I have two reasons for believing this. First, guidance technology development has tended to take stable organizational form. Cases of corrupt behavior that come to light tend to take place where there is genuine competition for contracts, and where anything that might give an edge—such as inside information—may seem worth paying for.[47] Although the contracts to produce guidance systems and their components are often open in that sense, the ones that are the center of my concern are those that involve the design of these systems and components. These, generally, have not been open. There has never been any serious doubt that the Draper Laboratory would retain its role as the chief guidance designer for every generation of U.S. submarine-launched ballistic missiles. Autonetics's role in Minuteman guidance was, similarly, never seriously threatened, and, as we saw, there was little doubt that MX would be guided by Draper's Advanced Inertial Reference Sphere. Why take the risk of illegal behavior if one is going to win the contract anyway, or stands no real chance of winning it even if one does?

Second, private economic gain and more legitimate motivations normally tend to run in the same direction rather than against each other. Military officers' careers will be advanced if the programs they run are seen as being successful, within

47. See, for example, the cases reported in *Aviation Week and Space Technology* in 1988 and 1989.

budget, and so on. Technologists can promote a pet concept or device out of technological enthusiasm: this seems to be accepted in the world of nuclear weaponry in the United States as a more-or-less legitimate motive. That their organizations, and probably the technologists personally, will benefit financially from its adoption cannot be admitted to publicly as motive, and need not be acknowledged in private.

This is, in a sense, simply one version of an elementary sociological point. Stable institutionalized reward systems can channel a wide variety of motives into one particular form of behavior. Scientists, for example, can be motivated by love of money, or fame, desire for the truth, the wish to help humanity, or the sheer pleasure of doing science, but all these motives are served by the publication of credible, interesting, novel results.[48] Similarly, the motivations of weapons technologists and program managers can be as different as a desire to enhance national security (as they perceive it), desire to serve the interests of the organization they work for, enthusiasm for their technology, or a wish to make money and advance their careers, without there being any great difference in how they behave. Ensuring the development of something that is theirs and negotiating its use in deployed systems, making their program a success; that will satisfy any and all of these goals.

The Separation of Politics and Technology

So the world of nuclear weaponry is lived, by those who are insiders to it, predominantly as ordinary, despite the extraordinary nature of its products. But I must now address squarely an issue that has been implicit so far in these conclusions. Is it two worlds or one: a world of technology and a world of politics, perhaps interacting but separate; or a world in which technology and politics are indistinguishable?

Again, this is one version of a more general issue, this time not such an elementary one: how to conceive of the relationship between technology and society. For a long time, the dominant view in the social sciences, with the partial exception of economics, was of technology as an external, autonomous force

48. See the discussion of the "cycle of credibility" in Bruno Latour and Steve Woolgar, *Laboratory Life: The Social Construction of Scientific Facts* (Beverly Hills, Calif.: SAGE, 1979), chapter 5.

exerting an influence on society.[49] From the 1960s onward, this technological determinism was increasingly rejected as scholars from history, sociology, and elsewhere found that technological change was itself shaped by the social circumstances within which it took place.[50]

By the mid-1980s, however, the terms of both the original thesis and the counter to it were being questioned, in the single most challenging argument to emerge from the "new sociology of technology" that came into being as sociologists of scientific knowledge turned their attention to technology.[51] Michel Callon, Bruno Latour, and John Law argued that just as there was no "technology" external to "society," so there was no "society" external to "technology." All there was were networks of varying degrees of durability linking together both human and

49. Economics was an exception because at least some economists felt that market forces could shape not just the adoption of technology but even invention itself. See Jacob Schmookler, *Inventions and Economic Growth* (Cambridge, Mass.: Harvard University Press, 1966). For a discussion of the thesis by two economists who believe it to be wrong, see David C. Mowery and Nathan Rosenberg, "The Influence of Market Demand upon Innovation: A Critical Review of some Recent Empirical Studies," in Rosenberg, *Inside the Black Box: Technology and Economics* (Cambridge: Cambridge University Press, 1982), 193–241, originally published in *Research Policy*, Vol. 8 (1979), 103–153. The trajectories of economics and sociology have been quite different in terms of the understanding of the relationship between technology and society. As sociology has moved away from technological determinism, some influential economists have moved some way toward it. For example, the currency of the metaphor of "technological trajectory," with its determinstic connotations, is largely within economics. For some insight into why this shift has taken place, see the reflections by a leader of it: Christopher Freeman, "Induced Innovation, Diffusion of Innovations and Business Cycles," in Brian Elliott, ed., *Technology and Social Process* (Edinburgh: Edinburgh University Press, 1988), 84–109.

50. In 1985, my colleague Judy Wajcman and I collected what we believed to be the clearest examples of this from the literature in our reader, *The Social Shaping of Technology*.

51. A key paper in the emergence of this field, though it does not make the particular argument I am referring to, was Trevor J. Pinch and Wiebe E. Bijker, "The Social Construction of Facts and Artefacts: Or How the Sociology of Science and the Sociology of Technology might benefit each other," *Social Studies of Science*, Vol. 14 (1984), 399–441. See also Wiebe E. Bijker, Thomas P. Hughes, and Trevor Pinch, eds., *The Social Construction of Technological Systems: New Directions in the History and Sociology of Technology* (Cambridge, Mass.: MIT Press, 1987).

non-human actors.[52] As historian of technology Thomas P. Hughes put it, the builders of technological systems such as Thomas Edison, "so thoroughly mixed matters commonly labeled 'economic,' 'technical,' and 'scientific' that his thoughts composed a seamless web." Only by the use of categories such as "network" or "system" that similarly refused these conventional distinctions could the nature of this activity be grasped.[53]

What we have found here is fully congruent with this idea. We have seen how the view of technology as an external, autonomous force cannot be sustained. We have seen how the political, organizational, economic, and legal circumstances—the "social" circumstances—of technological development shaped that development from its most general patterns to its most specific details. We have seen that this is true not only of technically "bad" decisions and designs, but also of technically "good" ones.[54] Minuteman II or Trident I, for example, would generally be regarded as technically excellent, barring the early troubles of the former, yet in each case we have seen the large variety of circumstances that helped shaped them.

Yet if technology is not an independent cause, is it not merely a dependent effect. I do not refer simply to the obduracy of the material world, though it is certainly true that the material world cannot simply be shaped at will. Of equal importance,

52. See, for example, Latour, *Science in Action;* Callon, "Society in the Making: The Study of Technology as a Tool for Sociological Analysis," in Bijker et al., *Social Construction,* 83–103; Law, "Technology and Heterogeneous Engineering: The Case of the Portuguese Expansion," in Bijker et al., *Social Construction,* 111–134. It is interesting to note that the second edition of Latour and Woolgar, *Laboratory Life: The Social Construction of Scientific Facts,* is entitled simply *Laboratory Life: The Construction of Scientific Facts* (Princeton, N.J.: Princeton University Press, 1986).

53. Thomas P. Hughes, "The Seamless Web: Technology, Science, et cetera, et cetera," in Elliott, *Technology and Social Process,* 9–19, quote on p. 13; this paper was first published in *Social Studies of Science,* Vol. 16 (1986), 281–292.

54. That social explanations should apply symmetrically and impartially to the technically "good" as well as the technically "bad" is a point of methodology inherited by the sociology of technology from the sociology of scientific knowledge. See, for technology, Pinch and Bijker, "Facts and Artefacts," and, for science, David Bloor, "Wittgenstein and Mannheim on the Sociology of Mathematics," *Studies in the History and Philosophy of Science,* Vol. 4 (1973), 173–191, also Bloor, *Knowledge and Social Imagery* (London: Routledge and Kegan Paul, 1976), chapter 1.

we have seen how mistaken it would be to see new technologies emerging because a socially-defined "need" for them existed. This is a standard way of responding to technological determinism, and a widely held view. Indeed, one could say it forms a third myth of invention to rival the "inspiration" and "applied science" myths.

But myth it is, at least for the inventions that constitute the larger breaks from existing practice.[55] Then, it is more common to find that "inventions are not made because they're needed. They're made and then you see what you can do with them." That remark, already quoted in chapter 4, was made by an interviewee in the context of a discussion of the invention of novel inertial sensors. But we have also found that it can be true for whole technological systems. Polaris is the clearest case where construction of the strategic "need" for the system was simultaneous with, not prior to, its technical development.

One could even go as far as to say that the development of Polaris caused the Navy to reassess where its interests lay. Having in the 1940s condemned a strategy based on the indiscriminate destruction of cities, the Navy embraced it, admittedly in somewhat different form, in the late 1950s. The cause was not a coarsening of ethical sensitivities, but a realization that Polaris could be used to forge a distinctive Navy contribution to national nuclear strategy, and, indeed, that the latter could very well be reshaped with Polaris's contribution in mind.

To summarize, what I would conclude is that technology in the nuclear world is not above politics as an autonomous determining factor, nor beneath it as a dependent effect, but part of it. Furthermore, as we enter the black box we find that the distinction between politics and technology becomes harder and harder to make. Is the diameter of a missile launch tube, for example, a political or a technical matter?

To that extent, therefore, I am wholly in sympathy with the argument that it is too weak a position even to see technology and politics as interacting: there is no categorical distinction to

55. Where technical change involves the identification and removal of bottlenecks or weak points in existing technologies, there is a sense in which it can be said to respond to "needs." But as Hughes, whose "reverse salients" model focuses on this, would emphasize, these needs cannot be seen as defined either solely socially, or solely technologically, but by the demands of a developing "system" that is both social and technical.

be made between the two. The web has no intrinsic seams. And yet we cannot stop with that observation.

For out of the seamless web, participants do construct relatively separate spheres of the "technical" and the "political." It is a distinction central to how they talk and, as I am about to argue, a distinction central to their success or failure. But if nothing absolute guarantees the distinction, how is it maintained?

In part, it is maintained through the division of labor. If you were a technologist in the Instrumentation Laboratory in the time when Charles Stark Draper was at his peak as a heterogeneous engineer, you could afford to think of what you did as just "technical." He did the persuading, got the contracts, and tied the laboratory's work on gyros to vital interests in the armed services. Similarly, the military officer quoted above enjoyed the luxury of not needing to talk in the office about the politics of nuclear strategy because both that office's director, and, interestingly, nowadays also the holder of the post of technical director, effectively devote themselves fulltime to managing the office's relations to the outside "political" world.[56]

There is also a pervasive way of thinking about technology that lends itself to the maintenance of separate spheres. This involves seeing politics as intruding on technology either when there is no single best technical solution, or where the best solution is not adopted. Let me give just two examples. In chapter 7, I quoted General Hepfer's comment on the Air Force's switch from conservative to optimistic estimates of the accuracy of its missiles: "the reason being political, they want to get the best accuracy they possibly can to make sure they stay better than submarines." In chapter 5, I mentioned one interviewee's account of the selection of a charge coupled device, rather than vidicon, stellar-sensor for Trident D5. The issue, he said, was "political"—the former gave you "not much greater accuracy" than the latter, but the Draper Laboratory was "trying to claw back more design responsibility" from Singer Kearfott, which had greater expertise in the vidicon.[57]

In both cases, then, the interviewees either disagreed with what had been done or considered there were other, equally

56. I draw this point from a 1987 interview by Graham Spinardi.
57. See chapter 7, p. 367 and chapter 5, p. 290.

technically valid, options that were not taken up. This made it possible for them to describe what happened as "political." They are not unique in this use of the technical/political dichotomy: other interviewees did the same, and similar ways of speaking are pervasive in science too.[58] Learning to avoid replicating in our analyses this unsymmetrical way of thinking has, indeed, been a major struggle for the social studies of science and technology.

More is involved, however, than the division of labor and this way of thinking. Perhaps the most important reason for the separation of technology and politics is that technologists and program managers work hard to maintain it. It is greatly in their interests, at least within U.S. culture, for their sphere of activity to be seen as one of purely technical work and technical decisions. Program managers, in particular, try to shape this work and those decisions so that the separation will be maintained.

The reason is quite straightforward. Outsiders—particularly those from the formal political system—taking anything other than a passive interest in the contents of a program can easily be either a source or a symptom of trouble. This is true especially if these outsiders control resources necessary for the program and if different sets of them have different preferences, for then it is impossible to opt for a "quiet life" by simply accommodating what the outsiders want. That in essence is what happened to the MMRBM and to MX, and what may be happening to the Small ICBM. The technical design of these missiles or their basing modes became "political," a matter of discussion and disagreement in Congress and other parts of the formal political system.

The way to avoid this trouble is for managers to keep their programs black boxes, to smoothly supply the technical output their environment demands, and thus to keep receiving the necessary input of resources from that environment without giving that environment reason to inquire into the details of the process. That, for example, is what the managers of the Fleet Ballistic Missile program in the U.S. Navy's Special Proj-

58. See, for example, Latour and Woolgar, *Laboratory Life*, second edition, 20–23; G. Nigel Gilbert and Michael Mulkay, *Opening Pandora's Box: A Sociological Analysis of Scientists' Discourse* (Cambridge: Cambridge University Press, 1984).

ects Office have almost always succeeded in doing. The major exception we have discussed, where stellar-inertial guidance became controversial in Congress, was not their doing, but rather the result of outsiders to the program recruiting allies in the formal political system to influence the contents of the black box.

One consequence for technological change of the need to keep the black box shut is a pervasive conservativism on the part of program managers, a willingness to accept only those goals they are sure they can meet, and a preparedness to use only those technical means they are sure will work. One manager told me: "there are two kinds of [program] managers: there are conservative ones and there are ones that get fired."[59]

Interestingly, though, conservativism may not always be the best strategy. Radical technological change is sometimes needed to create autonomy and a firmly closed black box. This we saw with the Navy intermediate-range ballistic missile, where program managers chose to avoid a marginally modified Jupiter missile, and opted instead for a dramatic combination of new technologies in Polaris. As noted in chapter 3, by doing so they removed their reliance on the Army and minimized their dependence on Admiral Rickover, father of the nuclear submarine, with his different interests and powerful allies in Congress and elsewhere.

Whether conservative or radical means are employed, however, the point is the same. Technology is shaped so as to maintain the separation between technology and politics. This is not universal. Little attempt, for example, seems to have been made to keep early postwar Soviet weapons programs black boxes in the eyes of the top political leadership.[60] There, a program could not be made successful simply by the leadership giving managers a budget and letting them get on with it. Detailed intervention by that leadership was needed to secure specific resources in short supply. However, in the American context, resource-rich but with a loose and conflict-ridden state structure, black-boxing programs appears to be an almost essential prerequisite of success.

59. Interview with Major Gregory Parnell (U.S. Air Force), Stanford, Calif., February 20, 1985.
60. See David Holloway, *The Soviet Union and the Arms Race* (New Haven, Conn.: Yale University Press, 1983), 149.

An example of a technology shaped so as to maintain the separation between technology and politics might be American stellar-inertial guidance. As we saw, what is distinctive about it is its unistar design, achieving mechanical simplicity at what some would argue to be a cost in terms of accuracy and flexibility. There is good evidence that it was unistar that made stellar-inertial guidance acceptable to program managers anxious to maintain the smooth black-boxing characteristic of the Fleet Ballistic Missile program, and tentative evidence that it may have arisen because of the search for that acceptability.[61]

If technology is sometimes shaped not directly by politics but by the exigencies of keeping politics and technology separate, then we have an interesting analogy with the much debated question of the relations of economics and politics. For it has been suggested there that it is mistaken to see politics as determined by economics, and unhelpful to see politics as autonomous, or even relatively autonomous, from economics. Rather, what characterizes capitalist societies is that within them the seamless web of social relations can be divided into separate spheres of economics and politics. What shapes politics is thus not economics but the separation of politics from economics, and efforts to maintain that separation.[62]

If technology is at least sometimes shaped in this indirect, structural way, then we have an interesting supplement to the simplest model of the social shaping of technology, which requires explicit technical alternatives connected to different "relevant social groups."[63] Technology may be shaped specifically so as to prevent this situation from arising!

Realizing the significance of the black-boxing of programs and of the maintenance of the separation of technology from politics is also of importance for the opponents of nuclear weapon systems. The managers of these realize how critical it

61. See chapter 5.
62. See John Holloway and Sol Picciotto, eds., *State and Capital: A Marxist Debate* (London: Arnold, 1978).
63. See Pinch and Bijker, "Facts and Artefacts," for "relevant social groups." For critiques of the model, see Stewart Russell, "The Social Construction of Artefacts: A Response to Pinch and Bijker," *Social Studies of Science,* Vol. 16 (1986), 331–346 and Stewart Russell and Robin Williams, "Opening the Black Box and Closing it After You," paper read to British Sociological Association Conference, Leeds, April 1987. There is a direct parallel with the issue of pluralism in political science.

is to their success to keep programs black boxes. Yet only some opponents of them realize the corresponding importance of opening the black box. To do so is not merely difficult. It also involves contact, even alliances, with "the other side," and lacks the moral clarity of outright opposition, that is, opposition to programs as black boxes. Yet if my analysis here is correct, it might on occasion be more effective.

Softening Hard Facts

There is a fourth, final cause of the apparent separation of technology from politics. The facts of politics are typically soft. That an assured destruction strategy "is simply not credible and therefore is ineffectual," that the Air Force is "smart and . . . ruthless . . . [in] the same way as the Communists," and even that there really is a "Soviet threat"; all these can be challenged, albeit with varying degrees of risk of disbelief.[64] But technology deals with facts that are much harder. That black-box navigation is "an entirely impossible undertaking," that "ball bearings simply do not cut the mustard in precision gyros," and that in stellar-inertial guidance "a single fix from an optimally located star will yield accuracy equivalent to a two-star fix"; these seem much harder to challenge.[65]

Indeed, they do not merely seem much harder to challenge. They are much harder to challenge. Only the first of them has been challenged successfully: the enormous, decades-long, resource-intensive effort to do so was the subject of chapter 2. There are a small minority who do not believe the second, but the belief it has commanded more generally has denied them the resources that would make a challenge possible. The third appears to have been challenged successfully in the Soviet Union but not in the United States.

Technologists' capacity to "speak for the facts" constitutes a vital resource in the area we have been studying. While the political and military elite can say "this weapon system is

64. 1980 Republican Party platform, as quoted in Scheer, *With Enough Shovels,* 128; Chief of Naval Operations Admiral Arleigh A. Burke, as quoted in chapter 3.
65. Maximilian Schuler, as quoted in chapter 2; John Slater, as quoted in chapter 4; the unistar principle, as expressed by Stephen F. Rounds and George Marmar, quoted in chapter 5.

needed" and economic advisers say "this weapon system cannot be afforded," the technologist can say "this can be done" or "this cannot be done." The setting of missile accuracy specifications, for example, always involved negotiation with guidance technologists at the very least. The only time in the United States where specifications were imposed unilaterally by military authorities was in the case of the early Air Force missiles.[66] But as we saw in chapter 3, there is reason to suspect no serious intention that a working missile actually be built. Otherwise, even where there was a strong desire for high missile accuracy, as there was, for example, with the Army's Jupiter missile, the Army "would wait for our [guidance engineers'] arguments whether it was possible."[67]

Drawing on work in the history and philosophy of science, the sociology of scientific knowledge has shown that the hardness of the facts of science is a construction: it follows neither from any unchallengeable correspondence to reality nor from the dictates of immutable logic. The same case can be made, as chapter 7 shows, for the facts of technology.

But how should analysts of science and technology deal with the way the in-principle challengeability of scientific-technical facts only rarely translates into their practical challengeability? Bruno Latour, in particular, has noted how the hardness of facts is made no less hard by not flowing from nature or logic alone. He concludes that unless there is controversy among scientists to undermine this hardness, analysts can go no further: "We do not try to undermine the solidity of the accepted parts of science. We are realists as much as the people [we study]."[68]

Latour's conclusion is, however, too pessimistic, at least if extrapolated to the facts of nuclear technology, for two reasons. The first is that uncertainty can exist, even where explicit controversy does not. Recall my speculative diagram (figure 7.2)

66. It may be that the accuracy specifications for the V-2 were imposed unilaterally (see chapter 2), though it may also be that the rocket engineers made optimistic promises to gain support for their project.
67. Dr. Walter Haeussermann, as quoted in chapter 3.
68. Latour, *Science in Action*, 100. "Realists" is used here not in its sense in the theory of international politics, but in the philosophical sense: "realists . . . believe that representations are sorted out by what really is outside, by the only independent referee there is, Nature" (ibid., p. 98).

of the distribution of uncertainty about the factual status of missile accuracies. The explicit controversy concerned those who were to be found on the right of the diagram, such as the proponents of the manned bomber, who were inclined to radical doubt as to whether missile accuracies were facts (or even whether missiles would work in the most basic sense of their warheads exploding). But those on the left of the diagram, those close to the heart of the production of knowedge of the technical characteristics of missiles, had uncertainties too, albeit less dramatic and of a somewhat different kind. General Hepfer's sense that "politics" affects how the Air Force calculates a missile's circular error probable, quoted above, is one example.

Uncertainties such as these would exist even if the public controversy over the factual status of missile accuracies had never taken place, and insiders such as Hepfer are certainly not to be counted as critics. Rather, the point is that nuclear facts are hardest, not for these insiders, but for those in "the certainty trough," the program loyalists and trusting readers of the "brochures" who are one or more stages removed from the production of technical fact.[69]

The analyst's role in gaining access to and generalizing the local uncertainties of the insiders is of some importance, for those uncertainties are evanescent in time and space. Even one office down the corridor, so to speak, a circular error probable can cease to be seen as a product of political negotiation, and could be entered into a nuclear targeting computer as hard fact. (It seems, though, that nuclear targeteers in the United States do have empirical, "safety-factor," formulae that they apply to circular error probable figures.[70]) When it comes to the general, operational rather than technical, advising the President or General Secretary in a crisis of the percentage of the opposing ICBM force that a preemptive strike against its silos will destroy, then the likelihood of any party to the con-

69. One possibility is that participants demonstrated their "insider" status to me as interviewer by pointing to sources of uncertainty. That may be so, but it does not affect the argument in the text, since these participants had "uncertainties" available to them to display that were not available to others. None of the "outsider" critics of the facticity of missile accuracies raised the point that accuracies could be seen as the outcome of political negotiation.
70. Interview.

versation having detailed knowledge of the processes leading to the construction of that percentage is small.

Crucial decisions bearing on matters of nuclear life and death have to be taken by those who are not "experts" in the sense of having access to the insiders' uncertainties. That is both right and probably inevitable: right because it is the only way that is at least partially compatible with democracy, inevitable because any significant decision will straddle more than one area of expertise, and there is no good reason to expect experts drawn from two areas to agree.[71] Given that uncertainty should incline to caution, any softening of the public image of nuclear facts that can be achieved is thus worthwhile. Giving voice to insiders' uncertainties is one way to help (in a small way, I have no illusions as to its great efficacy) in this process.

The other way in which nuclear facts can be softened has to do with what it takes to make a nuclear fact hard. Certain strategies for hardening these facts are simply not open to participants. They cannot fire a sample of ICBMs at their wartime targets, fire them with their warheads on board, or (at least in the United States) fire them out of their operational silos. This does not imply that the facts about these missiles are of necessity false, but it does mean that critics have a range of resources (all hinging around the argument that testing is not sufficiently like use) to deploy to undermine these facts. It may be that these constraints account for the extent of insider uncertainty, which is perhaps greater than in other fields of technology.

Constraints such as these affect not merely the certainty of technical knowledge, but also what artifacts are developed. Thus during the 1970s U.S. Navy Improved Accuracy Program, one option considered was starting to provide reentry vehicles with a means of homing in on their targets as they came down through the atmosphere. An argument deployed against this option was that such a vehicle could not be tested. "Its disadvantages," Rear Admiral Robert H. Wertheim told the Senate Armed Services Committee in 1978, "lie in the area of questions about the testability, our ability to conduct flight

---

71. Barry Barnes and David Edge, eds., *Science in Context: Readings in the Sociology of Science* (Milton Keynes, Bucks., England: Open University Press, 1982), part 5.

tests over land. Obviously this guidance scheme has to be able to recognize terrain features and correlate them with prepositioned information. We would certainly have to dramatically revise the way in which we now conduct our operational tests and demonstration shakedown tests, and introduce a whole new class of development risks and problems."[72]

Admiral Wertheim did not spell it out, but the "risks and problems" are as much political as technical. The population of the mainland United States is reckoned likely to be touchy about ballistic missiles being tested over their heads (and certainly litigious should something go wrong). This, for example, is why no U.S. ICBM has ever been fired from its operational silo. In 1974 the Air Force proposed trying to end some of the doubts about missile accuracy and reliability by firing eight Minuteman I ICBMs out of silos at Malmstrom Air Force Base, Montana. The plan was blocked by the ten senators from five northwestern states under or close to the proposed flight path.[73] The only ballistic missile testing in the United States that takes place outside of military reservations is on the remote (and relatively short) trajectory from Green River, Utah, into the White Sands Missile Range in New Mexico. The few residents under the flight path are sent registered letters enclosing checks to enable them, if they wish, to evacuate for the day of a test flight.[74]

72. Senate Armed Services Committee, *Department of Defense Authorization Hearings, Fiscal Year 1979, Part 9, Research and Development* (Washington, D.C.: U.S. Government Printing Office, 1978), 6684.

73. Interview with John Brett, Wayne, N.J., January 15, 1987; Joseph Albert Cernik, *Strategy, Technology, and Diplomacy: United States Strategic Posture, with Emphasis on the Tenure of James R. Schlesinger as Secretary of Defense, 1973–1975* (Ph. D. dissertation: New York University, 1982), 273.

74. Interview with Colonel Leonard R. Sugerman (U.S. Air Force, rtd.), Las Cruces, New Mexico, September 18, 1986. Colonel Sugerman told me that permission had been refused to open a longer range, from Idaho to White Sands, for testing the homing reentry vehicle of the intermediate range Pershing II. Cruise missiles are tested overland, it being argued that these tests are safe because "chase" aircraft tail the missiles and can take over control should anything go wrong. Even so, two of the test routes—one over northern Canada, and one from the Atlantic to inland Maine—have been the subject of considerable controversy. See Clyde Sanger, "Cruise Test Failure Raises Fears among Canadians," *The Guardian* (London) (January 24, 1986), 12; David Hughes, "Start of Tomahawk Flights Renews Controversy in Maine," *Aviation Week and Space Technology*, January 23, 1989, 28–29.

These restrictions on testing do not necessarily suffice to prevent facts being constructed. Booster rockets are used on the Green River to White Sands trajectory to make the conditions of reentry more like those at ICBM ranges.[75] Nor would the various negotiated prohibitions on ballistic missile flight-testing that are currently under consideration end fact construction. Even with a complete ban on all ballistic missile flight tests, it would still be claimed that computer simulations of those flights produced facts about matters such as accuracy.[76]

So such a ban could not be a simple "technical fix" for the technological arms race in ballistic missiles, as is sometimes imagined: its effects on the latter would be determined not by how valid computer simulation is but by how valid it is believed to be. But such a ban would give a further resource to those who sought to challenge nuclear facts. They would have available to them the argument that the gap between computer simulation and use is even larger than that between testing and use. A flight-test ban could thus reduce the credibility with which the nuclear ballistic missile could be presented as a surgically precise instrument of power, and help make it a weapon that could never be used rationally because the consequences of its unleashing are unknown. To ban ballistic missile flight tests could, therefore, potentially be an important adjunct to restrictions on the numbers of these missiles, because such a ban could soften facts.

One of the reasons the facts of science are hard is because "the layman is awed by the laboratory set-up, and rightly so. There are not many places under the sun where so many and such hard resources are gathered in so great numbers, sedimented in so many layers, capitalized on such a large scale . . . confronted by laboratories we are simply and literally impressed. We are left without power, that is, without resource to contest, to reopen the black boxes, to generate new objects, to dispute the spokesmen's authority."[77]

75. Sugerman interview.
76. For what such an argument might be based upon, see M. H. Andre and J. J. Czaja, "Minuteman Inertial Guidance Assessment: The Next Best Thing to Flight Tests," *Journal of Guidance and Control*, Vol. 6 (1983), 156–161.
77. Latour, *Science in Action*, 93.

Yet the inventors of accuracy have a problem. They possess all these resources. But, as we have seen, there is a sense in which they still require the world as their laboratory. And the world has yet to be persuaded that it should be so used. Acting wisely, we can prevent it ever being so.

Epilogue

Uninventing the Bomb

The manuscript of this book was completed in August 1989. By then East-West tension had already relaxed greatly. It still seemed possible, however, that what was going on was another temporary détente that events might reverse, especially if Gorbachev were replaced by a more hard-line Soviet leader.

Between then and now (July 1990), events moved on at a pace unprecedented in postwar history, sweeping aside the fossilized regimes of Eastern Europe, the Berlin Wall, the division of Germany, and with them the geopolitical framework of the cold war. The structure of the world system of states was altered decisively and irreversibly. No change of leadership in Moscow can now turn back the clock. In effect, the Soviet Union has abandoned its claim to being a superpower, permitting its previous satellites to find their own destiny. Secessionist movements threaten the Soviet Union's very survival in its current form.

If humanity is ever going to return the nuclear genie to its bottle, now is the moment to begin. Writing a year ago, the most that seemed achievable was to make use of the relaxation of tension to "uninvent" accuracy (see chapter 8). The dramatic changes of the past twelve months place on the agenda the alteration of social, political, and material conditions such that possession and use of nuclear weapons are never again possible.

There are at least three barriers to seizing this opportunity. The first is the substantial risk that, despite the end of the cold war, business-as-usual (perhaps at a somewhat reduced level) will continue as far as nuclear weapons are concerned. This book has emphasized that the division of the world into competing camps, each with a superpower at its center, has been the cause of the development of the nuclear arsenals in only an ultimate sense. The "ordinary world" of nuclear technology and politics, fueled as it is by career, bureaucratic, and corporate interests of a more mundane kind, will not simply change of its own accord to reflect the changes in the world system of states. No strategic nuclear system has yet been scrapped or canceled as a consequence of these changes. Nor has there been a perceptible increase in the pace of the long-standing strategic arms reduction talks.

Let me be optimistic and grant that the passage of time, plus determined political intervention (aided by economic and budgetary pressures), will force change in the nuclear arsenals to reflect changing East-West relations. A second barrier to taking full advantage of our current unique situation then becomes evident. East-West tensions are not the only ones that can fuel nuclear weapons development. Nuclear warhead and ballistic missile technology has already spread beyond its five traditional possessors, the United States, the Soviet Union, the United Kingdom, France, and China. In the Middle East, in particular, there are tensions in plenty to accelerate this process.

One likely response to this situation from the existing nuclear weapons states is to see in it a justification for their continuing possession of nuclear arsenals. If the threat from the East (or West) vanishes, there is always a potential threat from the South to consider. Yet the reassurance of having a nuclear arsenal to meet this danger, in case it should materialize, will be bought at a price. Third World states have always seen hypocrisy in the industrialized countries simultaneously arming themselves to the teeth while seeking to prevent the spread of these same weapons. To date, the existing nuclear weapons states have at least had the excuse of the need to deter each other. Without that excuse, the current nonproliferation regime will seem even more of a bastion of "Northern" privilege.

Stopping the proliferation of nuclear weapons is, however, an essential prerequisite of a safe world, and it can still be done. The South African nuclear arsenal, if it exists, can be dismantled along with the apartheid state that was its rationale. The certainly existing Israeli arsenal is vulnerable to determined pressure from the United States, especially if Israel can be reassured that its neighbors will be prevented from building nuclear weapons of their own. Other programs (even India's) are still far enough away from full operation to be prevented from ever becoming meaningful arsenals.

In the long run, all this is likely only if accompanied by the dismantling of the arsenals of the traditional nuclear weapons states. That would legitimate vigorous antiproliferation measures—and not just passive nonproliferation steps. It would also tie the interests of the existing nuclear weapons states and their allies firmly to stopping proliferation. The very anxieties generated by the loss of what is sometimes comfortingly thought of as a "nuclear umbrella" would serve as the strongest possible motive for preventing others from obtaining the very weapons that were being given up.

I do not underestimate the difficulties and risks involved in dismantling existing nuclear arsenals. The third barrier—even to it being seriously considered—is the feeling that these difficulties and risks will not be balanced by the long-term benefit of a world permanently free of nuclear weapons. Since we cannot reverse the invention of the bomb, so the argument goes, sooner or later someone will rebuild a nuclear arsenal.

This whole book argues against that kind of pessimism. Outside of the human, intellectual, and material networks that give them life and force, technologies cease to exist. We cannot reverse the invention of the motor car, perhaps, but imagine a world in which there were no car factories, no gasoline, no roads, where no one alive had ever driven, and where there was satisfaction with whatever alternative form of transportation existed. The libraries might still contain pictures of automobiles and texts on motor mechanics, but there would be a sense in which that was a world in which the motor car had been uninvented.

What exactly it would take to uninvent the bomb is not self-evident, and it is possible that a safe world can be achieved by measures less radical than the analogy suggests. To date, the issue has received remarkably little attention, with a few honorable exceptions. We have, however, had nearly half a century of "thinking the unthinkable"—pondering nuclear holocaust. The time has now surely come to think the other unthinkable, a feasible world permanently free of nuclear weapons.

Appendix A

Estimated Accuracies of American and Soviet Strategic Ballistic Missiles

Tables A.1 and A.2 contain estimates of the accuracies (circular error probable, or CEP) of American and Soviet ICBMs and submarine-launched ballistic missiles.[1] Unless otherwise stated, they refer to the year in which the missile was first deployed operationally. As figure 4.1 shows, the CEP of a missile tends to get smaller through time, even without hardware changes, as understanding of error processes develops.

The figures in these tables should, of course, be read with caution. The most important issues to bear in mind are those discussed in chapter 7. A CEP is not an unproblematic, self-evident attribute of a missile: it is the outcome of the complex processes described in there. The factual status of missile accuracy figures has been the subject of controversy.

There also are additional issues that arise from the classified nature of missile accuracy figures. Leaked or guessed accuracy figures appear in a variety of publications, most commonly *Aviation Week and Space Technology*. But the sources of these figures are not normally quoted, and they are frequently inconsistent, though seldom wildly so. Though I have not been wholly successful, in compiling tables A.1 and A.2 I have tried to avoid the arbitrary and ad hoc use of such figures.

For Soviet missiles this task is greatly eased by the existence of Barton Wright, *Soviet Missiles*, an attempt to compile effectively all the published estimates of the technical characteristics of Soviet missiles, and to collate these estimates to produce a

---

1. The CEP standardly is defined as the radius of the circle round the target within which 50 percent of warheads would fall in repeated firing. See chapter 7 for the contested nature of this definition.

Table A.1
Estimated Accuracies of U.S. and Soviet ICBMs

Year of first deployment	U.S. ICBMs	CEP[a]	Soviet ICBMs	CEP[a]
1958				
1959	Atlas D	1.8		
1960				
1961	Atlas E	1.0	SS-6	2.0
	Atlas F	1.0		
1962	Minuteman 1	1.1	SS-7	1.5
	Titan 1	0.65		
1963	Titan 2	0.65		
1964			SS-8	1.0
1965				
1966	Minuteman 2	0.26	SS-9	0.5
			SS-11 mod 1	0.75
1967				
1968				
1969			SS-13 mod 1	1.0
1970	Minuteman 3	0.21		
1971				
1972				
1973			SS-11 mod 3	0.6
			SS-13 mod 2	0.81
1974			SS-11 mod 2	0.6
			SS-18 mod 1	0.23
1975			SS-17 mod 1	0.24
			SS-19 mod 1	0.25
1976				
1977			SS-18 mod 2	0.23
			SS-18 mod 3	0.19
1978			SS-17 mod 2	0.23
			SS-19 mod 2	0.23
1979	Minuteman 3[b]	0.12	SS-18 mod 4	0.14
1980				
1981			SS-17 mod 3	0.20
			SS-19 mod 3	0.21
1982				
1983				
1984				
1985				
1986	MX	0.06	SS-25	?
1987			SS-24	?
1988				
1989			"SS-18 Follow-On"[c]	?

Notes: a. Unit = nautical miles.   b. Guidance improvements.
c. Flight tested in 1987. See text, p. 318.

Table A.2
Estimated Accuracies of U.S. and Soviet Submarine-launched Ballistic Missiles

Year of first deployment	U.S. SLBMs	CEP and range[a]	Soviet SLBMs	CEP and range[a]
1958			SS-N-4 (surface launched)	2.0 at 350
1959				
1960	Polaris A1	2.0 at 1,200		
1961				
1962	Polaris A2	2.0 at 1,500		
1963			SS-N-5	1.5 at 700
1964	Polaris A3	0.5 at 2,500		
1965				
1966				
1967				
1968			SS-N-6 mod 1	1.0 at 1,300
1969				
1970				
1971	Poseidon C3	0.25 at 2,500		
1972				
1973			SS-N-6 mod 2	1.0 at 1,600
			SS-N-8 mod 1	0.84 at 4,200
1974			SS-N-6 mod 3	1.0 at 1,600
1975				
1976				
1977			SS-N-8 mod 2	0.84 at 4,900
1978			SS-N-18 mods 1&3	0.76 at 3,500
1979	Trident C4	0.25 at 4,000[b]		
1980				
1981				
1982				
1983	Trident C4	0.12 at 4,000[c]	SS-N-20	0.5 at 4,500
1984				
1985				
1986				
1987			SS-N-23	? at 5,000
1988				
1989				
1990	Trident D5	0.06 at 4,000[d]		

Notes:
a. Unit = nautical miles.
b. Accuracy goal.
c. 1983 flight tests.
d. Accuracy requirement at 4,000 nmi range. As with C3 and C4, longer ranges are physically possible, but accuracy will degrade.

single "best estimate" for each missile.[2] Wright procedes not merely by averaging but by weighing the authority of different sources (some early intelligence estimates are now declassified, for example, and Wright gives considerable weight to these) and the plausibility of their estimates.

In only one case have I felt it necessary to depart from Wright's "best estimates." That case is the CEP of the SS-19, which Wright seems to underestimate, probably through being too influenced by a Defense Intelligence Agency figure for the CEP of the SS-19 mod 3 of 245 meters (0.13 nautical miles), which was released through a mistake in making security deletions in 1981 Congressional testimony.[3] Four years later, however, a March 1985 National Intelligence Estimate put the SS-19's CEP at 435 yards (0.215 nautical miles). Although the Defense Intelligence Agency dissented from this Central Intelligence Agency influenced figure, it appears to have defended not its 1981 estimate but only a more moderate 325 yards (0.16 nautical miles).[4] In table A.1, I use the 1985 National Intelligence Estimate figure, and have made corresponding upwards adjustments to Wright's CEP figures for the earlier versions of the SS-19.

The official sources used to construct my estimates of the accuracies of U.S. missiles are of two kinds: actual figures, now declassified, for the CEP of early missiles;[5] and statements of the CEP of one missile relative to another (such "relative" statements seem to be permitted under the security classification rules). The value of having both sources is clear: if a "chain" of relative statements can be extended back to a declassified absolute figure, then something approaching an official history of CEPs can be produced.

2. Barton Wright, *World Weapon Database, Vol.1, Soviet Missiles* (Lexington, Mass.: Lexington Books, 1986).
3. Ibid., 68.
4. Bill Keller, "Imperfect Science, Important Conclusions," *New York Times*, July 28, 1985, 4E, and Jeffrey T. Richelson, "Old Surveillance, New Interpretations," *Bulletin of the Atomic Scientists*, Vol. 42, No. 2 (February 1986), 18–23.
5. The most important single source is the now declassified presentation by General N. F. Twining, Chair of the Joint Chiefs of Staff, to the Defense Subcommittee of the House Committee on Appropriations, January 13, 1960 (Eisenhower Library, White House Office Staff Secretary Subject Series, Defense Department Subseries Box 11, Folders General Twining Posture Briefing). I am grateful to Graham Spinardi for a copy of this document.

Thus the original accuracy goal for Polaris A1/A2 has now been declassified: it was 4,000 yards or, roughly, two nautical miles.[6] A 1980 article by two successive directors of the Navy Special Projects Office and the president of the Charles Stark Draper Laboratory then enables us to set up a "chain." The objective for Polaris A3, they write, "was to improve accuracy . . . by a factor of about 4," relative to the original Polaris. For Poseidon C3, they continue, "accuracy needed to be improved, this time by about 50%," relative to the A3. The goal for Trident C4 was that accuracy should "be equivalent at 4000 nm [nautical miles] to that of Poseidon C-3 at 2000 nm."[7] Though the inference is not as precise as we might wish,[8] we can infer the following rough history of accuracies:

Polaris A1/A2   4,000 yards
Polaris A3   1,000 yards
Poseidon C3   500 yards
Trident C4   500 yards

Confidence in this "chain" is increased by the fact that the last of these figures, that for Trident C4, corresponds exactly to that in a secret Department of Defense report to Congress in 1982.[9]

That same leaked report states that in 1983 flight tests Trident C4 achieved a CEP twice as good as the goal: that is, 250

6. Berend Derk Bruins, *U.S. Naval Bombardment Missiles, 1940–1958: A Study of the Weapons Innovation Process* (Ph.D. dissertation, Columbia University, 1981), 285; this useful thesis draws extensively on material from U.S. Navy archives. The correctness of this figure was confirmed by an interview with a senior participant in the Fleet Ballistic Missile program.
7. Levering Smith, Robert H. Wertheim, and Robert A. Duffy, "Innovative Engineering in the Trident Missile Development," *The Bridge* (National Academy of Engineering), Vol. 10, No. 2 (Summer 1980), 11–12.
8. The major ambiguity concerns the starting point of the chain. Though, as indicated in appendix A, note 6, there seems little doubt that the accuracy goal for Polaris A1 was about two nautical miles, there was flexibility in this matter, as noted in chapter 3. Thus Levering Smith et al. (ibid., 10) write of Polaris A1: "a relatively simple missile carrying a single nuclear warhead to a maximum range of 1000–1500 nautical miles would satisfy the goal if it could have a circular probable error less than about four nautical miles." If we applied the "chain" to this latter figure, then clearly the figures for later CEPs would double. To my mind that would make them implausibly large, and accordingly I have chosen to apply it to the two nautical mile figure.
9. Quoted in William M. Arkin, "Sleight of Hand with Trident II," *Bulletin of the Atomic Scientists*, Vol. 40, No. 11 (December 1984), 6.

yards or roughly 0.12 nautical miles.[10] If we take this as the base for the official Navy statement that Trident D5 will be "at least twice as accurate" as Trident C4, we are led to infer an accuracy requirement for the D5 of 0.06 nautical miles, which is consistent with estimates commonly to be found in the literature.[11]

There is no single "relative" source for Air Force missiles as useful as the above cited article for Navy missiles. There is one very detailed and authoritative graph available of the relative guidance and control accuracies, year by year, of the three generations of Minuteman: I have reproduced it above as figure 4.1. Unfortunately, however, guidance and control is only one element (albeit a large one) of a missile error budget: reentry errors, deficiencies in geophysical and geodetic models and other factors also contribute. Furthermore, the declassified estimate to which the "chain" leads back is one of an interval, not a single figure: a January 1960 statement of the CEP goal of Minuteman I as 1.1 to 1.5 nautical miles.[12]

In constructing table A.1, I have thus had to make two contestible assumptions about the accuracy history of Minuteman. I have simply chosen the most straightforward: that overall CEP fell proportionately to guidance and control CEP; and that the 1960 "Autonetics contracted CEP requirement" in figure 4.1 corresponds to the midpoint of Twining's interval, 1.3 nautical miles. The result corresponds roughly to typical estimates in the literature, and yields a CEP for Minuteman III compatible with that implied by the leaked 1982 report.[13]

10. Ibid.

11. Testimony by Special Projects' Director Glenwood Clark to the Senate Armed Services Committee, *Department of Defense Authorization Hearings, Fiscal Year 1985* (Washington, D.C.: Government Printing Office, 1984), 3426. Again, I am grateful to Graham Spinardi for providing me with a copy of Admiral Clark's testimony. See, for example, the Trident D5 CEP of 400 feet (0.06 nautical miles) given in Thomas B. Cochran, William M. Arkin and Milton M. Hoenig, *Nuclear Weapons Databook, Vol. 1: US Nuclear Forces and Capabilities* (Cambridge, Mass.: Ballinger, 1984), 146.

12. Chart accompanying Twining, "Presentation."

13. See, e.g., Cochran et al., *Databook*, 114, 118; Colin S. Gray, *The Future of Land-Based Missile Forces*, Adelphi Paper no. 140 (London: International Institute of Strategic Studies, 1977), 5, 32. The 1982 report stated that "Trident I missile system operational accuracy is now about the same as Minuteman III operational accuracy," giving the former as 750 feet (0.12 nautical miles): Arkin, "Sleight of Hand," 6.

One further piece of relative data allows us to bring this history up to date. The 1987 debate over the MX Advanced Inertial Reference Sphere led to the publication, without a scale, of the flight test results for MX: see figure A.1. An upper bound to the scale can be inferred from the attempt by Air Force Brigadier General Charles May to defend the accuracy of MX. May told reporters that "if you take the outlier in that database, which is number 11, the accuracy on number 11 is more accurate than any missile system that we have in the inventory today or any missile save Peacekeeper [MX]."[14] If we follow the leaked Department of Defense report, the CEP of the most accurate US missiles apart from MX is 750 feet (roughly 0.12 nautical miles), and so the offset from the target of flight test 11 must be that or less.[15] The radius of the "current performance" circle in figure A.1 must thus be less than around 400 feet. That clearly does not define a single CEP figure, but rules out CEPs of more than 0.07 nautical miles (425 feet). Since I suspect May would have been more vigorous in his defense if the CEP had been much better than this, in table A.1 I choose a fairly conservative 0.06 nautical miles as my estimate, though would admit that there is a case for a less conservative 0.05 nautical miles.[16]

For U.S. ICBMs before the Minuteman series we have one official test-range accuracy figure: President Eisenhower's two (presumably statute) mile figure for Atlas D, quoted in chapter 3. For Atlas E and F, I have used the test-range result of 80 percent falling within 1.5 nautical miles of the target, as quoted in chapter 3, and assumed, for the purposes of a rough calculation of CEP, a circular normal distribution. For Titan I and II plausible CEP figures can be found in literature around

14. Transcript of briefing by General May, August 24, 1987, by Federal News Service, Washington, D.C.
15. Arkin, "Sleight of Hand."
16. Thus in 1984 Clarence Robinson of *Aviation Week,* a journalist with close Air Force connections, wrote that "in flight tests of the AIRS system with Mk. 12A reentry vehicles on the MX ICBM, USAF has achieved circular error probable accuracies of approximately 300 ft." or 0.05 nautical miles: Clarence A. Robinson, Jr., "Parallel Programs Advance Small ICBM," *Aviation Week and Space Technology* (March 5, 1984), 17. As will be seen from figure A.1, however, the early MX test results appear to be better than later ones.

UNCLASSIFIED

PEACEKEEPER FLIGHT-TEST RESULTS (U)
(WEAPON SYSTEM MEAN POINT OF IMPACT)

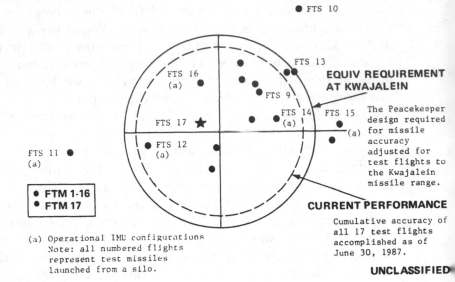

● FTS 10

FTS 13

EQUIV REQUIREMENT
AT KWAJALEIN

FTS 16
(a)

FTS 9

FTS 17 ★

FTS 14
(a)

FTS 15

The Peacekeeper
design required
for missile
accuracy
adjusted for
test flights to
the Kwajalein
missile range.

FTS 11 ●
(a)

FTS 12
(a)

● **FTM 1-16**
● **FTM 17**

CURRENT PERFORMANCE

Cumulative accuracy of
all 17 test flights
accomplished as of
June 30, 1987.

(a) Operational IMU configurations
Note: all numbered flights
represent test missiles
launched from a silo.

UNCLASSIFIED

Figure A.1
MX Flight-Test Results
Source: Subcommittees on Research and Development, and on Procure-
ment and Military, Nuclear Systems, of House of Representatives Armed
Services Committee, *The MX Inertial Measurement Unit: A Program Review*
(Washington, D.C.: Government Printing Office, 1987), 16.

the time of their deployment that are mutually consistent and
also consistent with later estimates.[17]

I have chosen nautical miles as the unit in tables A.1 and
A.2 because it is the most common in declassified documents,
and is also that used in the standard mathematics of attacks on

17. Anonymous, "Titan II to give USAF Well-Protected Fast-Reaction Strike
Force," *Aviation Week and Space Technology* (September 25, 1961), 138, says
that the accuracy of the inertially guided Titan II was "equal to that of its
predecessor," the radio-guided Titan I, and "industry speculation is that it
might be as low as two-thirds of a mile." J. Michael Fogarty, "Save the 70!,"
Aviation Week and Space Technology (January 14, 1963), 126, writes "Titan I,
fired under the most optimum conditions in its 1961 test series, chalked up
a circular error probable (CEP) of 0.65 mi." These figures may well be in
statute miles, and so there would be some justification for an estimate as low
as 0.55 nautical miles, but given the reservations expressed in each formu-

hard targets (see appendix B). The relevant conversion factors are:

$$
\begin{aligned}
\text{One nautical mile} &= 1{,}852 \text{ meters} \\
&= 2{,}025 \text{ yards} \\
&= 6{,}076 \text{ feet} \\
&= 1.151 \text{ statute miles.}
\end{aligned}
$$

lation, I have used a slightly more conservative 0.65 nautical miles in table A.1 More recent figures in the literature for the accuracy of Titan II prior to its retrofitting with a more modern, commercially available, guidance system range from 0.5 to 0.8 nautical miles. See Cochran et al., *Databook*, 112.

Appendix B

The Conventional Mathematics of Attacks on Point Targets

Let x and y be the errors around the target in any two mutually perpendicular horizontal directions. Assume

(1) x and y both normally distributed

(2) x and y independent

(3) standard deviation of x = standard deviation of y.
    Let the joint standard deviation be σ.

(4) mean of x = mean of y = 0.

That is, we are assuming a "circular" normal distribution of errors around the target, and no "bias."

Let $p(x)$ be the probability distribution of x, $p(y)$ the probability distribution of y, and $p(x, y)$ their joint probability distribution. Then

$$p(x)dx = \frac{1}{\sigma\sqrt{2\pi}} \exp - \left(\frac{x^2}{2\sigma^2}\right) dx$$

$$p(y)dy = \frac{1}{\sigma\sqrt{2\pi}} \exp - \left(\frac{y^2}{2\sigma^2}\right) dy$$

$$p(x, y)dxdy = p(x)dx.p(y)dy$$

$$= \frac{1}{2\pi\sigma^2} \exp - \left(\frac{x^2 + y^2}{2\sigma^2}\right) dxdy \tag{1}$$

To obtain the probability that the target will be destroyed, we need to find the probability that the distance between the target and the point of impact of the bomb will be less than or equal to the latter's "lethal radius" for a target of that type. Simplifying some complex issues, a dimensional argument establishes that the lethal radius r_k is a function of the one-third power of bomb's yield y. It is also inversely related to the

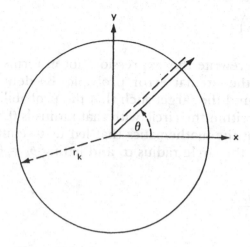

Figure B.1
Polar Coordinates

hardness h of the target, that is, the degree of pounds per square inch overpressure the target can withstand. The relation is complex mathematically, and is derived from studies of nuclear testing. Call the requisite function of hardness $g(h)$.

$$r_k = \frac{y^{1/3}}{g(h)} \tag{2}$$

Hence we need to integrate expression (1) over the circle with radius r_k. Transform into polar coordinates r, θ, as shown in figure B.1.

p_k, the probability of destruction of the target, or "single-shot kill probability," is given by

$$p_k = \int \int p(x, y)dxdy = \int_0^{r_k} \int_0^{2\pi} p(r \cos \theta, r \sin \theta) r d\theta dr$$

$$= \int_0^{r_k} \int_0^{2\pi} \frac{1}{2\pi\sigma^2} \exp - \left(\frac{(r \cos \theta)^2 + (r \sin \theta)^2}{2\sigma^2} \right) r d\theta dr$$

$$= \frac{1}{2\pi\sigma^2} \int_0^{r_k} \int_0^{2\pi} r \exp - \left(\frac{r^2}{2\sigma^2} \right) d\theta dr$$

$$= \frac{1}{\sigma^2} \int_0^{r_k} r \exp - \left(\frac{r^2}{2\sigma^2} \right) dr$$

$$= 1 - \exp - \left(\frac{r_k^2}{2\sigma^2}\right) \tag{3}$$

The final stage is to rewrite this expression not in terms of σ but in terms of σ_c, the circular error probable. By definition σ_c is that radius around the target such that the probability of the warhead falling within the circle with that radius is 0.5. All we need do is repeat the mathematics that led us to equation (3), integrating over the circle radius σ_c and letting $p_k = 0.5$.

This gives

$$0.5 = 1 - \exp - \left(\frac{\sigma_c^2}{2\sigma^2}\right)$$

$$\sigma_c^2 = \sigma^2 2 \ln 2$$

Substituting this expression and equation (2) into equation (3) gives

$$p_k = 1 - \exp - \left(\frac{y^{2/3} \ln 2}{\sigma_c^2 g^2(h)}\right) \tag{4}$$

or $p_k = 1 - 0.5^{(y^{2/3}/\sigma_c^2 g^2(h))}$ \tag{5}

Numerical values are conventionally obtained by expressing σ_c in nautical miles, y in megatons, and h in pounds per square inch. Then the standard public domain expression for $g(h)$, following Brode and Tsipis,[1] is

$$g(h) = h^{1/3}[0.1h^{-1} - 0.23h^{-1/2} + 0.068]^{1/3} \tag{6}$$

for $h \geqslant 300$ pounds per square inch.

Consider, for example, American targeteers using this conventional mathematics to work out the likelihood of an MX warhead destroying a very hard Soviet target: say a missile silo designed to withstand overpressures of 6,000 pounds per square inch.[2] Assume, perhaps optimistically (see appendix A), that σ_c for MX is 0.05 nautical miles. The W87 warhead for the MX reentry vehicles is around 300 kilotons in yield, so $y =$

1. See H. L. Brode, "Review of Nuclear Weapons Effects," *Annual Review of Nuclear Sciences*, Vol. 18 (1968), 180, and Kosta Tsipis, *Offensive Missiles*, Stockholm Paper No. 5 (Stockholm: Stockholm International Peace Research Institute, 1974).
2. See J. Richelson, "PD-59, NSDD-13 and the Reagan Strategic Modernization Programme," *Journal of Strategic Studies*, Vol. 6 (1983), 133.

0.3 megatons.[3] Substituting these figures into equations (5) and (6) produces a value of p_k of 0.90.

On the conventional mathematics, then, a single MX warhead has a 90 percent chance of destroying a very hard Soviet silo. If sufficient warheads of the right type are available to the targeteer it appears to be conventional to plan a two-on-one attack, where a second warhead from a different missile is programmed to attack the same target in the very short interval of time between the first explosion (and its immediate burst of radiation) and the sucking up of debris into the atmosphere. The resultant "two-shot kill probability," following the elementary mathematics of independent events, is $1 - (1 - p_k)^2$. Thus the two-shot kill probability for our example is $1 - (1 - 0.9)^2 = 0.99$. So two MX warheads targeted in this way have a 99 percent chance of destroying the silo. Only around 14 of the Soviet Union's 1,398 ICBM silos would, on this calculation, survive such an attack, were sufficient numbers of MX warheads available.

The actual calculus used in computer modeling of nuclear attacks will, of course, be a great deal more complex than this simple account, because it will correct for many secondary phenomena neglected here (such as the duration of the overpressure pulse). Furthermore, there can be argued to be many sources of practical uncertainty in counterforce attacks which are not captured by even a more sophisticated version of this calculus.[4]

3. T.B. Cochran et al., *Nuclear Weapons Databook*, Vol. 1: *U.S. Nuclear Forces and Capabilities* (Cambridge, Mass.: Ballinger, 1984), 126.
4. See chapter 7 above, and Matthew Bunn and Kosta Tsipis, *Ballistic Missile Guidance and Technical Uncertainties of Countersilo Attacks*, MIT Program in Science and Technology for International Security Report No. 9 (Cambridge, Mass.: MIT Program in Science and Technology for International Security, 1983).

Appendix C

List of Those Interviewed

In addition to those named, a certain number of interviews were carried out with members of the U.S. intelligence community under conditions of anonymity. Dr. Graham Spinardi and Dr. Wolfgang Rüdig also kindly gave me access to data from interviews they carried out, chiefly with participants in the Fleet Ballistic Missile program and the German work on inertial guidance and navigation in the 1930s and 1940s.

Military ranks are given when the interview chiefly concerned an interviewee's military service. All other titles (Professor, Dr., etc.) are omitted. Asterisks indicate that the interview was tape-recorded.

J. Edward Anderson*	Minneapolis, MN	March 9, 1985
Frederick Aronowitz*	Anaheim, CA	February 27, 1985
Benjamin Averbach*	Cambridge, MA	November 9, 1984
John Bailey*	Minneapolis, MN	March 7, 1985
Stewart Bailey and David Featherstone	Annapolis, MD	September 22, 1986
Anthony Bainbridge and David Dewar*	Bracknell, U.K.	March 21, 1986
Rollie Baldwin	Minneapolis, MN	March 8, 1985
J. W. Barnes*	Farnborough, U.K.	March 16, 1986
Richard Battin*	Cambridge, MA	October 11, 1984
Jack Becker*	Oak Creek, WI	March 11, 1985
Stuart Blain	Stevenage, UK	March 26, 1986
B. Paul Blasingame	Santa Barbara, CA	September 12, 1986
Norbert Bold	Valley Center, CA	September 10, 1986
Joseph Boltinghouse	Anaheim, CA	September 9, 1986

Philip Bowditch*	Cambridge, MA	October 11, 1984
John Brett	Wayne, NJ	January 15, 1987
Kenneth Brown*	Edinburgh, U.K.	July 17, 1985
Robert G. Brown*	Topsfield, MA	April 8, 1985
McGeorge Bundy*	New York, NY	September 26, 1984
Major General Aloysius Casey (U.S. Air Force), Mike Gorman, and Ed Rae*	San Bernadino, CA	February 25, 1985
Rear Admiral Glenwood Clark (U.S. Navy)*	Arlington, VA	March 28, 1985
Joseph D. Coccoli*	Cambridge, MA	November 19, 1984
Captain Steven R. Cohen (U.S. Navy)*	Arlington, VA	March 29, 1985
Richard Coleman*	Arlington, VA	March 29, 1985
Clifford Cormier*	Cambridge, MA	November 6, 1984
Daniel D. DeBra	San Francisco, CA	August 1, 1985
William Denhard*	Cambridge, MA	February 18, 1985
Andrew DePrete*	Arlington, VA	March 28, 1985
Paul Dow and Benedict Olson*	Cambridge, MA	October 5, 1984
Charles Stark Draper*	Cambridge, MA	October 2, 1984 and October 12, 1984
Robert Duffy*	Cambridge, MA	October 1, 1984
Chris Edwards	Stevenage, U.K.	March 26, 1986
Arthur Elias	Edinburgh, U.K.	April 24, 1984
Peter Entwhistle	Bracknell, U.K.	March 21, 1986
Harold Erdley	Los Angeles, CA	September 8, 1986
Frank Everest*	Stevenage, U.K.	March 26, 1986
Francis Everitt	Stanford, CA	February 19, 1985
David Ferguson*	Edinburgh, U.K; and Los Angeles, CA	June 6, 1985 September 10, 1986
Kenneth Fertig*	Cambridge, MA	November 12, 1984
Eric Firdman	Cambridge, MA (by telephone)	November 15, 1984
Carl Flom	Oak Creek, WI	March 11, 1985
John Foster	Cleveland, OH	September 24, 1986
Albert Freeman*	Cambridge, MA	October 3, 1985

Alton Frye*	Washington, DC	April 2, 1985
Hugh Galt	Anaheim, CA	September 9, 1986
Richard L. Garwin*	New York, NY	October 23, 1984
John S. Gasper and John Pinson	Anaheim, CA	September 9, 1986
Jerold P. Gilmore	Cambridge, MA	April 8, 1985
David Gold	Arlington, VA	April 2, 1985
Murray Goldstein, Ed Solov, and George Marmar	Wayne, NJ	January 15, 1987
Walter Haeussermann	Huntsville, AL	March 22, 1985
Edward J. Hall*	Cambridge, MA	October 4, 1984
Doug Harris*	Rochester, U.K.	March 18, 1986
Willis Hawkins*	Burbank, CA	March 4, 1985
Clifford V. Heer	Columbus, OH (by telephone)	March 27, 1985
Stephen Helfant	Cambridge, MA	April 8, 1985
Robert Henderson*	Cambridge, MA	October 12, 1984
Major General John Hepfer* (U.S. Air Force, rtd.)	Washington, DC	July 29, 1985
David G. Hoag*	Cambridge, MA	November 5, 1984
Tom Hutchings and Graham Martin*	Woodland Hills, CA	February 26, 1985
Norman Ingold	Holloman Air Force Base, NM	September 19, 1986
Aleksandr' Yu. Ishlinskii	Moscow, USSR	May 23, 1988
Martin Jaenke and Peter Zagone	Holloman Air Force Base, NM	September 19, 1986
John Jarosh	Los Angeles, CA	September 8, 1986
William Kaufmann	Cambridge, MA (by telephone)	April 5, 1985
Carl Kaysen*	Cambridge, MA	November 14, 1984
Joseph Killpatrick*	Minneapolis, MN	March 7, 1985
Thomas A.J. King*	Arlington, VA	April 2, 1985
Eric Kintner*	Newton, CN	September 25, 1984
Philip J. Klass*	Washington, DC	October 19–21, 1984
Howard Knoebel	Urbana, IL	September 3, 1986
Walter Krupick	Wayne, NJ	September 25, 1986

Henry Kudish and Wallace Lane	Farmington, CN	September 29, 1986
Morris Kuritsky, Murray Goldstein, and Ed Solov*	Wayne, NJ	March 18, 1985
J. Halcombe Laning*	Cambridge, MA	October 3, 1984
Stanley LaShoto*	Cambridge, MA	November 9, 1984
Polen Lloret	Paris, France	July 7, 1987
David Lynch	Goleta, CA	September 11, 1986
Edward J. MacCormack*	Cambridge, MA	November 13, 1984
Warren Macek	Great Neck, NY	April 5, 1985
John McHale	Silver Spring, MD	September 22, 1986
Robert S. McNamara*	Washington, DC	March 29, 1985
Arthur Metcalf*	Waltham, MA	November 6, 1984
Ernest Metzger*	Buffalo, NY	March 14, 1985
C.R. Milne and Sidney Smith*	Farnborough, U.K.	March 17, 1986
Robert Mitchell and Clifton Chappell*	Arlington, VA	March 28, 1985
John R. Moore*	Los Angeles, CA	February 25, 1985
Alice Moriarty*	Cambridge, MA	November 16, 1984
Fritz Mueller*	Huntsville, AL	March 23, 1985
Hugh Neeson, Norman Burnfield and Don McCallum*	Buffalo, NY	March 14, 1985
David Nisbet and James Holmes*	Edinburgh, U.K.	July 17, 1985
Bernard J. O'Connor*	Teterboro, NJ	March 19, 1985
Joseph O'Connor*	Cambridge, MA	October 2, 1984
Peter Palmer*	Cambridge, MA	October 10, 1984
Brad Parkinson*	Stanford, CA	February 20, 1985
Major Gregory Parnell (U.S. Air Force)*	Stanford, CA	February 20, 1985
Roland Peterson*	Woodland Hills, CA	February 26, 1985
Edward Porter*	Cambridge, MA	October 2, 1984
Ted Postol	Stanford, CA	February 19, 1985

Ralph Ragan*	Cambridge, MA	October 4, 1984
George Rathjens*	Cambridge, MA	November 12, 1984
Ronald G. Raymond*	Minneapolis, MN	March 8, 1985
Heinrich Rothe*	Huntsville, AL	March 23, 1985
Jack Ruina*	Cambridge, MA	November 9, 1984
Bernard de Salaberry	Versailles, France	July 21, 1987
Michele S. Sapuppo*	Cambridge, MA	October 10, 1984
Paul H. Savet	New York, NY	September 26, 1986
James Schlesinger*	Washington, DC	September 22, 1986
General Bernard Schriever (U.S. Air Force, rtd.)*	Washington, DC	March 25, 1985
Robert C. Seamans, Jr.*	Cambridge MA	October 9, 1984
Cal Senechal*	Minneapolis, MN	March 6, 1985
Hyman Shulman and Bruno W. Augenstein	Santa Monica, CA	January 13, 1987
Henry Singleton*	Los Angeles, CA	March 4, 1985
John Slater*	Fullerton, CA	February 22, 1985
Larry Smith*	Cambridge, MA	April 10, 1985
Rear Admiral Levering Smith (U.S. Navy, rtd.)*	San Diego, CA	February 23, 1985
E. V. Stearns*	By mail and tape from Evergreen, CO	May 25, 1985
Martin Stevenson	Goleta, CA	September 11, 1986
Guyford Stever*	Washington, DC	April 1, 1985
John Stiles	Wayne, NJ	September 25, 1986
Andrew Stratton	Farnborough, U.K.	March 17, 1986
Ralph Stretton	Datchet, U.K.	March 24, 1986
Colonel Leonard R. Sugerman (U.S. Air Force, rtd.)	Las Cruces, NM	September 18, 1986
Takeshi Takahashi*	Edinburgh, U.K.	February 27, 1986
Milton Traegeser	Cambridge, MA	November 12, 1984
Kosta Tsipis*	Cambridge, MA	November 2, 1984
John Walker	Minneapolis, MN	March 7, 1985

Rear Admiral Robert Wertheim (U.S. Navy, rtd.)*	Burbank, CA	March 4, 1985
Ron Whalley*	Bracknell, U.K.	March 24, 1986
Michael Willcocks	Slough, U.K.	March 19, 1986
Willis Wing	By telephone, and Glenhead, NY	March 26, 1985 September 26, 1986
Michael Wooley*	Bracknell, U.K.	March 24, 1986
Walter Wrigley*	Cambridge, MA	November 8, 1984
John M. Wuerth*	Fullerton, CA	February 22, 1985
Stanley Wyse, Jerome S. Lipman, and D. A. Ulrich*	Woodland Hills, CA	February 22, 1985

Index

*Readers seeking the meaning of a term should turn first to the entry or entries in **bold**.*

Printed in the United States
by Baker & Taylor Publisher Services

Printed in the United States
by Baker & Taylor Publisher Services